Advances in Science, Technology & Innovation

IEREK Interdisciplinary Series for Sustainable Development

Advances in Science, Technology & Innovation (ASTI) is a series of peer-reviewed books based on important emerging research that redefines the current disciplinary boundaries in science, technology and innovation (STI) in order to develop integrated concepts for sustainable development. It not only discusses the progress made towards securing more resources, allocating smarter solutions, and rebalancing the relationship between nature and people, but also provides in-depth insights from comprehensive research that addresses the **17 sustainable development goals (SDGs)** as set out by the UN for 2030.

The series draws on the best research papers from various IEREK and other international conferences to promote the creation and development of viable solutions for a **sustainable future and a positive societal** transformation with the help of integrated and innovative science-based approaches. Including interdisciplinary contributions, it presents innovative approaches and highlights how they can best support both economic and sustainable development, through better use of data, more effective institutions, and global, local and individual action, for the welfare of all societies.

The series particularly features conceptual and empirical contributions from various interrelated fields of science, technology and innovation, with an emphasis on digital transformation, that focus on providing practical solutions to **ensure food, water and energy security to achieve the SDGs.** It also presents new case studies offering concrete examples of how to resolve sustainable urbanization and environmental issues in different regions of the world.

The series is intended for professionals in research and teaching, consultancies and industry, and government and international organizations. Published in collaboration with IEREK, the Springer ASTI series will acquaint readers with essential new studies in STI for sustainable development.

ASTI series has now been accepted for Scopus (September 2020). All content published in this series will start appearing on the Scopus site in early 2021.

Muhammad Nawaz Tunio · Angeles Sánchez ·
Yasmin Moanis Latif Hatem · Ayman M. Zakaria
Editors

Sustainability in Creative Industries

Sustainable Entrepreneurship and Creative Innovations—Volume 1

A culmination of selected research papers from the International Conference on Sustainability in Creative Industries (SCI), Universitas Ciputra 2022

Springer

Editors
Muhammad Nawaz Tunio
Department of Business Administration
University of Sufism and Modern Sciences
Bhitshah, Pakistan

Angeles Sánchez
Department of Applied Economics
Faculty of Economics
University of Granada
Granada, Spain

Yasmin Moanis Latif Hatem
Department of Architecture
College of Engineering
Delta University for Science and Technology
Mansoura, Egypt

Ayman M. Zakaria
Department of Architecture
Faculty of Fine Arts
Minia University
Minia, Egypt

ISSN 2522-8714 ISSN 2522-8722 (electronic)
Advances in Science, Technology & Innovation
IEREK Interdisciplinary Series for Sustainable Development
ISBN 978-3-031-48452-0 ISBN 978-3-031-48453-7 (eBook)
https://doi.org/10.1007/978-3-031-48453-7

This Springer imprint is published by the registered company Springer Nature Switzerland AG
The registered company address is: Gewerbestrasse 11, 6330 Cham, Switzerland

Paper in this product is recyclable.

Scientific Committee

The Editors warmly thank all the Reviewers who have contributed their authority to the double-blind review process, to ensure the quality of this publication.

Preface

The massive impact of digital technology is transforming every industry. During the last 20 years, the narrower concept of creative industries (referred to arts and cultural industries) has evolved into the idea of creative economy. Creative economy envelops a multitude of fields, such as music and artistic expression, food and gastronomy, publishing, software, advertising and design, packaging, fashion and textile manufacturing, photography and videography. The ability to create and circulate intellectual capital has the potential to generate wealth and employment and, at the same time, promoting social inclusion, cultural diversity and environmental respect. That is, the creative economy can contribute effectively to the achievement of the Sustainable Development Goals.

Considering all these reasons, the celebration of the first International Conference "Sustainability in Creative Industries", organized by IEREK, has become an ideal forum to share ideas and experiences on this topic from different countries. This book includes a selection of papers presented at that conference, after passing a review process. The studies presented in this book volume have been carried out using a great variety of methods and techniques.

The book is structured in three parts. The first part "Sustainable Entrepreneurship: Characterization, Analysis, and Impacts" consists of five chapters. Chapter "The Key Characteristics of Sustainable Entrepreneurs" is focused on the study of the personal traits and skills that shape sustainable entrepreneurs. Key characteristics of sustainable entrepreneurs are identified: altruistic, innovativeness, the ability to balance values, probity and self-compassion. Chapter "Analysis of Factors Affecting Entrepreneurial Intention Among Undergraduates" shows that perceived desirability, locus of control, self-efficacy and entrepreneurial education are factors affecting entrepreneurial intention among undergraduates from Indonesia. The study in Chapter "The Influence of Entrepreneurial Alertness on New Venture Performance with Networking Capability as a Moderator at Start-Up in Surabaya", by using Partial Least Square, finds a significant association between entrepreneurial alertness and the new venture performance, where the moderating variable networking capability acts as a predictor of moderation. Chapter "Fashionpreneur: Sustaining Traditional Batik Craft Through Entrepreneurship Activity Among Students at Universiti Malaysia Kelantan (UMK)" presents the main results of the project "Fashionpreneur" in Malaysia with the objective of tracking how many students from this project will venture into the batik business after graduating. Finally, Chapter "Mascot and Brand Sustainability in Pandemic Era: Systematic Literature Review" aims at presenting the results of a systematic literature review on the utilization of mascots for place branding from 2020 to 2022.

In both creative and non-creative industries, probably the most valued skill of a worker is their creativity. To do this, from the academic field, activities and evaluation systems must be designed to encourage students' creativity. The second part of this book, with eight chapters, focuses on "Innovative Approaches to Teaching and Pedagogy in Creative Design Education". Chapter "Explorative Learning Space for Developing Motoric Skills in the Early Childhood Stage" shares the experience of the construction of a multi-function space designated for children to support the growth and development of their motoric skills. Following the Vitruvius design principles, this design allows integrating both activities of playing and learning in one space to encourage the spirit of learning in children. Chapters "Utilizing

Virtual Reality to Support Teaching the Design Principles of the Life-Safety System" and "Virtual Personal Branding Education Workshop for GenIUS School Students Using Framework for Innovation Participatory Design Method" discuss the advantages of using virtual learning methods in architecture studios and in a private school in the context of the pandemic, respectively. Following the survey method, Chapter "The Impact of a Connectivist Learning Environment on Indonesian Design Students' Learning Experiences Through MOOC" analyzes the satisfaction of undergraduate students in Indonesia with MOOC as a methodology to foster the learning process during the COVID-19 pandemic and, more importantly, for the after-pandemic period in an increasingly digitized education context. Results show that this method is very well evaluated by the students. Dealing with a radical vision of education for sustainability, Chapter "Competences, Capabilities, and Skills in Teaching and Learning Fashion Design for Sustainability" illustrates the definition of competencies, capabilities, and acquired skills in teaching and learning fashion design for sustainability through a holistic approach based on the sustainability pillars: environment, economy, society and culture. Chapter "Project-Based Learning (PBL): Student Creativities in the Upcycling Projects" presents the results of the project-based learning focused on increasing student's creativity in project upcycling for fabric waste. Positive outcomes of the project are improvements in student creativity in the products made, originality in describing problems, and creativity in finding solutions. Chapter "Upcycling the Abandoned Students Artwork with Bateson's Type of Learning in Entrepreneurship Course" describes the results of a practical experience with artworks by students from the entrepreneurship course in a private university in West Java, Indonesia. Specifically, they reused and recycled previous assets from their colleagues or themselves and explored the possibility of making them suitable for certain targets using Gutman's Means-End theory. Finally, Chapter "Student Perception and Behavioral Changes in Blended Learning Implementation" investigates students' behavior and attitude toward blended learning by comparing first-time and second-time users.

The third and last part of this book, with five chapters, is focused on "Technological Advancements and Sustainability-Based Innovations in Creative Industries". Chapter "The Collaboration Between Academic and Industry in Creative Industry and Sustainability-Based Programs: The Academic Perspective" highlights the advantages of the collaboration between academic and creative industry in Indonesia based on sustainability programs. Chapters "Sound Visualization Based on Font Modification Using the Sound of Angklung" and "The Implementation of Deep Learning Technique in Mobile Application as a Preservation and Learning Media of Javanese Letter" show the application of technological advances such as artificial intelligence to foster the conservation of local heritages. More specifically, Chapter "Sound Visualization Based on Font Modification Using the Sound of Angklung" presents the sound visualization-based font modification, using the Sound of Angklung difficulty. In Chapter "The Implementation of Deep Learning Technique in Mobile Application as a Preservation and Learning Media of Javanese Letter", mobile application is developed using deep learning technique with VGG-16 convolutional neural network architecture, as the cutting-edge method of artificial intelligence, to recognize Javanese letter and convert it into Latin alphabet. Chapters "Web-Based Human Resource Information System Design AT PT. Cakra Mandala Sakti Surabaya", "User Experience Toward Sustainable Choice: Case Study ZALORA Indonesia" and Chapter "Web-Based Human Resource Information System Design AT PT. Cakra Mandala Sakti Surabaya" conduct case studies of creative innovations that foster economic sustainability. Chapter "Web-Based Human Resource Information System Design AT PT. Cakra Mandala Sakti Surabaya" presents a Web-based human resource information system to optimize the operational management process. Chapter "User Experience Toward Sustainable Choice: Case Study ZALORA Indonesia" explains a technique of automotive modelmaking that uses waste wood. Finally, Chapter "User Experience Toward Sustainable Choice: Case Study ZALORA Indonesia" aims at studying the intervention needed on the user experience aspect that could encourage more conscious shopping through the Zalora Earth Edit platform.

Granada, Spain

Angeles Sánchez

Acknowledgments We extend our gratitude to the authors of the research papers which were selected to be included in this book. We also express our appreciation to the reviewers who generously shared their expertise and provided valuable feedback, contributing to the overall quality of the manuscript. We would like to acknowledge the editors of this book for their insightful organization of the volume and their dedicated efforts in editing it professionally. Additionally, we would like to thank the IEREK team for their support in facilitating the publication of the outstanding research papers submitted to the conference.

Contents

Sustainable Entrepreneurship: Characterization, Analysis, and Impacts

The Key Characteristics of Sustainable Entrepreneurs

Devanny Gumulya

Abstract

Sustainable entrepreneurs play an essential role in solving environmental degradation and social problems that continuously threaten our livelihood. They destroy existing traditional production methods, products, market structures and consumption patterns, and replace them with products and services that are environmentally friendly and ethically produced. Sustainable entrepreneurs promote environmental regeneration, responsible resources management, and practice good governance. However, little is known about the characteristics that shape sustainable entrepreneurs. Based on entrepreneurship characteristics theory, the study aims to explore and uncover the personal traits and skills that shape sustainable entrepreneurs. Our paper shows that six key characteristics of sustainable entrepreneurs are altruistic, innovativeness, the ability to balance values, economic motives, and perceived capabilities, probity, self-compassion, and holistic. The study contributes to a better understanding of the inherent characteristics that promote sustainable entrepreneurship. The findings of the study can be used to develop a sustainable entrepreneurship program in higher education.

Keywords

Sustainable entrepreneurs · Entrepreneurial characteristics · Creative social enterprises

1 Introduction

Industrialization has had bring extensive economic growth but also bring negative effects on natural environment such as ozone depletion, climate change, and destruction of biodiversity (UNDP, 2019). However, some scholars advocate for entrepreneurship as a means of addressing environmental issues, because entrepreneurship is considered to have a significant role to play in social and economic development for a developing nation (Schumpeter, 2008; Shane & Venkataraman, 2000). Entrepreneurship is good for the economy because it brings innovation in the business process through the establishment of small and medium enterprises. These enterprises increase income distribution and create job opportunities. Through entrepreneurship activities, society becomes more economically independent (Khovrak, 2018). Hence, in recent years, the term “Sustainable Entrepreneurship” has emerged from the integration of entrepreneurship, environmental, and prosperity economics literatures.

Sustainable entrepreneurship is defined as “*The discovery, creation, and exploitation of opportunities to create future goods and services that sustain the natural and/or communal environment and provide development gain for others*” (Shepherd and Patzelt, 2011, p. 632). Sustainable entrepreneurship aims to preserve nature, life support systems, and communities while exploring potential opportunities to develop future products, processes, and services for gain. The opportunity is broadly interpreted, encompassing both economic and non-economic benefits for individuals, the economy, and society as a whole (Shepherd and Patzelt, 2011).

The entrepreneur drives sustainable entrepreneurship, which is known as the sustainable entrepreneur. It is defined as, “*An individual who starts a business to serve both self-interests and collective interests by addressing unmet social and environmental needs* (Hoogendoorn et al., 2019, p. 1133)”. Another scholar also defines SEs are different than common entrepreneurs, “*An individual that is*

D. Gumulya (✉)
Product Design Department, School of Design,
Pelita Harapan University, Tangerang, Indonesia
e-mail: devanny.gumulya@uph.edu

M. Nawaz Tunio et al. (eds.), *Sustainability in Creative Industries*,
Advances in Science, Technology & Innovation, https://doi.org/10.1007/978-3-031-48453-7_1

fundamentally equivalent to the entrepreneur but differs by motivation. Driven by not only the financial/economical gains of his/her activity but also a broader gain in achieving environmental goals (i.e., improvement of the environment, pollution output, etc.*)*" (Chadha & Koellinger, 2012, p. 7). From these two definitions, we can conclude several characteristics of SE is a person who starts a business with the intention of meeting unmet social and environmental needs.

Sustainable entrepreneurs help to alleviate social and environmental problems by meeting current needs without putting the availability of resources for future generations or the well-being of society at risk. The term sustainable entrepreneurs are often used interchangeably with terms, such as, Ecopreneur, Green Entrepreneur, and Social Entrepreneur. Figure 1 depicts the key differences between each entrepreneur: a green entrepreneur focuses on the economy and the environment, a social entrepreneur focuses on the economy and the social, and a sustainable entrepreneur strives to balance three key aspects of sustainability: the economy, the environment, and the social (Yang et al., 2014).

2 Literature Review

In the first section of the literature review, the distinction between characteristic and personality is defined, because these two terms are often used interchangeably. The second section of the literature review, the classic theory of entrepreneurs' characteristic, is elaborated. The third section discusses previous research about the factors affecting SE's characteristic. The fourth section discusses the SE's characteristic and the challenges of becoming SEs, and finally, we compare the traditional and Sustainable entrepreneurs.

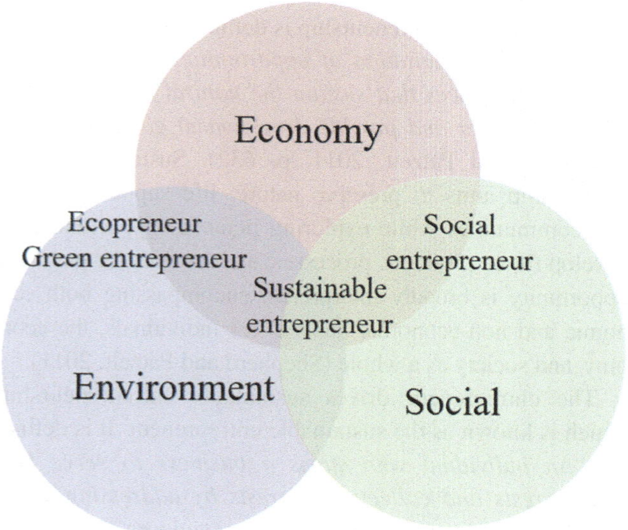

Fig. 1 Three dimensions of sustainability and their intersections. *Source* Adapted from Yang et al. (2014)

3 The Difference Between Entrepreneurs' Characteristics and Personality

In the entrepreneurship literature, the terms characteristic and personality often used interchangeably. Karimova and Nash (2018) defines the difference between the two concepts. First, character is the moral qualities and features that comes from one's core beliefs. It is distinctive and typical to each individual. While personality is a person inherent quality that comes from individual's distinctive character that make people behave in every situation. According to Banicki (2017), the concept of personality is frequently associated with a significant superficiality and the shallowness of just a appearances, whereas the concept of character is frequently perceived as more in-depth. Based on this definition, the study focuses on entrepreneur's characteristic.

4 Entrepreneur's Characteristics

The role of an entrepreneur is a leading force in the success of an entrepreneurial enterprise. Entrepreneurs need specific characteristics to make them more successful. Characteristic of the entrepreneur defined as "personal traits and skills that constitute the competence of the entrepreneur needed to be successful." (Abdulwahab & Al-damen, 2015).

Scholars have mentioned several entrepreneur's characteristics that commonly included in the study of traditional entrepreneurs (Caird Chye Koh, 1996; Gartner, 1989; Hansemark, 2003; Hwee Nga & Shamuganathan, 2010; Karimi et al., 2012; McCelland, 1971; Paunonen & Jackson, 1996; Sally, 1991):

1. Risk Propensity defined as "a personality trait involving the willingness to pursue decisions or courses of action involving uncertainty regarding success or failure outcomes" (Paunonen & Jackson, 1996). It is suggested that risk-bearing is a key factor that distinguishes entrepreneurs from managers, and others suggested that the entrepreneurial function primarily involves risk measurement and risk-taking. Risk-taking is considered a very important characteristic of entrepreneurship (Mill, 2004).
2. Need for achievement can be defined as "behavior toward competition with a standard of excellence" (McCelland, 1971). In other words, "the expectations of doing something better or faster than anybody else or better than the person's own earlier accomplishments" (Hansemark, 2003). People with a high "need for" are

believed to have a strong wish to succeed and are thus more likely to be entrepreneurial achievement.

3. Need for independence can be defined as “the need to do and say as one likes despite conventional expectations” (Caird Sally, 1991). For entrepreneur, the need of independence is so important that they want to run their own businesses, be their own bosses, be independent and flexible, and they don’t want to work for any company.
4. Therefore, many people are leaving their job and become entrepreneurs. They are tired of working for an employer. Therefore, they developed their own business. Entrepreneurs find themselves motivated by freedom and flexibility. They want to be their own boss and be independent.
5. Entrepreneurial self-efficacy is “individuals’ self-perceptions of their skills and abilities to accomplish a specific course of action within a given domain or achieve a desired outcome” (Bandura, 1997). According to Markman et al. (2002), it is “the perceptions of self-efficacy, rather than objective ability that motivate individuals to demonstrate entrepreneurial behavior”. Previous study has suggested that empirical studies in the entrepreneurship literature have found entrepreneurs to have a higher level of self-efficacy than non-entrepreneurs (MacKo & Tyszka, 2009).
6. “Locus of control” is defined as the degree to which individuals believe they can control or influence events in their lives (Boag & Tiliopoulos, 2011). While people with an internal control locus believe they can control life events, people with an external control locus believe that the events of life derive from external influences such as fortune, chance, and destiny. It also have been suggested that locus of control could distinguish successful entrepreneurs from unsuccessful ones (Carsrud & Brännback, 2009). Generally, entrepreneurs are believed to prefer internal control rather than external factors (Mitton, 1989).
7. Tolerance of ambiguity: When there is insufficient information to make a decision, it becomes ambiguous. It is defined as “a person’s level of uncertainty and tolerance toward ambiguity” (Chye Koh, 1996). A person who has high tolerance of ambiguity is one who finds ambiguity stimulating and enjoys facing the trying and unfamiliar tasks. Entrepreneurs are not deterred by the uncertainty of the marketplace, but willingly seek out and manage risk. It is believed that those who are entrepreneurially inclined show more tolerance of ambiguity than others (Fellnhofer, 2017).
8. Self-confidence: Given the general notion of an entrepreneur as a person who prefers to engage in his/her own business, an entrepreneur must be convinced that he/she can achieve the objectives he/she sets. An entrepreneur, in other words, should be aware of his/her self-esteem and skill that are aligned with his/her business affairs (Robinson et al., 1991). It has been suggested that autonomy is a necessary entrepreneurial characteristic and is linked to the psychological features such as internal control, risk-taking propensity and ambiguity tolerance. Empirical studies in the literature of entrepreneurship have found that entrepreneurs are more confident than the non-entrepreneurs (Chye Koh, 1996; Dinis et al., 2013).
9. Innovativeness: It is defined as “perceiving and acting on business activities in new and unique ways” (Robinson et al., 1991). It involves the perception and performance in new and unique forms of business activities (Robinson et al., 1991). Evidence from literature on entrepreneurship shows that entrepreneurs are much more innovative than the non-entrepreneurs (Otilia, 2019).

5 Factors Affecting Entrepreneur’s Characteristics

In this section, several factors that influence the development of entrepreneur’s characteristic are discussed.

- **Organization Culture**: Selamat et al. (2018) state that entrepreneur characteristics are influenced by organizational culture, which is a collection of values, expectations, and practices that guide and advise the actions of all members within a company, and it affects the company’s overall performance.
- **Demographic**: Trihopoulou (2005) states that the characteristic of human population like income, age, race, ethnicity, gender, marital status, income, education, and employment affect entrepreneurs’ characteristic.
- **Capital Endowments**: According to Kerr et al. (2017), a person’s entrepreneurial traits are influenced by their knowledge, skills, education, formal, informal, and business experience.
- **Entrepreneurial Regions**: In addition, Kerr et al. (2017) also state that region affects entrepreneurship’s characteristics because some regions have certain legacies that favor or hinder entrepreneurship.
- **Entrepreneurial Orientation**: According to Sari and Rahmantika (2018), the entrepreneurial orientation influences how entrepreneurs define success. Entrepreneurial orientation defines how executives develop strategies to pursue new opportunities, and thus influences the characteristics of entrepreneurs.

6 Sustainable Entrepreneur's Characteristic

Based on a review of the literature on sustainable entrepreneurship, I propose that sustainable entrepreneurs possess six inherent qualities that come from their unique situation of starting a business to serve both self-interests and collective interests by trying to address unmet social and environmental needs.

1. **Altruistic**

The importance of helping others and surroundings as well as care for the environment.

Previous research has highlighted the importance of being altruistic in driving behavior toward solving social problems and positive environmental attitudes (Dees, 2012; Kai Hockerts, 2015). In addition, being altruistic is a form of self-transcendence. Individuals with strong self-transcendent values have a more altruistic orientation and are more willing to engage in pro-environmental behaviors; whereas individuals with high self-enhancement values are more self-interested and are less inclined to participate in pro-environmental behaviors (Schultz, 2002). Another study found that people with self-enhancement are more likely to engage in pro-environmental behaviors if they see personal benefits, such as self-improvement, from doing good to the environment (De Dominicis et al., 2017).

2. **Innovativeness**

The societal and environmental issues are frequently difficult, and the customary solutions do not appear to be effective. The issues are complicated because they are frequently overlooked, neglected, or resolved ineffectively by governments, incumbent businesses, or civil society organizations (Hockerts & Wüstenhagen, 2012). As a result, SEs need to have a strong willingness to transform the industry in which they operate by being inventive and creative to bring innovation to the market (Kai Hockerts, 2015).

3. **The Ability to Balance Values, Economic Motives, and Perceived Capabilities**

SEs have triple goal setting in sustainable entrepreneurship. Shepherd and Patzelt (2011, p. 137) formulate this so-called multiple bottom line as "the preservation of nature, life support, and community in the pursuit of perceived opportunities to bring into existence future products, processes, and services for gain, where gain is broadly construed to include economic and non-economic gains to individuals, the economy, and societies". SEs earn a living and want to be successful while also wanting to have a positive influence on societal and environmental studies, even if it means foregoing personal economic gain. Because SEs must create triple value, they must have a strong belief that they possess the necessary skills and knowledge to become entrepreneurs. This is defined as perceived capability (Vuorio et al., 2018). According to Hoogendoorn et al. (2019), SEs seek both economic and non-economic gains, so they are more willing to take risks than regular entrepreneurs.

4. **Probity**

There are numerous conflicts between value creation (for society) and value capture (for private gain). For example, if SEs provide a solution to a deprived community, the capacity for value capture is clearly limited because the communities typically lack the means to pay for the value creation. To tackle this kind of dilemma, SEs must have a strong moral compass, as well as decency and honesty (Chadha & Koellinger, 2012). Furthermore, in order to achieve their intended entrepreneurial outcomes, SEs must leverage both their formal and informal relational resources in order to solve the complex issues of sustainability. SEs can leverage their relational resources if they maintain their reputation, probity, and credibility.

5. **Self-compassion**

It is a three-part inner process that begins by noticing other people's suffering, progresses to developing affective feelings toward the suffering, and ends with action (Dutton et al., 2014). As Engel et al. (2019) explain that when compassion is stimulated, selfish concerns seems rather less significant, and the rewards of taking actions to alleviate the suffering of others are more highly valued. Another study by Yitshaki et al. (2022) claims that SEs' compassion arises from their own experiences, as well as a deep understanding of what people with similar encounters feel and need. Hence self-compassion plays a critical role in encouraging SEs to perform prosocial behavior.

6. **Holistic**

SEs don't see people, profit, planet as three separate things, but seeing them as one whole. A profit should bring benefit to the planet and society as well (Kai Hockerts, 2015). In addition, the holistic characteristic enables SEs to recognize and evaluate all relevant (sub)systems across multiple domains (people, planet, profit), as well as their boundaries (Ploum et al., 2018).

7 Challenges in Becoming Sustainable Entrepreneurs

SEs capitalize on opportunities created by unaddressed social and environmental issues. They combine self-interest with group interests. Their goals are broader than those of most entrepreneurs. This adds to the complexities that SEs must deal with; I present four challenges that SEs may face in their business processes.

1. **Trade-Offs**

In order to remain competitive in the market, SEs sometimes need to some compromises and trade-offs. For instance, selecting organic base material will result in an increase in the cost of the product and will make it more difficult for you to compete in the market. Therefore, in order to be a sustainable entrepreneur, it is critical to have the necessary sense, ethics, and experience. They help SEs to strike a balance in day-to-day business decisions between people, profits, and the environment (Muñoz, 2018).

2. **Lack of Support**

Sustainable entrepreneurs present a unique opportunity to effect social change and bring about a sustainable society from the bottom-up rather than the top-down. For more entrepreneurs to scale the heights of the sustainable entrepreneurship model, they need external support especially in funding. They need funding to achieve collective effectiveness through collaborative partnerships and cooperation, and they also need to gain support from both national and local governments, as well as research institutes (Jegede et al., 2006). The lack of financial support can be attributed to the low number of social investors in developing countries, particularly in Indonesia, as well as the low number of SE enablers (Siregar et al., 2015). In addition to lack of financial support, another barrier that may hinder SEs is lack of community support. SEs to create impact, they need to work with certain community if they don't support and trust SEs it will be hard for SEs to do their business.

3. **Lack of Innovation Capability**

As mentioned earlier, innovativeness is important in SEs, because environmental and social problem are complex issues that require the ability to innovate (Bosma et al., 2015). Unfortunately, innovation capability in firms developing countries is specified by insufficient knowledge, lack of organizational facilities, absence of R&D activities, and poor cooperation with key actors, and lack of funding for innovation (Tesfaye & Kitaw, 2018). In regard to this barrier, Tesfaye and Kitaw (2018) in their research suggest that SEs in developing country should focus on developing three key constructs which are knowledge accumulation, knowledge application and fund generation.

5. **Lack of Knowledge**

Innovation capability is driven by knowledge. SEs work under circumstances of market imperfections and complex environmental and social issues, so a broad knowledge base is needed. To overcome these challenges, sustainable entrepreneurs should invest in the acquisition of external knowledge, collaboration, and the establishment of internal knowledge. SEs should possess more advanced networking skills and a greater reliance on social capital to acquire knowledge than regular entrepreneurs (Hoogendoorn et al., 2019).

From the existing studies, it is found that the study about the sustainable entrepreneur's characteristic is still very limited. It is worth noting that all studies have been conducted in, mainly, Western countries; no studies have so far tried to explain the sustainable entrepreneurs' characteristics in the Indonesian context. Our paper's aim is to fill this gap. Hence, we raise the questions of, what are the personal traits of sustainable entrepreneurs that drive them to discover, create, and exploit opportunities from unmet social and environmental needs? What are the creative skills of sustainable entrepreneurs that allow them to create future goods and services that help to sustain the natural environment while also providing development gain for others?

8 Comparing Traditional and Sustainable Entrepreneurs' Characteristic

From the literature review, we can see that SEs have some traditional entrepreneurs' characteristics such as innovativeness and risk propensity. Therefore, I argue that when compared to regular entrepreneurs, SEs must be more innovative and have a higher risk propensity than traditional entrepreneurs because they are operating under conditions of market failure regarding environmental and societal challenges.

SEs, in addition to being more innovative and willing to take risks, have additional qualities related to their personal values such as altruism and probity, which allow them to be more compassionate about the challenges in their surroundings and more ethical in trying to balance concerns for environmental and economic sustainability. I also argue that SEs' ability to make sustainable business decisions is also shaped by their holistic thinking feature. They see people, profit, and the planet as a unified system that is inseparable from one another and do not exist independently.

9 Conclusion and Implication

Overall, our paper summarized and detailed traditional entrepreneur characteristics, as well as the factors that influence the entrepreneurs' characteristics. After understanding the traditional context, I conducted additional research on the distinct characteristics and characteristics of sustainable entrepreneurs. They have these characteristics because they serve both personal and collective interests by addressing unmet social and environmental needs. Furthermore, I identified the challenges of becoming a sustainable entrepreneur and, finally, I compared the characteristics of traditional and sustainable entrepreneurs. In our view, mapping the characteristics of sustainable entrepreneurs is a useful issue, but it remains largely vague and unresearched, with little theoretical and empirical evidence. This is clearly a new and promising field for researchers. Further research should find empirical evidence on each trait and support more descriptive data.

As a result, it appears that our findings and outcomes—the six characteristics of sustainable entrepreneurs—can be used to develop a university-based sustainable entrepreneurship program. Because becoming a successful sustainable entrepreneur does not happen overnight, knowing what characteristics to cultivate allows teachers to stimulate and foster these traits in various types of learning activities. We propose lectures should use project-based learning to foster these characteristics in their entrepreneurship programs. Students should be challenged to start a business with limited resources and tools. Businesses that address sustainability issues within the university and its surroundings. Throughout the project, they will learn how to be innovative, debate and discuss putting their probity and altruist moral values into practice and take action to demonstrate their compassion for the mission. By the end of the project, the students should have a better understanding of what it takes to be a successful entrepreneur.

Acknowledgements We thank our Faculty of Design Dean Dr. Martin L. Katoppo S.T.M.T., and the Chairman of Center for Research and Community Development (CRD) Dr.-Ing.Ihan Martoyo, S.T., M.Sc. This article is a research publication registered in Center for Research and Community Development (CRD) of Pelita Harapan University under number 129/LPPM-UPH-VI/2022.

References

Abdulwahab, M. H., & Al-damen, R. A. (2015). The impact of entrepreneurs' characteristics on small business success at medical instruments supplies organizations in Jordan Amman Arab University Business Administration Department Amman Arab University. *International Journal of Business and Social Science, 6*(8), 164–175.

Bandura, A. (1997). Theoretical perspectives: The nature of human agency. In *Self-efficacy: The exercise of control* (p. 3). http://search.ebscohost.com/login.aspx?direct=true&db=a9h&AN=9703260522&site=ehost-live

Banicki, K. (2017). The character–personality distinction: An historical, conceptual, and functional investigation. *Theory and Psychology, 27*(1), 50–68. https://doi.org/10.1177/0959354316684689

Boag, S., & Tiliopoulos, N. (2011). Personality and individual differences: Theory, assessment, and application. In *Personality and individual differences: Theory, assessment, and application.*

Bosma, N., Schøtt, T., Terjesen, S., & Kew, P. (2015). Global Entrepreneurship monitor special topic report: Social entrepreneurship. *Global Entrepreneurship Monitor* (pp. 1–44). www.gemconsortium.org

Carsrud, A. L., & Brännback, M. (2009). Understanding the entrepreneurial mind. *Understanding the Entrepreneurial Mind.* https://doi.org/10.1007/978-1-4419-0443-0

Chadha, A., & Koellinger, P. (2012). Characterizing sustainable entrepreneurship. In *BSc. International bachelor economics & business economics (IBEB)* (Vol. 305972). https://thesis.eur.nl/pub/11872/BA-thesisSA.Chadha-Final.pdf

Chye Koh, H. (1996). Testing hypotheses of entrepreneurial characteristics: A study of Hong Kong MBA students. *Journal of Managerial Psychology, 11*(3), 12–25. https://doi.org/10.1108/02683949610113566

De Dominicis, S., Schultz, P. W., & Bonaiuto, M. (2017). Protecting the environment for self-interested reasons: Altruism is not the only pathway to sustainability. *Frontiers in Psychology, 8*(June), 1–13. https://doi.org/10.3389/fpsyg.2017.01065

Dees, J. G. (2012). A tale of two cultures: Charity, problem solving, and the future of social entrepreneurship. *Journal of Business Ethics, 111*(3), 321–334. https://doi.org/10.1007/s10551-012-1412-5

Dinis, A., do Paço, A., Ferreira, J., Raposo, M., & Rodrigues, R. G. (2013). Psychological characteristics and entrepreneurial intentions among secondary students. *Education and Training, 55*(8–9), 763–780. https://doi.org/10.1108/ET-06-2013-0085

Dutton, J. E., Workman, K. M., & Hardin, A. E. (2014). Compassion at work. *Annual Review of Organizational Psychology and Organizational Behavior, 1*, 277–304. https://doi.org/10.1146/annurev-orgpsych-031413-091221

Engel, Y., Ramesh, A., & Steiner, N. (2019). Powered by compassion: The effect of loving-kindness meditation on entrepreneurs' sustainable decision-making. *Journal of Business Venturing.* https://doi.org/10.1016/j.jbusvent.2019.105986

Fellnhofer, K. (2017). Drivers of innovation success in sustainable businesses. *Journal of Cleaner Production, 167*, 1534–1545. https://doi.org/10.1016/j.jclepro.2017.08.197

Gartner, W. B. (1989). Some suggestions for research on entrepreneurial traits and characteristics. *Entrepreneurship Theory and Practice, 14*(1), 27–38. https://doi.org/10.1177/104225878901400103

Hansemark, O. C. (2003). Need for achievement, locus of control and the prediction of business start-ups: A longitudinal study. *Journal of Economic Psychology, 24*(3), 301–319. https://doi.org/10.1016/S0167-4870(02)00188-5

Hockerts, K. (2015). Determinants of social entrepreneurial intentions. *Entrepreneurship: Theory and Practice, 41*(1), 105–130. https://doi.org/10.1111/etap.12171

Hockerts, K., & Wüstenhagen, R. (2012). When David meets Goliath: Sustainable entrepreneurship and the evolution of markets. *Social Innovation.* https://doi.org/10.1057/9780230367098_12

Hoogendoorn, B., van der Zwan, P., & Thurik, R. (2019). Sustainable entrepreneurship: The role of perceived barriers and risk. *Journal of Business Ethics, 157*(4), 1133–1154. https://doi.org/10.1007/s10551-017-3646-8

Hwee Nga, J. K., & Shamuganathan, G. (2010). The influence of personality traits and demographic factors on social entrepreneurship start up intentions. *Journal of Business Ethics, 95*(2), 259–282. https://doi.org/10.1007/s10551-009-0358-8

Jegede, O. O., Jegede, O. E., & Mustapha, N. (2006). Informal sector measurement of openness, collaboration and innovation: The case of Otigba hardware microenterprises cluster in Africa. In *Business history* (Vol. 48, Issue January). http://www.oxfordhandbooks.com/oso/public/content/oho_business/9780199286805/toc.html

Karimi, S., Biemans, H. J. A., Lans, T., Arasti, Z., Chizari, M., & Mulder, M. (2012). Application of structural equation modelling to assess the effect of entrepreneurial characteristics on students' entrepreneurial intentions. *SSRN Electronic Journal.* https://doi.org/10.2139/ssrn.2152932

Karimova, H., & Nash, J. (2018). *Personality & character traits: The good, the bad and the ugly*. https://positivepsychology.com/character-traits/

Kerr, P. S., Kerr, W. R., & Xu, T. (2017). Personality traits of entrepreneurs: A review of recent literature. In *Harvard business school* (Vol. 9, Issue 33).

Khovrak, I. (2018). Social entrepreneurship and its role in sustainable development: Regulation, funding and trust building. *Studia i Materiały*. https://www.ceeol.com/search/article-detail?id=775116

MacKo, A., & Tyszka, T. (2009). Entrepreneurship and risk taking. *Applied Psychology, 58*(3), 469–487. https://doi.org/10.1111/j.1464-0597.2009.00402.x

Markman, G. D., Balkin, D. B., & Baron, R. A. (2002). Inventors and new venture formation: The effects of general self-efficacy and regretful thinking. *Entrepreneurship Theory and Practice, 27*(2), 149–165. https://doi.org/10.1111/1540-8520.00004

McCelland, D. (1971). *The achieving society* (p. 6).

Mill, J. S. (2004). *Principles of political economy (Abridged): With some of their applications to social philosophy*. http://books.google.nl/books?id=kxSw72qiV7YC%5Cn, http://files/5969/Mill-2004-PrinciplesofPoliticalEconomy(Abridged)WithSomeofTheirApplicationstoSocialPhilosophy.pdf

Mitton, D. G. (1989). The complete entrepreneur. *Entrepreneurship Theory and Practice, 13*(3), 9–20. https://doi.org/10.1177/104225878901300303

Muñoz, P. (2018). A cognitive map of sustainable decision-making in entrepreneurship. *International Journal of Entrepreneurial Behavior and Research*. https://doi.org/10.1108/IJEBR-03-2017-0110/full/html

Otilia, M. P. E. (2019). Entrepreneurship, a determinant factor in the sustainable development of small businesses. *Jurnal Perspektif Pembiayaan Dan ....* http://online-journal.unja.ac.id/JES/article/view/6134

Paunonen, S. V., & Jackson, D. N. (1996). The Jackson personality inventory and the five-factor model of personality. *Journal of Research in Personality, 30*(1), 42–59. https://doi.org/10.1006/jrpe.1996.0003

Ploum, L., Blok, V., Lans, T., & Omta, O. (2018). Toward a validated competence framework for sustainable entrepreneurship. *Organization and Environment*. https://doi.org/10.1177/1086026617697039

Robinson, P. B., Stimpson, D. V., Huefner, J. C., & Hunt, H. K. (1991). An attitude approach to the prediction of entrepreneurship. *Entrepreneurship Theory and Practice, 15*(4), 13–32.

Sally, C. (1991). Testing enterprising tendancy in occupational groups. *British Journal of Management, 2*(July), 177–186.

Sari, I. A., & Rahmantika, D. N. (2018). The effect of firm characteristics and entrepreneur characteristics on the success of Msme business in Central Java Province. *Jurnal Manajemen: Dayasaing, 23*(1), 26–35.

Schultz, P. W. (2002). Inclusion with nature: The psychology of human-nature relations. *Psychology of Sustainable Development*, 61–78. https://doi.org/10.1007/978-1-4615-0995-0_4

Schumpeter, J. A. (2008). *Capitalism, socialism, and democracy*. Harper Perennial Modern Thought.

Selamat, F., Tunjungsari, H. K., Chairy, & Utama, D. W. (2018). Entrepreneurial characteristics amongst different professional backgrounds: Evidence from Indonesia. *International Journal of Business, 2*(1), 25–32. https://doi.org/10.32924/IJBS.V2I1.30

Shane, S., & Venkataraman, S. (2000). The promise of enterpreneurship as a field of research. *The Academy of Management Review, 25*(1), 217. https://doi.org/10.2307/259271

Shepherd, D. A., & Patzelt, H. (2011). The new field of sustainable entrepreneurship: Studying entrepreneurial action linking "what is to be sustained" with "what is to be developed." *Entrepreneurship: Theory and Practice, 35*(1), 137–163. https://doi.org/10.1111/j.1540-6520.2010.00426.x

Siregar, H., Tampubolon, N., & Yulius. (2015). *The art of sustainable giving*. http://image-src.bcg.com/The-Art-of-Sustainable-Giving-May-2015_tcm93-40480.pdf

Tesfaye, G., & Kitaw, D. (2018). An innovation capability development process for firms in developing countries: A theoretical conceptual model. *Journal of Entrepreneurship, Management and Innovation, 14*(3), 87–110. https://doi.org/10.7341/20181434

Trihopoulou, K. S. A. (2005). Female entrepreneurs' personal characteristics and motivation: A review of the Greek situation. *Women in Management Review, 20*(1), 24–36. https://doi.org/10.1108/09649420510579559

UNDP. (2019). *Creative industries and sustainable development* | Sustainable Development Goals Fund. https://www.sdgfund.org/creative-industries-and-sustainable-development

Vuorio, A. M., Puumalainen, K., & Fellnhofer, K. (2018). Drivers of entrepreneurial intentions in sustainable entrepreneurship. *International Journal of Entrepreneurial Behaviour and Research, 24*(2), 359–381. https://doi.org/10.1108/IJEBR-03-2016-0097

Yang, M., Vladimirova, D., Rana, P., & Evans, S. (2014). Sustainable value analysis tool for value creation. *Asian Journal of Management Science and Applications, 1*(4), 312. https://doi.org/10.1504/ajmsa.2014.070649

Yitshaki, R., Kropp, F., & Honig, B. (2022). The role of compassion in shaping social entrepreneurs' prosocial opportunity recognition. *Journal of Business Ethics, 179*(2) (Springer, Netherlands). https://doi.org/10.1007/s10551-021-04860-x

Analysis of Factors Affecting Entrepreneurial Intention Among Undergraduates

Justin Wijaya and Carolina Novi Mustikarini

Abstract

Learning about entrepreneurship is very widespread in the world of education, one of which is at the level of higher education, especially in Indonesia. This can be seen from the existence of one of the Independent Learning Campus (MBKM) programs initiated by the Ministry of Education and Culture (Mendikbud) regarding entrepreneurship. This program encourages campuses to integrate entrepreneurship courses into their curricula. During the process of forming this curriculum, campuses need to know the factors that encourage the desire to be entrepreneurial. This study aims to find out what factors encourage entrepreneurial intentions in undergraduate students. The number of respondents in the study amounted to 30 undergraduate students from province of South Sulawesi and East Java, Indonesia. This research method is a quantitative approach which is analyzed using PLS. Results of this study show that perceived desirability, locus of control, self-efficacy, and entrepreneurial education (universities courses support) are factors affecting entrepreneurial intention among undergraduates from province of South Sulawesi and East Java, Indonesia.

Keywords

Entrepreneurial intention · Perceived desirability · Self-efficacy · Entrepreneurial education · Locus of control

J. Wijaya (✉) · C. N. Mustikarini
Sekolah Tinggi Ilmu Ekonomi Ciputra, Makassar, Indonesia
e-mail: justin.wijaya@ciputra.ac.id

C. N. Mustikarini
e-mail: cmustikarini@ciputra.ac.id

1 Introduction

The fourth-most populous country in the world is Indonesia with the population which is estimated at 272 million people. From that population, the number of labor force (age between 15 and 60 years) is estimated to be more than 144 million people (Kusnandar, 2022). Statistics Indonesia (BPS) data from February 2022 shows the unemployment rate of 5.83%, down 0.43% from February 2021.

One of the ways to reduce unemployment rate is developing entrepreneurial intention since the early stage. Entrepreneurial intention can be upgraded through entrepreneurship education. Learning about entrepreneurship is very widespread in the world of education, one of which is at the level of higher education, especially in Indonesia. According to Liñán (2004), there are three types of entrepreneurship education program, which is entrepreneurship basic education, entrepreneurship dynamic education, and continuous education for entrepreneurs. Entrepreneurship education in Indonesian universities has not been linked to the basic principles of economic management as mandated in the 1945 Constitution of the Republic of Indonesia. The spirit contained in Articles 33 and 34 is related to entrepreneurship, namely prioritizing the interests of all parties based on the spirit of mutual cooperation and compassion in human relationships with humans and the environment. Increasing welfare and prosperity to achieve a sustainable quality of life is the goal of national economic development (Witjaksono, 2016).

To overcome these problems, it is necessary to introduce an entrepreneurship education learning model that can stimulate student interest in entrepreneurship. Indonesian Ministry of Education and Culture through the Directorate of Learning and Student Affairs and the Directorate General of Higher Education always striving to develop and increase the number of entrepreneurial students. The government realizes the importance of student entrepreneurship programs in higher education and is realized through

M. Nawaz Tunio et al. (eds.), *Sustainability in Creative Industries*,
Advances in Science, Technology & Innovation, https://doi.org/10.1007/978-3-031-48453-7_2

the Independent Learning Independent Campus Policy (MBKM) which encourages the development of student entrepreneurial interests with appropriate learning activity programs. The program encourages campuses to include courses on entrepreneurship in their curriculum.

The purpose of this study is to show the results of an exploratory study to find the factors that influence entrepreneurial intentions conducted on undergraduate students in Indonesia from two provinces, namely South Sulawesi and East Java.

2 Literature Review

2.1 Definitions of Entrepreneurship, Entrepreneur, Intention, and Entrepreneurial Intention

The term "entrepreneurship" was initially proposed by a French economist named Cantillon in 1755 (Hisrich et al., 2005). The verb Entreprendre, which means to begin or to carry out, is the root of the word (Parker, 2009). Since then, other academics have attempted to define the word in a variety of ways from various angles. More than 70 alternative definitions of "entrepreneurship" were found by Morris et al. (1994). The majority of definitions provided have to do with starting a new business or working for oneself. Other versions place more emphasis on a variety of tasks, such as developing organizations, combining resources in novel ways, investigating opportunities, and accepting risks. Additionally, "entrepreneurship" is a creative activity that entails utilizing readily available resources to generate profit in a novel way, according to Drucker (1985). "Entrepreneurship" is the spirit of creation and innovation, according to Schumpeter (1934). According to Bygrave and Timmons (1992), "entrepreneurship" is the act of generating or grasping chances and pursuing them despite available resources. According to Shane and Venkataraman (2000), "entrepreneurship" refers to the identification, seizing, and utilization of opportunities that lead to the development of novel goods, novel services, or novel production processes, novel organizational forms, novel strategies, and novel markets for goods and inputs that have not previously existed. From the perspectives of entrepreneurship, it can be concluded that entrepreneurship is the beginning of a new business process. The procedure calls for a great deal of creativity and innovation to create or grasp business chances, utilizing available resources, and taking risks by engaging in commercial initiatives with the goals of achieving money, independence, or supplying some values to society.

An entrepreneur is characterized as a person who carries the inventiveness to collect resources in original manners and ready to take the risk and/or uncertainty related to the activity (Hisrich et al., 2012). An entrepreneur is portrayed as somebody who expresses creative thinking and initiative, turns resources and situations to practical account with his or her ability to organize social and economic mechanisms, and accepts the associated risk and failure (Hisrich, 1990).

Parker (2004) defines "intention" as a particular propensity shared by all people to carry out an activity or series of activities. A conscious thought process that can direct conduct produces intention. According to Bird (1998), "intention" can be characterized as a mental state that focuses attention and behaviors toward a certain objective or method of achieving something.

The researchers cited above take a variety of perspectives when looking at the term "entrepreneurship." Consequently, it follows that the phrase "entrepreneurial intention" has a wide range of applications. The majority of studies, however, view "entrepreneurial intention" as a mental state that shapes entrepreneurial activities before they take place. In 1993, Krueger defined "entrepreneurial intention" as the determination to launch a firm. This commitment reveals a person's capacity or propensity to launch a new company in the future. Kruger noted that entrepreneurial intention can be seen as a cognitive process that is recognized just before the act of beginning a firm in another study up until the year 2005. Thompson (2009) further suggested that entrepreneurial intention should be seen as a deliberate identification of mind that aims to promote the behaviors required to launch a firm.

2.2 Perceived Desirability

"Perceived Desirability" was defined by Shapero and Sokol in 1982 as the allure of beginning a business with each topic, or the degree of interest of an individual, or the attitude of an individual toward start-ups.

2.3 Entrepreneurial Education (Support of University Courses)

Liñán (2004) identified four types of entrepreneurship education programs. The first type is entrepreneurship education, which aims to broaden knowledge about entrepreneurship and influence relationships that can influence intentions. The second is start-up training, usually designed for people with an entrepreneurial idea and have a need to solve practical questions about how to become an entrepreneur. Third, the Entrepreneurship Dynamics training is aimed for people that are already an entrepreneur and want to promote dynamic behavior after the launch phase. The fourth "Lifelong Learning for Entrepreneurs" describes a lifelong learning program and is aimed at experienced entrepreneurs. Along with various types of entrepreneurship education, the entrepreneurship education curriculum includes (i) emphasis on the role of entrepreneurship programs for individuals and society, (ii) development of a multimedia environment or curriculum, (iii) study of content and implementation of entrepreneurship programs, (iv) focus on the individual need of participants in entrepreneurship programs (Bechard et al., 2002). According to Ratten and Usmanij (2021), entrepreneurship education is an important way government can encourage economic growth in their regions. More resources have poured into entrepreneurship education programs particularly those focusing on early stage business start-ups. However, the rate of new business success is low and could be improved with more target policy initiatives. For the sustainable entrepreneurship in the region, the role of universities that educate, train, and prepare graduates to become masters of their own careers is vital. Despite the several challenges and hurdles in the entrepreneurial process, the educated youths' inspiration to be self-reliant and independent is highly valuable for their unmatched endeavors into sustainable entrepreneurship in developing regions (Tunio et al., 2021).

2.4 Entrepreneurial Self-efficacy

The foundation of self-efficacy is social cognitive theory (Bandura, 1997). Human conduct is "a consequence of the interaction of intrapersonal factors, the activity individuals engage in, and the environmental forces that press upon them," according to Bandura (2012). One's perceptions about one's capacity to successfully carry out a certain conduct in a particular environment and one's expectations regarding the results of the behavior are shaped by the interplay between these aspects (Bandura, 1999). Self-efficacy influences how people complete their current task and determines how they will accomplish their future task goals. It is both the cause and the effect of an action decision (Bandura, 2000). Self-efficacy beliefs have a significant impact on how people choose an activity despite having alternatives, how much effort they put in to carry it out, how persistent they are in the face of obstacles and problems, and how successful they are at carrying out the action (Bandura, 1997; Dwyer & Cummings, 2001). According to Bandura (2012), self-efficacy has the greatest influence on behavior since it affects it both directly and indirectly through its effects on other processes and elements like goal-setting, outcome expectancies, and perceptions of environmental facilitators and barriers. Self-fundamental efficacy's influence on human behavior has inspired scholars to use the idea in a variety of contexts, including business.

2.5 Previous Empirical Studies

The first study was conducted by Pihie and Bagheri (2013) and used Bandura's structural path model for components to examine the relationship between entrepreneurial self-efficacy, self-regulation, and entrepreneurial intention. 722 Malaysian university students from the public and private sectors made up the research sample. The findings showed that the most important and beneficial factor influencing someone's decision to become an entrepreneur is their level of entrepreneurial self-efficacy. Particularly, both directly and indirectly, students' entrepreneurial inclination is significantly influenced by entrepreneurial self-efficacy. The association between students' entrepreneurial intention and entrepreneurial self-efficacy is also partially mediated by self-regulation. These findings' implications for entrepreneurial research and instruction are highlighted.

The second study, by Bui et al. (2020), tries to pinpoint the factors that influence graduates of the National Economics University's entrepreneurial intention. They conducted a questionnaire survey utilizing a convenience sampling technique on 250 full-time third-year undergraduates at the National Economics University in order to determine the sample size. 150 of the 250 undergraduates were studying business administration, and 100 were studying economic management. 215 replies—93.07% of the total—were used for the data analysis process after few invalid responses were discarded for lack of information or poor information. According to research findings, the element "Perceived desirability" has the greatest influence on a person's intention to launch a business. The next element was "Perceived Feasibility," which found that "ability to search and prepare for start-up" had a larger impact on undergraduates' intention to start their own business than "leadership and capacity to overcome adversity."

The third study, by Thuo et al. (2016), sought to define the entrepreneurial intention of students graduating from Ethiopian colleges and to pinpoint the variables that affect students' entrepreneurial behavior. 665 final-year university students from five different universities were randomly chosen to participate in a survey. A self-evaluation survey questionnaire was used to obtain the quantitative data, which was then examined using Pearson correlation and Logistic Regression. Additionally, information from open-ended questions was used to identify the elements that affect students' entrepreneurial behavior. The findings showed that

Ethiopian university students in their last year had strong entrepreneurial intentions but had low propensity to start a business within two years of graduation. In a Logistic Regression Analysis, perceived desirability, perceived feasibility, and support for the course were found to be substantial predictors of entrepreneurial intention as well as perceived locus of control to some extent, although gender and past work experience in a firm had no effect.

Fourth study is from Setyawati et al. (2021) aimed to determine student entrepreneurship interest by means of entrepreneurship education based on entrepreneurship MBKM. This research is a quasi-experimental type, and the research participant are students who take entrepreneurship courses. As a result, it is proven that MBKM-based entrepreneurship education can stimulate student interest in entrepreneurship. Students find it useful to reach their potential in the spirit of cooperation and develop awareness of caring for others and the environment.

3 Research and Analysis Method

Research method for this study is quantitative research. All data used in this study is primary data. Collecting data in this study using questionnaires that are distributed directly to the respondents, measuring the questionnaires using Likert scale and analyzed statistically using Confirmatory Factor Analysis (CFA) method. The population of this study was representatives of undergraduates from few universities in South Sulawesi and East Java Province in Indonesia. The research sample was 30 people who were determined by using convenience sampling technique.

Analysis method used in this research is Confirmatory Factor Analysis (CFA). Figure 1 shows various factors that will be tested in this study. All the analysis and tests in this study is done by using SmartPLS 3.0 software program as a medium for confirmatory variable analysis using second-order confirmatory factor analysis.

Fig. 1 Research model

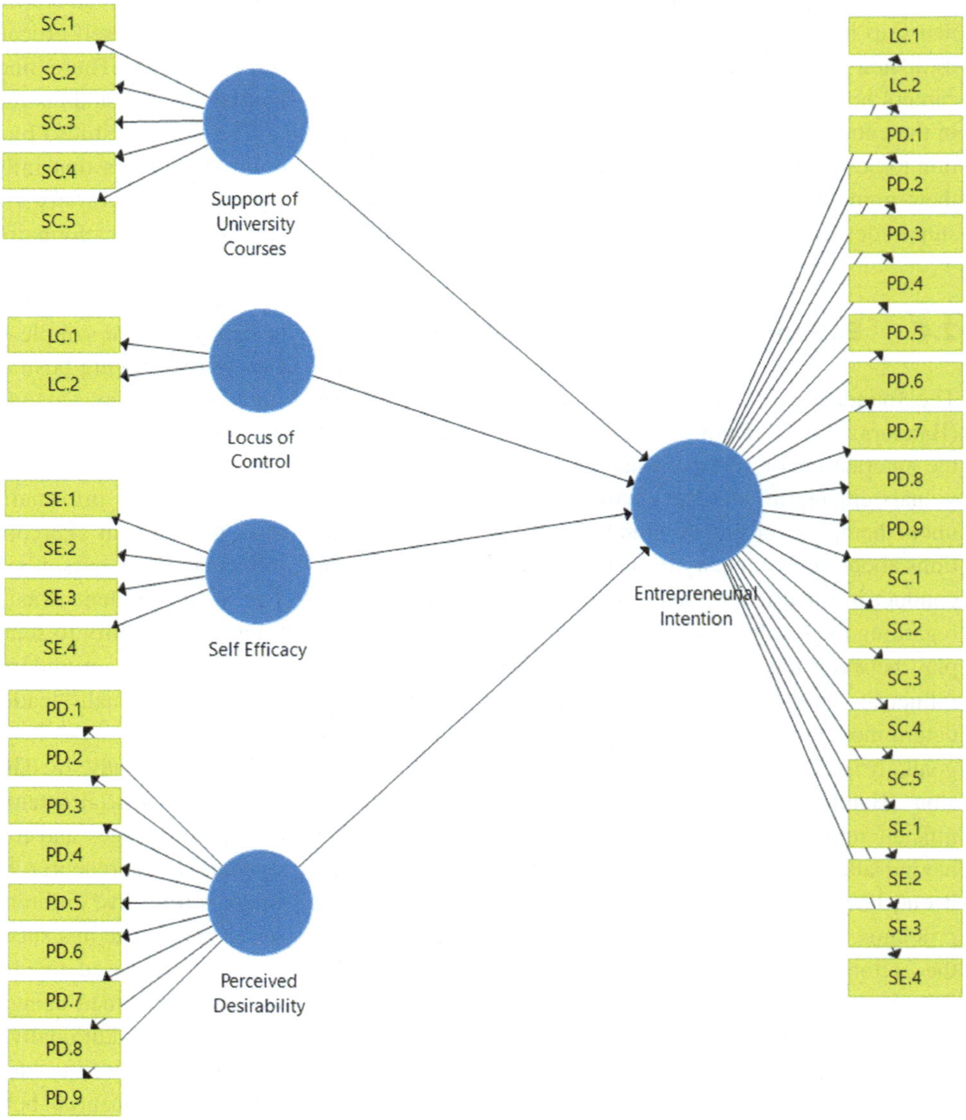

4 Result and Discussion

4.1 Respondent's Data

From data on Table 1, this research's respondent is dominated by women respondents (73%). The characteristic-based place of origin is dominated by student from South Sulawesi respondents (73%). Based on their graduation status, this research is dominated by undergraduate student (57%). Based on the year entered university, the respondent is dominated by students from the year of 2020 batch (33.33%) which coincidentally the first year of the implementation of the MBKM program by the Ministry.

4.2 Data Analysis

From data on Fig. 2, it is necessary to carry out further data processing by eliminating several indicators from Entrepreneurial Intention and Perceived Desirability variable to meet the minimum Average Variance Extracted (AVE) value of 0.5. Data in Table 2 is an analysis of the reliability and validity of which the indicators have been reduced to the AVE (Average Variance Extracted) criteria above 0.5. The following figures and table will show the results after the adjustment.

Data from Fig. 3 show a good results with all of the indicators have reached the Average Value Extracted (AVE) minimum value of 0,5. There are one indicator deleted from Perceived Desirability variable and five indicators deleted from the Entrepreneurial Intention variable. The adjustment also has an effect when the validity and reliability test was conducted again, as can be seen from the data given in Table 3, all of variables have the AVE value above 0.5.

Table 1 Sample's demographic

No	Characteristics		Total	Percentage (%)
1	Gender	Male	8	27
		Female	22	73
2	Place of origin	South Sulawesi	22	73
		East Java	8	27
3	Status	Alumni	13	43
		Undergraduates Students	17	57
4	Year	2011	1	3.3
		2012	3	10
		2013	1	3.3
		2016	3	10
		2017	1	3.3
		2018	3	10
		2019	5	16.67
		2020	10	33.33
		2021	3	10

Source Data processed

First Order Construct

The first validity test is the loading factor test with the recommended value above 0.70. The loading factor test will improve the model if there are indicators that have numbers below 0.70. The next step is to test the AVE with the recommended value above 0.50 and cross loading with the recommended value above 0.70. The AVE test results in this Table 3 indicate an acceptable AVE value, namely all variables have values above 0.50.

Next is to do a Composite Reliability Test with a recommended value above 0.70 to be declared valid. The results of the Composite Reliability tests show that all of the variables have Composite Reliability value above 0.70.

Table 4 results of the *T*-Statistics test show that all indicators have a value above 1.96, so the first order construct test can be accepted.

Second-Order Construct

The test for the second-order construct has several parts that must be considered carefully, such as the loading factor value which can be seen from the path coefficient value with a suggested value above 0.70, a high coefficient of determination and a reflective hypothesis test with a value above 1.96. The test results show that all variables have a loading factor value higher than 0.70, then all variables are acceptable. The value of R^2 or the coefficient of determination will show the value of the indicator forming the main variable. Table 5 gives results of the R^2 test which has a value of 0.995 which means that 99.5% of the indicators can form the main latent variable. The *t*-statistic results value above 1.96 indicates the hypothesis is acceptable.

4.3 Discussion

Based on the data that has gone through the analysis process, it can be concluded that all the variables that used in this study can encourage students' interest in entrepreneurship which is perceived desirability, entrepreneurial education (support of university courses), locus of control, and self-efficacy.

Self-efficacy variable is the variable ranked first as a factor that reflects as main driver of entrepreneurial intention. Self-efficacy is both the antecedent and the consequence of an action choice and affects the way in which individuals perform their current task and direct their future task accomplishments (Bandura, 2000). Confidence in oneself

Fig. 2 Analysis results

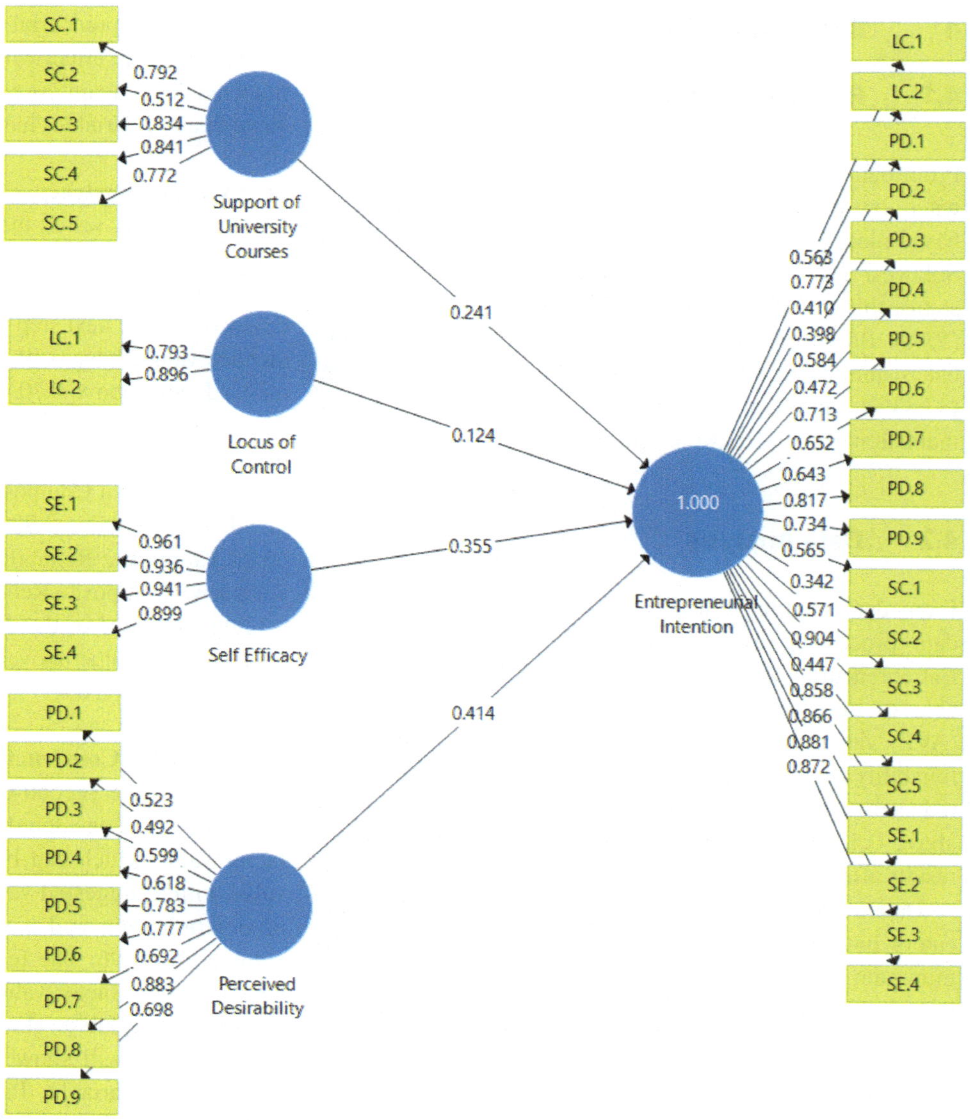

Table 2 Measurement model test

Variable	Composite reliability	Average variance extracted (AVE)
Entrepreneurial intention	0.940	0.458
Locus of control	0.834	0.716
Perceived desirability	0.885	0.469
Self-efficacy	0.965	0.873
Entrepreneurial education	0.869	0.578

Source Data processed

could see opportunities is one of the important things driving interest in entrepreneurship. Respondents who believe in their ability to see opportunities tend to have higher entrepreneurial intentions. Confidence in yourself to be able to start and build relationships with important people is also very necessary for an entrepreneur, because all business processes require important people who will contribute to these processes. Confidence in oneself can organize work to achieve the desired results. This is supported by the opinion of Tumasjan and Braun (2012) that self-efficacy gives entrepreneurs ability to overcome the difficulties during entrepreneurship process such as opportunity recognition, collecting resources, and improving the performance of new business.

The locus of control variable is the ranked second for variable that reflects the driver of entrepreneurial intention. According to Phares in Annisa et al. (2021), locus of control is defined as a belief that every event that occurs to oneself is caused by factors that exist in his or herself. In this variable, there are two indicators, first is diligence and hard work usually lead to success and second is I am confident of my skills and abilities to start a business. People

Fig. 3 Research model adjusted

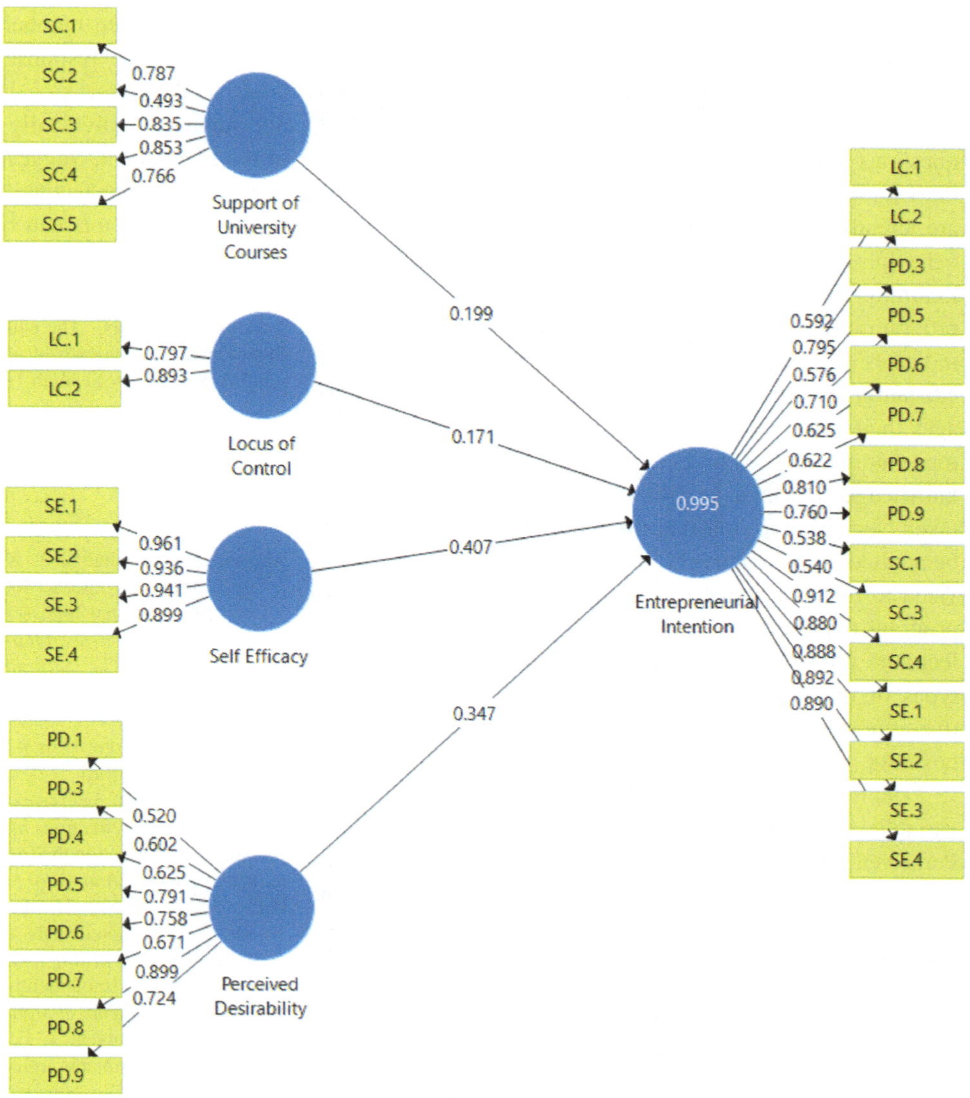

Table 3 Measurement model test adjusted

Variable	Composite reliability	Average variance extracted (AVE)
Entrepreneurial intention	0.948	0.560
Locus of control	0.835	0.717
Perceived desirability	0.887	0.501
Self-efficacy	0.965	0.873
Entrepreneurial education	0.887	0.663

Source Primary data, processed

Table 4 Path coefficient

	T-Statistics
Perceived desirability → entrepreneurial intention	9.965
Self-efficacy → entrepreneurial intention	6.690
Entrepreneurial education → entrepreneurial intention	6.255
Locus of control	5.130

Source Primary data, processed

Table 5 *R*-Square

	R-Square
Entrepreneurial intention	0.995

Source: Data processed

that want to open their own business must have diligence and confidence in their own business to operate their business and never give up pursuing their dreams.

The entrepreneurial education variable (support of university courses) is the third-ranked variable that reflects the driver of entrepreneurial intention. Liñán (2004) identified four types of entrepreneurship education programs. The first type is entrepreneurship education, which aims to broaden knowledge about entrepreneurship and influence the relationships that can influence intentions. The second is Start-up Training, usually designed for people with an entrepreneurial idea and have a need to solve practical questions about how to become an entrepreneur. Third, the Entrepreneurship Dynamics training is for people that are already an entrepreneur and want to promote dynamic behavior after the launch phase. The fourth "Lifelong Learning for Entrepreneurs" describes a lifelong learning program and is at experienced entrepreneurs. This variable indicates the response of scholars to the entrepreneurial education they have received and whether learning encourages their entrepreneurial intention. Based on the 5 indicators that are owned, first is undergraduates gain sufficient knowledge to run a business after following the lessons taken, second is to find that the input from the lessons can be applied in the real world, third is students are more motivated to run a business after completing the undergraduate program than before taking it, fourth through learning my lecturers prepare me to enable me to recognized potential types of industrial/business areas to enter, and fifth creative atmosphere of university inspires me to work out new business.

Perceived Desirability variable is the fourth ranked for variable as a factor that drive entrepreneurial intention. Perceived Desirability is defined as the attractiveness of starting a business with each subject, or the level of interest of an individual, the attitude of an individual toward start-ups (Shapero & Sokol, 1982). The desire to create new job opportunities, work flexibility, be creative, gain experience, gain higher control, and achieve vision are indicators that encourage interest in entrepreneurship. There are 3 indicators that are removed because the AVE value is below 0.5, namely, to get financial benefits, to become independent worker and to get a higher social position. The result that perceived desirability as a driver of entrepreneurial intention was found by Bui et al. in 2020 and Thuo et al. in 2016 also.

5 Conclusion

The purpose of this study was to examine what factors encourage entrepreneurial intention in undergraduates in the provinces of South Sulawesi and East Java. This study's results indicate that the factors of perceived desirability, self-efficacy, locus of control, and entrepreneurial education (support of university courses) advocates entrepreneurial intention in undergraduates.

There are several limitations in this study, such as first, the number of respondents who tend to be minimal and dominated by certain campuses, so if you want to get more comprehensive results, a study with more research samples with a more even number can be carried out. The second limitation is the response time of the respondents which tends to be rather long because the researcher sends the questionnaire individually and adjusts to the response time of the respondent. Suggestions for future research can also take several other factors by conducting exploratory studies so that other factors can be found that encourage entrepreneurial intention in undergraduates.

Acknowledgements The authors gratefully acknowledge the support shown by all respondents during data collection phase and the financial support provided by Sekolah Tinggi Ilmu Ekonomi Ciputra Makassar.

References

Annisa, D. N., Tentama, F., & Bashori, K. (2021). The role of family support and internal locus of control in entrepreneurial intention of vocational high school students. *International Journal of Evaluation and Research in Education, 10*(2), 381–388.

Bandura, A. (1997). *Self-efficacy: The exercise of contro*. W.H. Freeman and Company.

Bandura, A. (1999). A social cognitive theory of personality. In L. Pervin & O. John (Eds.), *Handbook of personality* (2nd ed., pp. 154–196). Guilford.

Bandura, A. (2000). Exercise of human agency through collective efficacy. *Current Directions in Psychological Science, 9*, 75–78.

Bandura, A. (2012). On the functional properties of perceived self-efficacy revisited. *Journal of Management, 38*, 9–44.

Bechard, J., Gregoire, D., & Grégoire, D. (2002). Entrepreneurship education revisited: The case of higher education. *Academy of Management Learning and Education, 4*(September), 840–853.

Bird, B. (1998). Implementing entrepeneurial ideas: The case for intention. *Academy of Management Review, 13*, 442–453.

Bui, T. H. V., Nguyen, T. L. T., Tran, M. D., & Nguyen, T. A. T. (2020). Determinants influencing entrepreneurial intention among undergraduates in universities of Vietnam. *The Journal of Asian Finance, Economics and Business, 7*(7), 369–378.

Bygrave, W. D., & Timmons, J. A. (1992). *Venture capital at the crossroads*. Harvard Business School Press.

Drucker, P. F (1985). *Innovation and entrepreneurship: Practice and principles*. Harper Business.

Dwyer, A. L., & Cummings, L. C. (2001). Stress, self-efficacy, social support and coping strategies in university students. *Journal of Counselling, 35*, 3.

Hisrich, R. D. (1990). Entrepreneurship/intrapreneurship. *American Psychologist, 45*(2), 209–222.

Hisrich, R., Peters, M., & Shepherd, D. (2005). *Entrepreneurship* (6th ed.). McGraw–Hill.

Hisrich, R. D., Peters, M. P., & Shepherd, D. A. (2012). *Entrepreneurship* (9th ed.). McGraw-Hill Education.

Kusnandar, V. B. (08 June, 2022). *Berapa Jumlah Angkatan Kerja Indonesia 2022?*. Katadata. https://databoks.katadata.co.id/datapublish/2022/06/08/berapa-jumlah-angkatan-kerja-indonesia-2022

Krueger, N. F. (1993). The impact of prior entrepreneurial exposure on perceptions of new venture feasibility and desirability.*Entrepreneurship Theory and Practice, 18*(1), 5–23.

Liñán, F. (2004). Intention-based models of entrepreneurship education. *Piccola Impresa/small Business, 2004*(3), 11–35.

Morris, M. H., Lewis, P. S., & Sexton, D. L. (1994). Reconceptualizing entrepreneurship: An input—output perspective. *Advanced Management Journal*, Winter, 29–30.

Parker, S. C. (2004). *The economics of self-employment and entrepreneurship*. Cambridge University Press.

Parker, S. C. (2009). *The Economics of Entrepreneurship. (n.p.).* Cam bridge University Press.

Pihie, Z. A. L., & Bagheri, A. (2013). Self-efficacy and entrepreneurial intention: The mediation effect of self-regulation. *Vocations and Learning, 6*(3), 385–401.

Ratten, V., & Usmanij, P. (2021). Entrepreneurship education: Time for a change in research direction? *The International Journal of Management Education, 19*(1), 100367.

Schumpeter, J. (1934). *The theory of economic development: An inquiry into profits, capital, credit, interest, and the business cycle*. Harvard University Press.

Setyawati, Y., Sugiharto, F. B., Rosyanafi, R. J., Cahyanto, B., Rini, T. A., & Yusuf, A. (2021). Pengaruh MBKM Kewirausahaan Terhadap Minat Berwirausaha Mahasiswa. *MOTORIC, 5*(2), 311–318.

Shane, S., & Venkataraman, S. (2000). The promise of entrepreneurship as a field of research. *The Academy of Management Review, 25*(1), 217–226.

Shapero, A., & Sokol, L. (1982). Social dimensions of entrepreneurship. *Encyclopedia of entrepreneurship* (pp. 72–90). Prentice-Hall.

Thompson, E. (2009). Individual entrepreneurial intent: Construct clarification and development of an internationally reliable metric. *Entrepreneurship Theory and Practice, 33*(3), 669–694.

Thuo, M., Abo, T., & Toma, S. (2016). Entrepreneurial intentions of university students: Insights for entrepreneurial education in Ethiopia. *European Journal of Business and Management, 8*(22), 25–35.

Tumasjan, A., & Braun, R. (2012). In the eye of the beholder: How regulatory focus and self-efficacy interact in influencing opportunity recognition. *Journal of Business Venturing, 27*(6), 622–636.

Tunio, M. N., Chaudhry, I. S., Shaikh, S., Jariko, M. A., & Brahmi, M. (2021). Determinants of the sustainable entrepreneurial engagement of youth in developing country—An empirical evidence from Pakistan. *Sustainability, 13*(14), 7764.

Witjaksono, M. (2016). Analisis Kritis dan Pragmatis Caring Economics sebagai Paradigma baru Kajian Ekonomi. *Jurnal Ekonomi Dan Studi Pembangunan, 8*(2), 214–242.

The Influence of Entrepreneurial Alertness on New Venture Performance with Networking Capability as a Moderator at Start-Up in Surabaya

Carolina Novi Mustikarini and Putu Arisintha Putri Pratiwi

Abstract

Entrepreneurial alertness is important for start-ups to have in improving new venture performance (Adomako et al., 2018). One of the factors that can influence the success of a start-up is networking capability. Start-ups that can empower the network, and they have the opportunity to compete one step ahead compared to their competitors (Anwar et al. in Int J Emerg Markets 13:998–1025, 2018). This study aims to examine the effect of entrepreneurial alertness on new venture performance and examine the effect of moderating networking capability on the effect of entrepreneurial alertness on new venture performance at start-up in Surabaya, by replicating and modifying research by Adomako et al. (2018). This research is quantitative research. Respondents in this study were 70 respondents. Respondents in this study are start-up owners who are registered in the Tokopedia marketplace. This study uses data analysis namely Partial Least Square (PLS). The results showed a significant influence between entrepreneurial alertness on new venture performance, where the moderating variable networking capability acts as a predictor of moderation.

Keywords

Entrepreneurial alertness · Networking capability · New venture performance · Start-up business

C. N. Mustikarini (✉)
Sekolah Tinggi Ilmu Ekonomi Makassar, Makassar, Indonesia
e-mail: cmustikarini@ciputra.ac.id

P. A. P. Pratiwi
Universitas Ciputra Surabaya Makassar, Makassar, Indonesia
e-mail: parisintha@student.ciputra.ac.id

1 Introduction

Indonesia is currently facing the digital era of the industrial revolution 4.0 with the use of very intensive technology. The development of technology through the use of the Internet has become a basic need of the Indonesian people in various daily activities. This is supported by an increase in the number of Indonesian Internet users from year to year (Statistas, 2022). The penetration of Internet users in 2022 in Indonesia has increased to 64.8% of the total population of Indonesia, equivalent to 264.16 million people. The overall penetration growth of Internet users in Indonesia has experienced an average increase of 23.6% from 2021 to 2022.

According to Fig. 1, the number of Internet users in Indonesia in 2022 grew by 10.4% from 2021 to 5065,23 million. The growth trend of internet penetration is one of the factors that accelerates the flow of digitalization in various sectors in Indonesia. Figure 1 also shows that in 2022, around 224 million people access the internet in Indonesia, meaning that with more than 204 million Internet users, Indonesia is one of the largest online markets in the world. As of July 2021, online penetration in the country is around 70%. Popular online activities include mobile messaging and social media. The most popular social network in Indonesia is WhatsApp, with around 89% of the online population using the platform. Mobile Internet usage is experiencing double-digit growth rates and currently stands at more than 64% among the population. The digital industry is a sub-sector of the creative industry which shows significant growth as a contributor to the national economy (Saputra, 2015). The rapid growth of the digital industry is shown by the presence of start-ups that appear in various regions in Indonesia (Bekraf, 2018). According to Gallop and David (2019), start-up is an institution established to provide new and innovative products or services in conditions of high uncertainty. Start-up is synonymous with relatively small-scale businesses with innovative ideas and has

M. Nawaz Tunio et al. (eds.), *Sustainability in Creative Industries*,
Advances in Science, Technology & Innovation, https://doi.org/10.1007/978-3-031-48453-7_3

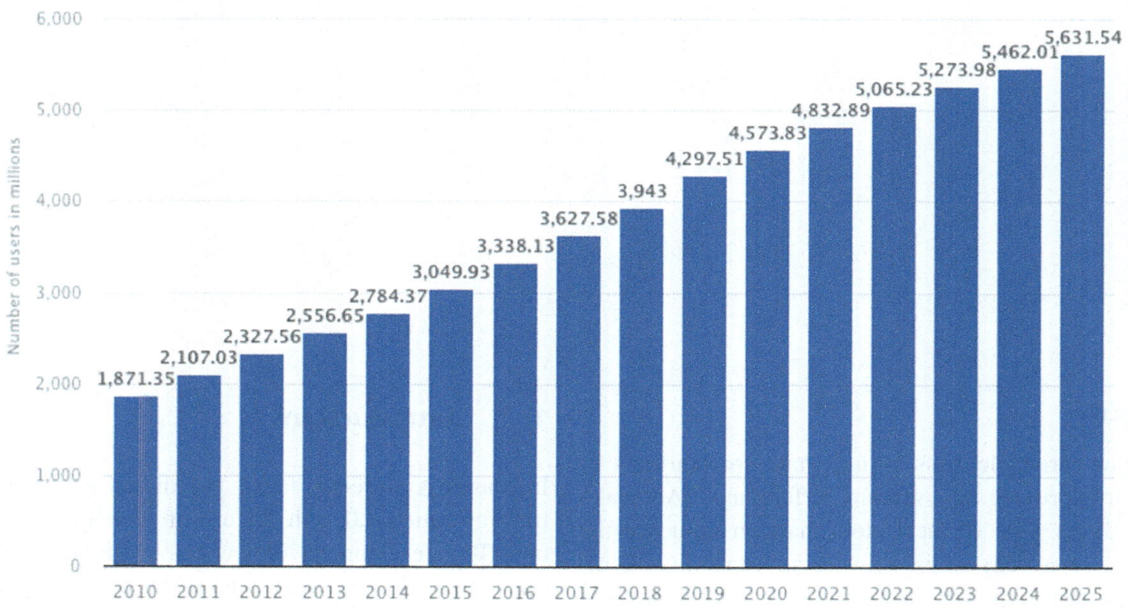

Fig. 1 Data of Internet users in Indonesia from 2010 to 2022. *Note* Figure 1 explains the data of a significant increase in the number of Internet users in Indonesia in the last twenty years from 2010 to 2022. Copyright 2022 by Statista.com

a relatively young business age (Anwar et al., 2018). The presence of start-ups in Indonesia in 2018 has experienced growth in several regions.

The opportunity for developing start-ups in Indonesia is increasingly supported by advances in technology and increasingly easy Internet connectivity. According to law number 20 of 2008, there are categorizations of start-ups in Indonesia based on turnover. Start-ups are categorized based on the turnover they have (1) the micro-scale category generates a turnover of 50–300 million per year (2) the small-scale category generates turnover reaching 300 million–2.5 billion per year (3) the medium-scale category reaches a turnover of 2.5–50 billion per year and (4) large-scale categories reached more than 50 billion. The increasingly rapid digital flow makes new start-up opportunities develop faster into larger scales. There are 5 start-ups that have developed rapidly and have been categorized as Indonesian unicorns with a valuation of over 1 billion USD (Rahman, 2019).

Gojek became the first start-up from Indonesia which has unicorn status. Gojek provides a one-stop service application that is in high demand by Indonesians and now Gojek has a valuation of 10 billion USD. The second place is Tokopedia, an online marketplace service provider that holds unicorn status after 10 years standing with a valuation of 7 billion USD. The third position is Traveloka through an online ticket sales service that holds unicorn status after 5 years of standing with a valuation of 2 billion USD. Fourth place, a start-up in the e-commerce sector Bukalapak received a unicorn title after receiving funding from the Emtek Group with a valuation of 2.5 billion USD. OVO, an electronic payment service provider that is now in unicorn status, is ranked fifth. OVO has a valuation of 2.9 billion USD and is a very fast-growing start-up with unicorn status in just 2 years.

The achievements of OVO and the four other start-ups do not require a long process to be in their current position. One of the successful start-ups owned by Indonesia after 9 years of existence is the Tokopedia marketplace. PT Tokopedia was founded by Wiliam Tanuwijaya and Leontinus Alpha Edison on February 6, 2009. This start-up has a mission to achieve digital economic equality (Tokopedia, 2019). The mission was created to encourage new start-ups to develop online by providing a place for business continuity to start-up. Now, the number of online start-ups on Tokopedia has reached 7.2 million partners (Tokopedia, 2019).

The development of high online start-ups with technological advances and digitalization now increases the potential of small-scale start-ups to grow and develop faster. Increasing competition with the growth of new start-ups, demands that start-ups continue to innovate and have sensitivity in capturing opportunities in a dynamic environment (Firdha, 2019). According to Sharma (2018) in a dynamic environment, start-ups must have cognitive abilities to have sensitivity in capturing opportunities. Cognitive abilities help start-ups in seeing and identifying opportunities through the information process they have. Start-up partners who join Tokopedia are examples of business people who have cognitive abilities who can identify the great potential of the development of online start-ups to grow and develop in the digital age. Tokopedia partners can be said to be start-ups who have sensitivity because they are able to apply their cognitive abilities one step further using marketplace such as Tokopedia as a digital business development tool to increase the success of their start-ups.

According to Adomako et al. (2018), the driving factor for a successful start-up is to have sensitivity in capturing opportunities that are often also called entrepreneurial alertness. Entrepreneurial alertness is the ability of start-ups to find out opportunities before they are discovered by competitors (Kirzner, 1999; in Sharma, 2018). Conceptually, the utilization of opportunities identified by start-up through alertness affects the level of new venture performance (Andersén, 2019; Tang et al., 2012). Entrepreneurial alertness helps start-ups make decisions quickly and nimbly so that it impacts the performance of start-ups. According to Karami and Tang (2019), enhancing new venture performance requires networking capabilities, often called networking capabilities.

Start-ups that develop network capabilities are able to utilize the resources they have (Ebbers, 2014). Networking skills help start building relationships with outsiders in agreeing on opportunities. Networking leads to start-up relationships with consumers, suppliers, and business partners. Networks supported by new companies can provide the latest market information and times that will be used to explore new companies (Boso et al., 2013). The ability to network connect start-ups with new markets, suppliers, and new customers so that it will improve the performance of new businesses from increased sales and profits (Anwar et al., 2018).

Based on this background, it inspired the research to examine the effect of entrepreneurial alertness on new venture performance with networking capability as moderating. This is very relevant to the research model conducted by Adomako et al. (2018). This research will replicate and adapt the research model. The purpose of this research is to examine the influence of entrepreneurial alertness on new venture performance and examine the moderating influence of networking capability on the influence of entrepreneurial alertness on new venture performance at start-up in Surabaya.

2 Literature Review

2.1 Entrepreneurial Alertness

According to Kirzner (1983: 48; in Adomako et al. 2018), alertness is identified as an entrepreneurial characteristic which means the ability to know without looking for opportunities sought by competitors. Roundy et al. (2018) state that start-ups who have entrepreneurial alertness can access a high sensitivity to opportunity. Alertness makes start-up decisions making more accurate, innovative, and agile which impacts on improving start-up performance.

According to Sharma (2018), entrepreneurial alertness is formed on several main components, namely information searching and sensing, knowledge and experience, personality factors, social networks, cognitive abilities, and the environment. Entrepreneurial alertness helps start-ups identify many opportunities to be turned into reality in creating new values and successful start-ups (Zhao et al., 2020). Adomako et al. (2018) state that the level of entrepreneurial alertness affects the success of a start-up. There are 3 indicators of entrepreneurial alertness based on Tang et al. (2012; in Adomako et al., 2018), namely:

1. Scanning and search: scan the environment and find information.
2. Association and connection: associate or collect information that was not previously connected.
3. Evaluation and judgment: make evaluations and research on the commercialization of ideas.

2.2 Networking Capability

According to Anwar et al. (2018), networking capabilities possessed by start-ups can facilitate access to many resources and information that are not easily available. The higher the network level that is owned will affect the continuity of a start-up (Karami & Tang, 2019). According to Adomako et al. (2018), networking capability consists of 2 things, namely social networking capability and business networking capability. Fang et al. (2014) states that social networking capability refers to the ability possessed by start-ups to mobilize the resources available in the structure of social relations. Based on the research of Adomako et al. (2018), there are 3 indicators of social networking capability, namely:

1. Get information about the industry from contacts that are owned faster than competitors.
2. Have a professional relationship with influential people in one industry.
3. Informal activities with influential people in one industry.

Adomako et al. (2018) state that business networking capabilities are the ability of start-ups to build buyer, supplier, and competitor relationships. Anwar et al. (2018) state that business networking capability is needed in establishing relationships with suppliers, buyers, and competitors. Without a relationship with these 3 actors, a start-up can't survive. Business networking capability has 3 indicators based on Boso et al. (2013) in Adomako et al. (2018):

1. Customers: build good relationships with buyers.
2. Supplier: build good relationships with suppliers.
3. Competitors: build good relationships with competitors.

2.3 New Venture Performance

According to Luk et al. (2008), new venture performance is defined as the results of the company's achievements in a certain period. Performance can also be reflected as the ability of start-up to produce achievements in the dimensions that have been determined about the target (Anwar et al., 2018). The performance of start-ups can be measured through available financial data, but it is difficult to obtain for start-ups that are still relatively new (Dibrell et al., 2014). The measurement of start-up performance in this study uses a self-reported approach. Measurement of the self-reported approach involves assessing the perspective of the owner or manager of start-ups because they have responsibility for the strategic planning and performance of the company (Tajeddini & Mueller, 2012). The use of self-reported approaches is used because of three bases, namely: (1) objective financial data not publicly available for new businesses (2) the literature suggests that subjective performance measurement has been used in previous studies (3) subjective performance measurement is used to avoid problems associated with measurement objective performance in developing countries such as financial reports that are not by standards (Adomako et al., 2018). According to Dess and Robinson (1984; in Adomako et al., 2018), measurements through self-reporting have a strong correlation with objective performance indicators.

Based on Adomako et al. (2018), there are 7 indicators of new venture business, namely:

1. Profitability: evaluation of the company's profit level compared to the previous period.
2. Profit margin: evaluation of the company's profit margin level compared to the previous period.
3. Return on investment: evaluation of the level of investment returns of the company compared to the previous period.
4. Market share: evaluates the company's market share compared to the previous period.
5. Return on assets (return on assets): evaluation of the rate of return on company assets compared to the previous period.
6. Profitability growth: evaluates the rate of growth of the company's profit compared to the previous period.
7. Sales growth: evaluates the company's sales growth rate compared to the previous period.

The measurement of start-up performance in this study uses a self-reported approach, where the owner is asked to provide an assessment of the company's performance based on three things, namely profitability growth, sales growth, and market share of the company compared to a certain period. The selection of the three indicators is due to the limited information regarding the accessibility of start-up financial data that is relatively new, so the three indicators were chosen to facilitate the owner to provide a subjective assessment related to the company's performance.

2.4 Previous Empirical Studies

Adomako et al. (2018) aims to examine the effect of entrepreneurial alertness on new venture performance with the moderating effect of networking capabilities. This study uses Hierarchical Regression analysis techniques and uses random sampling techniques. Primary data was collected from 203 new businesses operating in Ghana. The results of this study are entrepreneurial alertness significantly and positively correlated to new venture performance and networking capability as a moderating force for that relationship. The relevance of the study is that there is a model that is adopted and replicated with the variables studied, namely entrepreneurial alertness, new venture performance, and networking capability as moderation variables.

Roundy et al. (2018) aim to explore whether entrepreneurial alertness has a direct influence on strategic change decisions and organizational performance. This study uses Hierarchical Multiple Regression analysis techniques and sample data used by 231 companies in Arlington within a 15-mile radius of Cowboys Stadium. The findings of this study are the direct influence of entrepreneurial alertness on strategic change decisions and organizational performance. The relevance of this study is the similarity in the variables studied, namely entrepreneurial alertness and performance so that it is relevant to be a reference for this research.

Sharma (2018) aims to integrate research variations on entrepreneurial alertness, identify key components, and develop an understanding of the concept of entrepreneurial alertness. This study uses a systematic review of secondary research as an analysis technique and the sampling technique uses purposive sampling. This study found that identification which became the main component in alertness was information searching and sensing, knowledge experience, personality factors, social networks, cognitive abilities, and environment. The relevance of this study is that there are similarities in the variables studied, namely entrepreneurial alertness so that it is relevant to be a reference for this research.

Karami and Tang (2019) aims to broaden the knowledge of entrepreneurial orientation (EO) by testing EO in the context of the international performance of SMEs. This research also adds to the understanding of the foreign markets of SMEs by mediating networking capability and experiential learning. The analysis technique used in this study is the Structural Equation Model (SEM) with 164 international MSME samples in New Zealand. The results of this study indicate that EO influences international performance and the mediating role of experiential learning and networking capability in this relationship. The relevance of this study is the similarity in the variables studied, namely performance and networking capability so that it is relevant to be a reference for this research. Mitrega et al. (2017) aim to determine the networking capability in supplier relationships and how they affect product innovation and firm performance. This research uses FIMIX analysis and general sampling frame as the sampling technique. The sample used was 156 suppliers from the Iranian automotive industry. The results of this study are the positive influence of networking capability on product innovation and firm performance. The relevance of the research is the similarity in the variables studied, namely performance and networking capability so that this research is a reference.

Anwar et al. (2018) aim to examine the effect of different types of networking, namely business networking, financial networking, and political networking on new venture performance and looking for types that constitute a competitive advantage. This research uses the Structural Equation Model analysis technique and sample technique that is convenience-based sampling. The results of this study indicate that business networking, financial networking, and political networking have a significant effect on new venture performance and competitive advantage. Competitive advantage is a strong mediator between financial networking and new venture performance as well as business networking and new venture performance. The relevance of this study is the similarity in the variables studied, namely new venture performance and networking capability so that it is relevant to be a reference for this research.

3 Research Methods and Data Analysis

This type of research is quantitative research. The population in this study is start-ups who have partnered with Tokopedia in the Indonesian. The sample used in this study was 70 start-ups in the Tokopedia Surabaya "BaruMulai" database

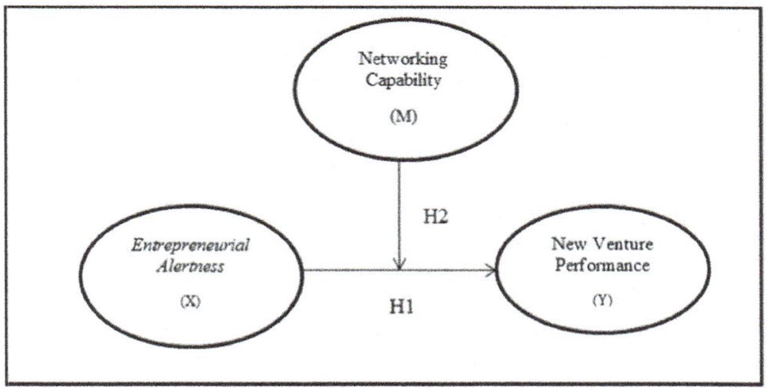

Fig. 2 Research model. *Note* Figure 2 is a research model adapted from previous research. From "Entrepreneurial alertness and new venture performance: facilitating roles of networking capability", by Adomako et al. (2018), *International Small Business Journal: Researching Entrepreneurship*, *36*(5), 453–472

accessed on February 15, 2020. The sampling technique used a convenience sampling technique. This study obtained primary data sources by distributing questionnaires to start-up owners and measuring indicators using a Likert scale.

The method used in this study is the Partial Least Square (PLS) method using the SmartPLS 3.0 program. SmartPLS is one of the computer programs used in estimating the PLS-SEM (Structural Equation Modeling) model. According to Ghozali (2006), PLS is a soft modeling analysis method which is assumed that the data is not with a certain scale measurement, meaning that the data can use a small sample size (under 100 people). Moderation analysis that is used in this research is according to Baron and Kenny (1986) theory. The model of this research is replicated and modified by Adomako et al. (2018) research model as can see in Fig. 2.

Based on Fig. 2, the research seeks to test the following hypotheses:

H1: Entrepreneurial alertness has a significant influence on new venture performance
H2: Networking capability moderating the influence of entrepreneurial alertness on new venture performance

4 Result and Discussion

4.1 Respondent Data

Based on Table 1, it is known that from 70 respondents, the owner of start-ups that are male is more in number with a percentage of 84.3% or equivalent to 59 respondents. The owner of the start-up who is a woman has a percentage of 15.7% or the equivalent of 11 respondents much smaller than the male gender. Start-ups listed in the database are more dominantly owned by male owners, although this does

Table 1 Sample demographic

No	Characteristic		Total	Percentage (%)
1	Gender	Man	59	84.3
		Woman	11	15.7
2	Start-up established duration	1–3 year (s)	16	22.9
		4–5 years	54	71.1

Note Data collected by Author in August 2022

not become a barrier to start-ups to have good performance. Based on established duration, Table 1 also gives that 54 companies have been established for 4–5 years, equivalent to 71.1% of the total 70 start-ups. The remaining 16 companies stood between 1 and 3 years, equivalent to 22.9% of the total start-up. These results indicate that many of the start-up companies found in the Tokopedia Surabaya database #BaruMulai are young companies in the range of 1–5 years. In this age range, relatively young companies still have time to develop, improve company performance, which is more stable.

4.2 Measurement Model Test

Based on Fig. 3 shows that all indicators of the variable entrepreneurial alertness (X), new venture performance (Y), and networking capability (M) have values above 0.7 and this indicates all indicators have met the requirements of the thumb of rule value of the convergent validity test seen from the value of outer loading is greater than 0.7. All indicators have met re requirements of the thumb of rule value of AVE is greater than 0.7. Cronbach Alpha value of the entrepreneurial alertness variable (X), new venture performance (Y), networking capability (M), and moderating effect in this study showed a value>0.8. Based on these results, it can be stated that the item measuring instrument used in this study is classified as very good and reliable to use.

4.3 Structural Model (Inner Model)

This study used a moderation variable, the main thing to consider before conducting the inner model test is testing the main effect. The main effect test is performed to determine the relationship between the independent variable (X) and the dependent variable (Y). According to Baron and Kenny (1986) in Abdillah and Jogiyanto (2015), a new inner model test can be carried out if the independent variable and the dependent variable have a significant effect.

Based on Fig. 4, show that the t-statistic value between the variables of 10.701 indicates that there is a significant relationship between the entrepreneurial alertness variable (X) and the new venture performance variable (Y). The relationship can be said to be significantly seen from the *t*-statistic value from the testing of the main effects worth more than 1.96 (Abdillah and Jogiyanto, 2015). Testing significant main effects, the inner model testing can be continued by adding a moderating variable (networking capability) to the research model.

Figure 5 shows the *t*-statistic value of the relationship variable networking capability (M) and new venture performance (Y) of 2.222 significant because *t*-statistic > 1.96 and the *t*-statistic moderating effect value of 0.963 can be interpreted that the networking capability variable (M) weakens the relationship between the entrepreneurial alertness variable (X) to the new venture performance (Y) because of the *t*-statistic < 1.96.

Based on Table 2, it shows R^2 value of the new venture performance variable (Y) is 0.704. The results of the R^2 value illustrate that the new venture performance variable can be explained by 0.704 (70.4%) by the entrepreneurial alertness and networking capability variable, while the remaining 29.6% is explained by other variables outside this research model.

According to Table 3, the effect of entrepreneurial alertness (X) on new venture performance (Y) after the moderation variable was included in the inner model test remained significant (*t*-statistic = 3.687). The main requirement for

	Y1	Y2	Y3	M 1	M 2	M 3	M 4	M 5	M 6	X1	X2	X3	X4	X5	X6	X7	X8	X9	X10	X11	X12	Moderating effect
Outer loading	0,906	0,912	0,900	0,819	0,724	0,881	0,812	0,880	0,905	0,891	0,816	0,772	0,862	0,877	0,868	0,855	0,868	0,774	0,767	0,836	0,795	1,917
AVE	0,821			0,704						0,694												1,000
Cronbach Alpha	0,891			0,914						0,960												1,000

Fig. 3 Measurement model test. *Note* Figure 3 shows measurement of all indicators of the variables. From the results of PLS 4.0 data processing

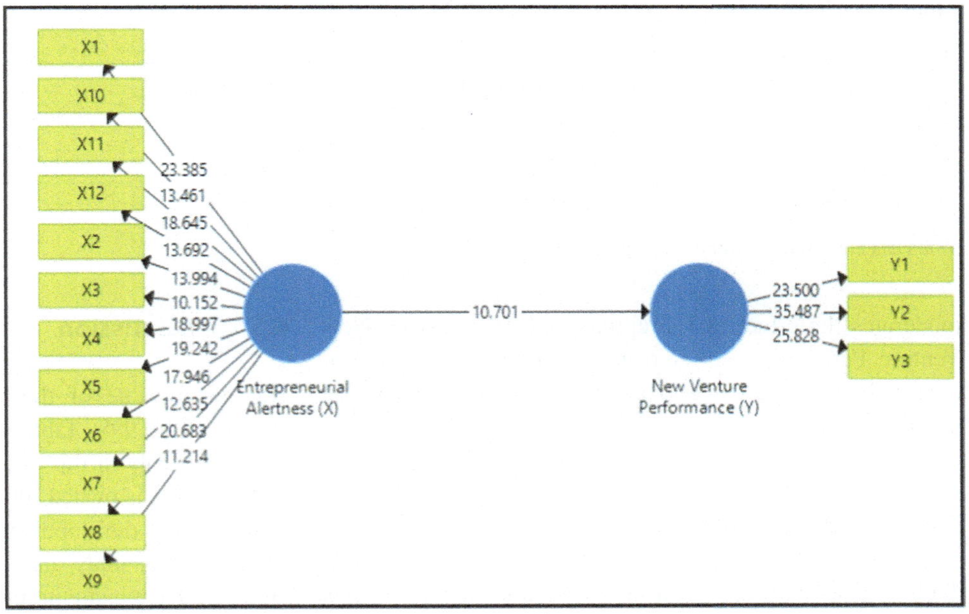

Fig. 4 Inner model. *Note* Figure 4 is the result of statistical data processing based on the inner model. From the results of PLS 4.0 data processing

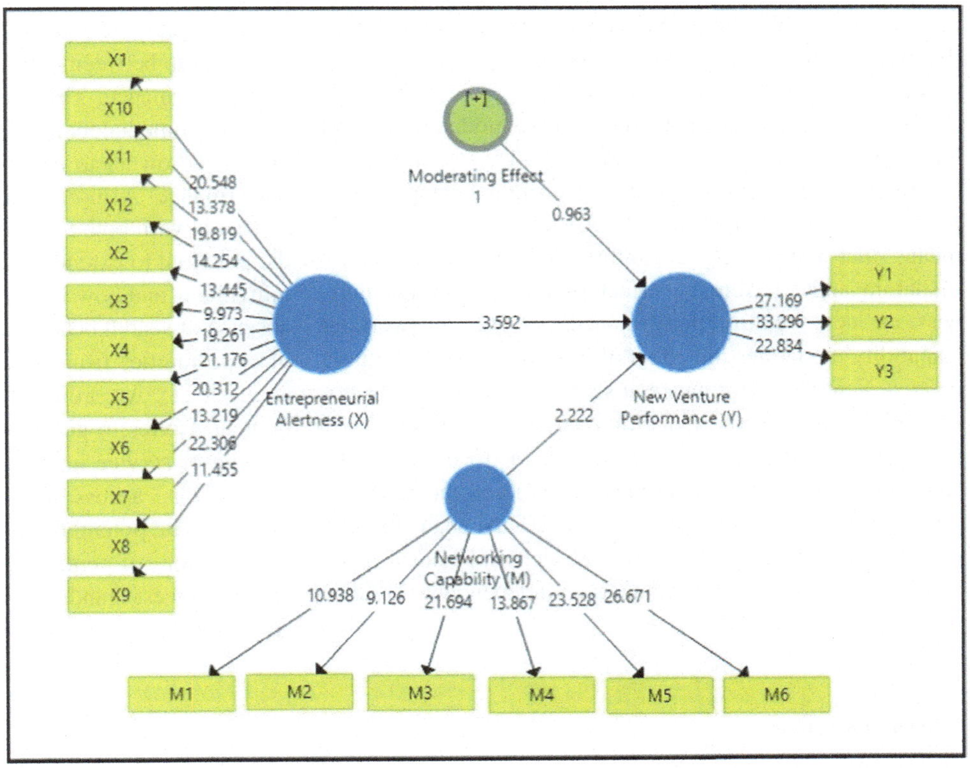

Fig. 5 Inner model. *Note* Figure 5 is the result of statistical data processing based on the inner model along with the results of the moderation test. From the results of PLS 4.0 data processing

Table 2 *R* square

R square	
Variable	R square
New venture performance (Y)	0.704

Source Primary data, analyzed
Note Data collected by Author in August 2022

moderation testing is that the influence of variable X on Y must be significant. Therefore, the results of the study prove that H1 "Entrepreneurial Alertness has a significant effect on new venture performance at start-up in Surabaya". The effect of the moderating effect on new venture performance (Y) did not have a significant effect because *t*-statistics showed a value of 0.941 (*t*-statistics < 1.96). The results of this study prove that H2 "Networking Capability

Table 3 Path coefficient

Path coefficient				
Variable	Original sample	Sample mean	Standard deviation	*T* statistic
X–Y	0.557	0.604	0.151	3.687
M–Y	0.285	0.265	0.122	2.329
Moderating effect-Y	– 0.077	– 0.076	0.082	0.941

Note Data collected by Author in August 2022

moderates the effect of Entrepreneurial Alertness on New Venture Performance at start-up in Surabaya" is not supported, but can be categorized as a predictor moderation model.

4.4 Moderation Test

Table 4 gives that the significance test in testing the effects of moderation. The moderation effect is not only done to test the direct effect of the independent variable on the dependent variable, but also to see the relationship between the independent variable and the moderating variable on the dependent variable (indirect effect).

Table 4 gives that the path between the moderating effect and new venture performance does not have a significant *t*-statistic value of 0.941 (<1.96). The original sample value possessed by the moderating effect with Y is – 0.077. This means that networking capability does not significantly moderate and weaken the effect of entrepreneurial alertness on new venture performance, so H2 "Networking Capability moderates the effect of Entrepreneurial Alertness on New Venture Performance at start-up in Surabaya" is not supported. The type of moderation variable in this study is a predictor of moderation, where the variable M has a significant effect on the dependent variable (Y) and the moderating effect has no significant effect on Y. This means that the networking capability (M) variable is possibly not a moderation variable but stands alone as an independent variable.

4.5 Discussion

The mean value of the overall entrepreneurial alertness indicator is 3.998. Table 5 gives that the owners of start-up respondents in this study agree that the entrepreneurial alertness that is owned influences in achieving good performance in the development of the business being run. The results of this study also show that the start-up owners in Surabaya have had entrepreneurial alertness shown by how they use Tokopedia to be one of the means to market their products in the digital era like today. This is supported by Adomako et al. (2018) which describes the strong influence relationship between entrepreneurial alertness and new venture performance. The higher ability to read opportunities and can immediately apply them into a business opportunity allows the owner to make a successful business with good performance. A more inclusive concept of entrepreneurial alertness, with both virtuous and intrinsic aspects, will have the potential to extend theories of how alertness develops in individuals and how it influences opportunity seeking in the absence of search behavior, or the creation and development of opportunities from purely subjective origins (Chavoushi et al., 2021). According to Tang et al. (2021), if the entrepreneur aspires to get involved in activities leading to new product or service innovation, they must actively seek new information, access alertness, and combine information then evaluate potential future business opportunities.

Indicator of entrepreneurial alertness variable that has the lowest standard deviation is the ability to connect any information obtained, it shows that the answers from

Table 4 Total effect

Total effects				
Variable	Original sample	Sample mean	Standard deviation	*T* statistic
X–Y	0.557	0.604	0.151	3.687
M–Y	0.285	0.265	0.122	2.329
Moderating effect-Y	– 0.077	– 0.076	0.941	0.941

Note Data collected by Author in August 2022

Table 5 Standard deviation, mean, and total mean

Variable	Indicator	Standard deviation	Mean	Total mean
New venture performance	Y1	0.768	4.157	3.948
	Y2	0.809	3.786	
	Y3	0.831	3.900	
Networking capability	M1	0.715	4.057	3.867
	M2	0.736	4.029	
	M3	0.754	3.871	
	M4	0.879	3.643	
	M5	0.769	3.743	
	M6	0.761	3.857	
Entrepreneurial alertness	X1	0.732	4.086	3.998
	X2	0.819	4.014	
	X3	0.850	3.814	
	X4	0.754	3.943	
	X5	0.773	4.129	
	X6	0.799	4.071	
	X7	0.695	3.943	
	X8	0.717	3.971	
	X9	0.774	3.839	
	X10	0.746	3.986	
	X11	0.759	4.100	
	X12	0.754	4.057	

Note Data collected by Author in August 2022

respondents to the statement of the item are consistent. The indicator that has the highest standard deviation is browsing the internet every day which shows the diverse responses of respondents. Based on these results, not all of the start-up owners access the Internet every day. The existence of employees such as admin who has had the task of checking the conditions of sale, this does not require the owner to check at any time and every time because it will be reported by the admin in charge.

Data results of respondents' answers produced an overall mean value of the networking capability variable of 3.867. Table 5 gives that start-up owners in Surabaya agree that networking capability can help the development and performance of their businesses with their networking capabilities. The indicator of networking capability variable which has the lowest standard deviation is the acquisition of information faster than competitors with the contact network they have, and this shows that respondents' responses are consistent with the statement. The statement of the item illustrates that with a contact network owned by the start-up owner can get information faster than its competitors. The indicator with the highest standard deviation is building a good relationship with consumers, this high value indicates that respondents' responses are more diverse. Based on these results, statement items establishing good customer relations have diverse views from the owner start-up. A good relationship with consumers is one of the best ways for companies to maintain their business performance. Start-ups need ways and efforts to innovate in their business models because they face relatively more obstacles such as strategic capabilities and limited resources, poor network capabilities, and have limited capacity to grow and survive so network capabilities are needed by start-up to improve performance of the business (Lew, et al. 2022). According to Anwar et al. (2018), maintaining relationships with consumers is important in the sustainability of the business being run and has a direct impact on business performance. The mean value of all indicators of the new venture performance variable is 3948. The mean value means that the start-up in the Tokopedia database shows good performance in terms of self-reported from the owner of each start-up.

The results of this study indicate that the start-up respondent has sought to improve the company's performance in terms of sales, profits, and market share in terms of self-reported owner start-up valuation. Indicator of new venture performance variable with the lowest standard deviation is the company's sales performance which has increased compared to the last 1 year. This shows the respondents' answers related to the statement items are consistent. The indicator with the highest standard deviation is that the company's market share has increased compared to the past 1 year which shows the respondents' answers varied for this statement item. Market share is one of the benchmarks used in measuring a company's financial performance. This is supported by Anwar et al. (2018) which explains that measuring company performance can be measured based on self-reported information from the owner regarding market share, profits, levels of ROA, ROE, ROI, and others.

The mean of the highest variable X indicator is 4.129 and the mean of the highest variable Y indicator is 4.157. This shows that having entrepreneurial alertness and this will affect the performance of their business. The higher the ability to read opportunities, higher chance business for growth and success (Adomako et al., 2018). The results of this study are also supported by the research of Roundy et al. (2017) entrepreneurial alertness will speed up decision-makers to make decision choices and reduce uncertainty which will only waste time. Usually sensitive decision-makers have high sensitivity, accuracy, persistence, and innovative views which will make them quickly modify, update, and perfect their operations. Based on that attitude will positively affect the performance of the company itself.

The standard deviation of the lowest indicator X is 0.695 and the standard deviation of the lowest Y indicator is 0.850. The data can be interpreted that the ability possessed by the owner that is connecting all information in making decisions will impact on company performance. All information that has been obtained will be considered in making strategic decisions, the right decision will certainly affect the company's performance. This is supported by research conducted by Adomako et al. (2018) company performance is a common thing for a company to achieve to realize these targets requires entrepreneurial alertness in a company either from the owner himself or the decision-maker. The ability to read these opportunities can help companies read the market and respond to market opportunities. Entrepreneurial alertness will have an impact on company performance when business people can directly imply the opportunities that have been identified as new business opportunities in the market. This situation will make the company one step ahead in competing with competitors in one field.

This research shows networking capability cannot be a moderating variable in the effect of entrepreneurial alertness on new venture performance in this study. The results of the study show that the perspective of networking ability that is formed on two aspects of social and business networking as a moderating variable does not have a positive moderation effect on the relationship between entrepreneurial alertness to new venture performance. According to Fang et al. (2014) looking for opportunities to enable businesspeople to build new businesses and develop successful businesses such as entrepreneurial efforts will more or less influence new venture performance depending on the networked effort being undertaken. Networking efforts by companies differ, and the level of capability networking they have will affect the company's performance. According to Adomako et al. (2018), the higher the networking capabilities, the higher the level of alertness will certainly improve the company's performance. Conversely, if the ability to network is low, the level of alertness is low, the company's performance will decrease. This research shows that networking capability has a role as a predictor of moderation.

This research shows that the networking capability moderation variable is more likely to act as an independent variable that has a direct relationship to the dependent variable. The moderating variable of networking capability can also weaken the relationship between entrepreneurial alertness and new venture performance. Based on this, it can be concluded that the ability to network the company does not have a significant influence on increasing the sensitivity of the company to the company performing optimally. The network capabilities of a company directly influence the company's performance depending on how the company's ability to empower its network resources. The company can optimize its network so that it can access things that cannot be accessed directly by the company, with this capability can provide benefits for the company and can improve company performance.

5 Conclusion and Suggestion

Based on the results of research that has been done, several conclusions can be drawn as follows:

1. Entrepreneurial alertness has a significant effect on new venture performance at start-ups in Surabaya.
2. Networking capability cannot support the moderation of the effect of entrepreneurial alertness on new venture performance at start-up in Surabaya.

The alertness by the owner of the start-up is needed in the conditions of this dynamic business environment. An understanding of environmental conditions and finding opportunities among these conditions will allow the owner to excel over his rivals. After finding opportunities, it is also necessary to implement these opportunities into new business opportunities for the owner to create his new business. The development of a business with innovations will improve the performance of the businesses being managed.

The contribution of his study found that the role of networking capability owned by companies did not significantly strengthen the relationship of entrepreneurial alertness to new venture performance. The findings from this study provide insight for start-ups that the company's network capabilities do not affect the company's sensitivity to company performance. To improve company performance, high alertness is required. Additionally, the level of networking ability owned by the company will affect the level of performance of the company itself on how companies can leverage their network to improve company performance. The strategy that can be implemented by start-ups based on the results of this research is to improve business performance through the use of entrepreneurial alertness and strengthen the relationships that the company has. First, taking advantage of entrepreneurial alertness by employing strategic managers who have high scores in alertness is known through a pre-job offer assessment test. Second, the utilization of the company's internal and external relationships or networks to support the company's operational and strategic activities.

Future studies can use different variables as moderating variables such as competitive advantage, entrepreneurial orientation, and several other variables that might influence and strengthen the relationship between entrepreneurial alertness to new venture performance. Future studies are expected to use a wider scope and population that is different from the population in this study to get accurate results related to variables that influence the new venture performance. It is expected that further research on studies related to new venture performance in a long or longer period so that it can make deeper and consistent observations about the performance of a start-up from time to time.

References

Abdillah, W., & Jogiyanto. (2015). *Partial least square (PLS) Alternatif structural equation modeling (SEM) dalam Penelitian Bisnis* (edisi 1). ANDI.

Adomako, S., Danso, A., Boso, N., & Narteh, B. (2018). Entrepreneurial alertness and new venture performance: Facilitating roles of networking capability. *International Small Business Journal, 36*(5), 453–472.

Andersén, J. (2019). Resource orchestration of firm-specific human capital and firm performance—The role of collaborative human resource management and entrepreneurial orientation. *International Journal of Human Resource Management, 0*(0), 1–33. https://doi.org/10.1080/09585192.2019.1579250

Anwar, M. (2018). Business model innovation and SMEs performance—Does competitive advantage mediate? *International Journal of Innovation Management, 22*(7), 1–31. https://doi.org/10.1142/S1363919618500573

Anwar, M., Rehman, A. U., & Shah, S. Z. A. (2018). Networking and new venture's performance: Mediating role of competitive advantage. *International Journal of Emerging Markets, 13*(5), 998–1025. https://doi.org/10.1108/IJoEM-07-2017-0263

Baron, R. M., & Kenny, D. A. (1986). The moderator-mediator variable distinction in social psychological research: Conceptual, strategic, and statistical considerations. Journal of personality and social psychology, 51(6), 1173.

Bekraf. (2018). Mapping dan database startup Indonesia 2018. In *Badan Ekonomi Kreatif.*

Boso, N., Story, V. M., & Cadogan, J. W. (2013). Entrepreneurial orientation, market orientation, network ties, and performance: Study of entrepreneurial firms in a developing economy. *Journal of Business Venturing, 28*(6), 708–727. https://doi.org/10.1016/j.jbusvent.2013.04.001

Chavoushi, Z. H., Zali, M. R., Valliere, D., Faghih, N., Hejazi, R., & Dehkordi, A. M. (2021). Entrepreneurial alertness: A systematic literature review. *Journal of Small Business and Entrepreneurship, 33*(2), 123–152.

Dess, G. G., & Robinson, R. B. (1984). Measuring organizational performance in the absence of objective measures: The case of the privately-held firm and conglomerate business unit. *Strategic Management Journal, 5*(3), 265–273. https://doi.org/10.1002/smj.4250050306

Dibrell, C., Craig, J. B., & Neubaum, D. O. (2014). Linking the formal strategic planning process, planning flexibility, and innovativeness to firm performance. *Journal of Business Research, 67*(9), 2000–2007. https://doi.org/10.1016/j.jbusres.2013.10.011

Ebbers, J. J. (2014). Networking behavior and contracting relationships among entrepreneurs in business incubators. *Entrepreneurship: Theory and Practice, 38*(5), 1159–1181. https://doi.org/10.1111/etap.12032

Fang, S. R., Chang, E., Ou, C. C., & Chou, C. H. (2014). Internal market orientation, market capabilities and learning orientation. *European journal of marketing, 48*(1/2), 170–192.

Firdha, S. (2019). Bagaimana Perusahaan Startup Mencapai Level "Unicorn" bahkan "Decacorn"? *Eproc*, 1. Retrieved from https://www.eproc.id/berita/detail/bagaimana-perusahaan-startup-mencapai-level-unicorn-bahkan-decacorn

Gallop, & David, L. (2019). The lean startup: How today's entrepreneurs use continuous innovation to create radically successful businesses. *Defense A R Journal, 26*(2), 168

Ghozali, I. (2006). *Aplikasi Analisis Multivariate dengan Program SPSS* (Edisi Ke 4). Badan Penerbit Universitas Diponegoro.

Karami, M., & Tang, J. (2019). Entrepreneurial orientation and SME international performance: The mediating role of networking capability and experiential learning. *International Small Business Journal: Researching Entrepreneurship, 37*(2), 105–124. https://doi.org/10.1177/0266242618807275

Kirzner, I. M. (1983). *Perception, opportunity, and profit*. Chicago, IL, USA: Chicago University Press.

Kirzner, I. M. (1999). Creativity and/or alertness: A reconsideration of the Schumpeterian entrepreneur. *Review of Austrian Economics, 11*(1–2), 5–17. https://doi.org/10.1023/A:1007719905868

Kurniawan, A. W., & Puspitaningtyas, Z. (2016). *Metode Penelitian Kuantitatif* (Cetakan Pe). Pandiva Buku.

Lew, Y. K., Zahoor, N., Donbesuur, F., & Khan, H. (2022). Entrepreneurial alertness and business model innovation in dynamic markets: International performance implications for SMEs. *R&D Management.*

Luk, C. L., Yau, O. H. M., Sin, L. Y. M., Tse, A. C. B., Chow, R. P. M., & Lee, J. S. Y. (2008). The effects of social capital and organizational innovativeness in different institutional contexts. *Journal of International Business Studies, 39*(4), 589–612. https://doi.org/10.1057/palgrave.jibs.8400373

Mitrega, M., Forkmann, S., Zaefarian, G., & Henneberg, S. C. (2017). Networking capability in supplier relationships and its impact on product innovation and firm performance. *International Journal of Operations & Production Management, 37*(5), 577–606.

Neuman, W. L. (2015). *Metodologi Penelitian Sosial: Pendekatan Kualitatif dan Kuantitatif.* PT Indeks Permata Puri Media.

Oebaidillah, S. (2019, October). Pemerintah Dukung Pengembangan Startup di Indonesia. *Media Indonesia*, 1. Retrieved from https://mediaindonesia.com/read/detail/268323-pemerintah-dukung-pengembangan-startup-di-indonesia

Priyatno, D. (2014). *SPSS 22 Pengolahan Data Terpraktis*. CV. Andi Offset.

Rahman, A. F. (2019, October). Ovo Jadi Startup Unicorn ke-5 Indonesia. *Detik.Com*, 1. Retrieved from https://inet.detik.com/business/d-4733416/ovo-jadi-startup-unicorn-ke-5-indonesia

Roundy, P. T., Brockman, B. K., & Bradshaw, M. (2017). The resilience of entrepreneurial ecosystems. *Journal of Business Venturing Insights, 8*, 99–104.

Roundy, P. T., Harrison, D. A., Khavul, S., Pérez-Nordtvedt, L., & McGee, J. E. (2018). Entrepreneurial alertness as a pathway to strategic decisions and organizational performance. In *Strategic Organization* (Vol. 16). https://doi.org/10.1177/1476127017693970

Saputra, A. (2015). Peran Inkubator Bisnis dalam Mengembangkan Digital Startup Lokal di Indonesia. *Jurnal Ilmiah.*

Sharma, L. (2018). A systematic review of the concept of entrepreneurial alertness. *Journal of Entrepreneurship in Emerging Economies, 11*(2), 217–233.

Tajeddini, K., & Mueller, S. L. (2012). Corporate entrepreneurship in Switzerland: evidence from a case study of Swiss watch manufacturers. *International Entrepreneurship and Management Journal, 8*, 355–372.

Tang, J., Baron, R. A., & Yu, A. (2021). Entrepreneurial alertness: Exploring its psychological antecedents and effects on firm outcomes. *Journal of Small Business Management, 8*, 1–30.

Tang, J., Kacmar, K. M. M., & Busenitz, L. (2012). Entrepreneurial alertness in the pursuit of new opportunities. *Journal of business venturing, 27*(1), 77–94.

Zhao, W., Li, J., Li, X., & Schøtt, T. (2020). Implications of network diversity for venture growth: The mediation effect of entrepreneurial alertness. *Sustainability, 12*(22), 9762.

Fashionpreneur: Sustaining Traditional Batik Craft Through Entrepreneurship Activity Among Students at Universiti Malaysia Kelantan (UMK)

Hanisa Hassan, Norhasliyana Hazlin Zainal Amri, Mohd Zaimmudin Mohd Zain, and Nurulahda Sulaiman

Abstract

Local craft such as batik is on a regressive trend in Malaysia. Statistic showing the decline of batik makers during the past 10 years is worrying. Most artisans are veterans in the craft business and often have problems passing down their legacy due to the lack of interest among youngsters. The labour-intensive work of batik-making and the competitive market may discourage them from venturing into the batik business. Therefore, the 'Fashionpreneur' project was introduced. The word 'Fashionpreneur' is a combination of fashion and entrepreneurship disciplines. The project was targeted to combine craft and entrepreneurship, targeting younger generations to venture into local batik business. The project was done during the 2020/21 COVID-19 pandemic, which aimed to stimulate interest in batik among the final-year students at Universiti Malaysia Kelantan (UMK), in the state of Kelantan, Malaysia. The overall project took about 14 weeks to finish, from designing toile until producing the final prototypes. As this project was done during the period of semi-lockdown, half of the semester was done at the Faculty's premise whilst the remaining weeks were done at a local batik workshop. The project's objective is to track how many students from this project venture into the batik or fashion business after graduating. Previous literature shows that those who had family business in batik have higher chances of venturing into similar business. Prior research also suggested that lack of experience or support from family plus lack in funding were the contributing factors to the setback. A set of survey was given to all eight students involved in the project. Seemingly, the theories regarding batik business are more sustainable when it is run by family with proper funding are justified. After a year of graduating, none of the sampled students ventured into batik or fashion-related business. Even though, the project was unsuccessful in achieving its target to produce younger entrepreneurs in fashion/batik, it has initiated a new strategy to do similar projects in future. Targeting students from batik-related families in future may have a higher chance of producing batik 'fashionpreneur'. Sustaining the craft must start at an early stage so that the interest and passion in batik can be developed later in the future.

Keywords

Batik · Craft · Entrepreneurship · Fashionpreneur

1 Introduction

Batik making in Kelantan has suffered a regressive trend throughout the years. The number of batik stamping (*cop/sarang*) makers also showed a worrying decline that indicates the shrink of the industry (Faradiba Liana Naser, 2020). One of the declining factors contributors is a massive flood in 2014 that swept many batik-makers' workshops along the riverbank (Hassan, 2018). After the flood, batik makers ventured to other types of jobs to sustain their income. Another contributing factor to the decline of batik makers is that they cannot cope with the competing rivals of imported textiles, which are cheaper and faster to produce (Noor & Baharin, 2019).

H. Hassan (✉) · N. H. Z. Amri · M. Z. M. Zain · N. Sulaiman
Faculty of Creative Technology and Heritage, Universiti Malaysia Kelantan, Kota Bharu, Malaysia
e-mail: hanisa@umk.edu.my

N. H. Z. Amri
e-mail: hazlin.za@umk.edu.my

M. Z. M. Zain
e-mail: zaimmudin@umk.edu.my

N. Sulaiman
e-mail: nurulahda.s@umk.edu.my

M. Nawaz Tunio et al. (eds.), *Sustainability in Creative Industries*,
Advances in Science, Technology & Innovation, https://doi.org/10.1007/978-3-031-48453-7_4

The decline is also contributed by the fact that many young adults have lost interest in batik making and prefer to work in a more conducive working environment that provides a steady income (Razali, 2021a, 2021b). The fact that younger generations are the ones who will determine the sustainability of the craft makes the situation worrying. The lack of interest among young adult in batik business forecast further decline of Kelantanese batik producers.

Many case studies had mentioned that most batik entrepreneurs run within the family business. Not many started their own batik business despite the hundreds of textile design graduates produced yearly. The labour-intensive process in batik making might not seem appealing to the younger generation unless they had been exposed to such experience since childhood. Therefore, early exposure in batik, whether in their family business or family working within the field, contributes to the sustainable of the craft.

From annual surveys (from 2021 to 2022 data) conducted by the Student Affairs Department at Universiti Malaysia Kelantan (UMK) also revealed that not many students pursue their business after graduating; despite all the guidance and prior knowledge given. Working within their area of study also derailed in most cases. Therefore, this study tries to identify whether instilling batik projects as future business during their study could draw their interest in becoming 'fashionpreneur' in batik.

A high impact grant entitled 'fashionpreneur' that was funded by Ministry of Higher Education Malaysia was aimed to aid the batik initiative. The project was led by four researchers from the Faculty of Creative Technology and Heritage (FTKW) of UMK. The grant conducted is aimed to understand the reasons why younger generation are uninterested in venturing batik-related businesses, which lead to the declining numbers of batik entrepreneurs in Kelantan.

A simple survey was conducted during the student's graduation day, and the data was used to formulate hypotheses related to the main aim of this study as stated earlier. Next, the researchers aimed to identify the measures to overcome the problem. It should be noted that since only 12 students took part in the survey, their opinions cannot be generalised, but the data was sufficient enough to be used for future research relating to area of study, interest and choice of profession after graduating. All the students had studied fashion and textile for three years and continued with another year of attachment at the actual industry. This study tries to investigate whether a real project with the batik industry can instil their interest of batik business after graduating.

From a simple survey upon the students' graduation, information was gathered to determine whether the hypothesises suggested are accurate. If so, the question will be, what should be the measures to overcome the problem? Nevertheless, since only 12 students took part in the survey, their opinions cannot be generalised; but enough to be used for future research relating to area of study, interest and profession after graduating. All the students had studied fashion and textile for three years and continued with another year of attachment at the actual industry. This study tries to understand whether a real project with the batik industry can instil their interest of batik business after graduating.

1.1 Literature Review

Three hypotheses constructed to understand this case study as mentioned below:

H1: Most batik entrepreneurs inherit batik-making skills from their families. Those who are not born in the batik-related business rarely venture into the batik-making business.

Even though batik has been made in Kelantan since the late 1920s and 1939 in Terengganu, it has been monopolised approximately 86% by microenterprises (Noor Haslina Mohamad Akhir, 2015). Most batik entrepreneurs inherit batik-making skills from their family and continue their business. Those who are not born in the batik-related family rarely venture into the batik-making business.

In Indonesia, batik companies managed by Chinese families, such as in Lasem are usually run by a family firm model. Such companies are able to sustain their business for the next generation. Quite a number successfully passed on their business until its 3rd generation, unlike in the West, only 14% of family business lasts to their 3rd generation (Budhiyanto, 2019).

In another study by Sobirin on batik in Pekalongan, the success of a family run business in batik relies on its succession planning among their family members. Findings from four families of batik makers revealed that they do not have any written agreement on batik succession, but they already equipped their children inheriting the business as a part of the responsibility for the family. Their family taught them the love for batik since they were small, so that they would not be reluctant to continue the business (Achmad Sobirin, 2013).

In many family oriented businesses relating to batik making in Indonesia, e.g. the trade secret is kept around the chosen family members whom the head of family choose to continue the legacy. The colour mixing formula or the mixture of the wax is something they keep as family secret and legacy which they will never reveal to others. Such action is meant to keep the family business intact, therefore it was never passed to outsiders. Recent study by Bingbing Ge and Giovanna Campopiano (2021) mentioned that,

> In particular, managing knowledge can prove a key success factor for handing over the business to the next generation members. Indeed, this is not a unidirectional process where all knowledge is passed over from the incumbent to their children during succession, but it is a bidirectional process (Woodfield and Husted, 2017), which involves a complex and time-dependent set of activities and practices to succeed. (Le Breton-Miller et al., 2004)

Until today, the people of Girilayu still hold the teaching of Mangkunagara I to possess, to sustain and to be brave enough to act on useful matters (Desy Nurcahyanti, 2020).

H2: Most graduates did not work in their field of study. In this case, even though the students were exposed to batik making during their degree course, many might opt for an easier job with fixed salary.

An article entitled *Current Challenges of The Batik Industry in Malaysia and Proposed Solutions* (2021) mentioned that the lack of skilled workers from young people should also be considered when many young people voiced that they are no longer interested in working in the batik industry. This may be because the labour-intensive batik-making process may not appeal to the younger and educated population. The study summarised that the shortage of workers in Terengganu's batik industry contributed to 76.2% of the batik production problems. Despite many institutions offer batik courses at certificate, diploma, and degree levels, the shortage of workers in the industry is still upsetting (Razali, 2021a, b). Most of them did not pursue their practice in batik after graduation, which may due to the fact that other sectors offer less labour-intensive work with a steady income. A study by Kent University (Yoon Jin Ma, 2013) revealed in its descriptive analyses on merchandising and design-major students (using SPSS 21.0); that the most important expectations from design students are job security (4.56) and a feeling of accomplishment (4.66). Job security means assurance that you will be able to work in your job as long as you please and will not become unemployed (HarperCollins, 2022).

- H3: Producing 'fashionpreneur' will not be easy without fundamental support such as funding for a start-up business in fashion or batik.

'An entrepreneur without funding is a musician without an instrument' (Jr, 2012). For a graduate student with a very limited entrepreneurship experience, venturing into a new business is a scary thought. Most students rushed to get a job to cover their basic needs rather than starting a business, which might risk their only savings. About 80% of UMK students are in the B40 income group, making it too risky to start their own business. Therefore, they rushed to grab any job opportunity that comes as long as they got paid; often underpaid. According to the Graduate Employability report, among students of UMK in 2020/21, less than 20% are getting salaries above RM2000 (USD500). Therefore, funding is essential for them to start-up a new business under their label. It is even more crucial when the person has no family support. 'A business without a funding source will crumble under the weight of its own loans. Funding is the foundation on which every start-up thrives. One can escort to different ways and methods to attain funding, and more than one option can always be used' (Shivam, 2017).

2 Methodology

A case study approach was conducted to investigate whether the chosen final-year students from the Faculty of Creative Technology and Design were interested in batik business. 12 students volunteered to venture into this project called 'fashionpreneur', whereby an industry expert mentored in the fashion and batik field at his workshop. They must produce at least three completed fashion styles using batik for their final project.

During the first to the seventh week, the students developed their sketches and design concepts with their lecturers. After getting approval for their final three designs, they had to produce a set of toiles for fitting so that they would know the amount of fabric needed to make the collection. Three students made a fashion collection consisting of a fashion style, another three made batik sarong, and the remaining two made batik scarf.

Unfortunately, on week eight (8), there was another curfew round due to the increased number of covid-19 cases in Malaysia. All schools and higher institutions were instructed to close down and to be conducted online. Therefore, another seven weeks needed to be done outside the university's compound as all institutions were ordered to close.

Consequently, a new location is needed for the project. All the students involved in the project were moved to a batik workshop owned by Rumah Gahara at Penambang in Kelantan. They were stationed there for the remaining seven weeks and stayed at the house belonging to the workshop owner. The premise also has a batik workshop behind the house, so the students could continue with their projects there (Figs. 1 and 2). Sewing machines were also transported from the university to the house to ensure that students could finish their projects on time. All supplies of food rations were also given to them so that they could survive during the lockdown without going out to buy food. During that time, the owner of the batik workshop, Mr. Nik Faiz, also occasionally came to check on the students' progress and monitor their batik-making process. He also taught the students the batik business trade since

Fig. 1 Students at work during the lockdown

Fig. 2 Batik colouring process at the workshop

he has been a batik and fashion entrepreneur for more than 15 years. All the students completed their project within the given time frame and made a photo shoot for their project portfolio (see Figs. 3, 4, 5 and 6).

3 Analysis

From the 12 students who had participated in this project, only eight responded to the survey. All the eight respondents graduated in November 2021. The students were contacted a year later to answer five simple questions regarding their industrial attachment under this project at Rumah Gahara's workshop. The aim of this survey is to understand whether the project had been able to produce 'fashionpreneur' after the students' graduation.

Fig. 3 Office wear in batik

Fig. 4 Casual wear in batik

i. Do you have the ambition to become a fashionpreneur while studying?
ii. After completing your industrial attachment (at Rumah Gahara) do you still have interest in becoming a fashionpreneur in batik?

Fig. 5 Batik sarong on cotton

Fig. 6 Batik scarf

iii. Does your current job relate to fashion or batik?
iv. Why are you not pursuing your job as fashion/batik entrepreneurs?
v. What is your view on fashion/batik business after you had graduated?

4 Discussion

Based on the survey result as indicated in Figs. 7, 8, 9, 10 and 11, all the students involved in this project did not have prior knowledge or exposure to batik-making until they entered UMK. All of them came from the B40 (lower income group) family, earning less than RM4850 (USD1091) median income monthly. As per discussed in H1 regarding family business (see Budhiyanto, 2019), it is not a surprise to see that they did not continue with their batik business after graduation, even though all of them had registered a company under Companies Commission of Malaysia (SSM) during their final year of study. As stated by Noor Haslina Mohamad Akhir, 2015, majority batik entrepreneurs inherit batik-making skills from their family and continue their business. Therefore, the result showing alumni not venturing in batik business upon graduating is not surprising. Since they neither have family support nor sustainable income, it was unlikely for them to choose the industry as career path after graduation.

It also appears that job stability and fixed salaries are still their main goals even if they have to settle for a lesser pay as graduates. In Malaysia, the basic minimum salary is RM1500 (approx. USD341) with a minimum high school education. Bachelor graduates should get a minimum wage of RM2000, but in most cases, they sadly have opted for income which is below than what they should receive. Recent report extracted from the Graduate Tracer Study Report (SKPG) for UMK in 2022 revealed that only about 13.7% of the alumni get paid more than RM2000.00 (Approx. USD455). When the survey is zoomed to the faculty (FTKW), only 20% work in similar field of study. This may be due to other areas that may offer better salary. It is also discovered that only 10.16% works independently as small entrepreneurs or working in their family business. The rest of 60% opt for fixed salary profession (UMK, 2022).

With reference to this study, the students' interest in becoming a batik entrepreneur was quite high, i.e. the response was 75%, before they went for practical training at the actual batik workshop. Unfortunately after learning the real hardship or the process or batik making and finishing the final prototypes, their interest level dropped significantly to 37.5% (as survey responded 'interested' and 'very interested' combined). Their attachment with industry partners at actual batik workshop helped them learn more effectively (62.5%), and the other 37.5% strongly agreed. Unfortunately, after their first-hand experience with the labour-intensive process of making batik, their interest in the batik business declined to 62.5% of unsure answers, whereby the remaining 25% are still very interested, and 12.5% are interested. Further discussion with the students revealed that working in a non-conducive environment weakened their spirit to venture into such business. As mentioned in the prior review, those who are not exposed to batik making since childhood are more likely to surrender after learning its demanding processes.

Next, H2 stated that even though the students were exposed to batik making during their degree course, they still

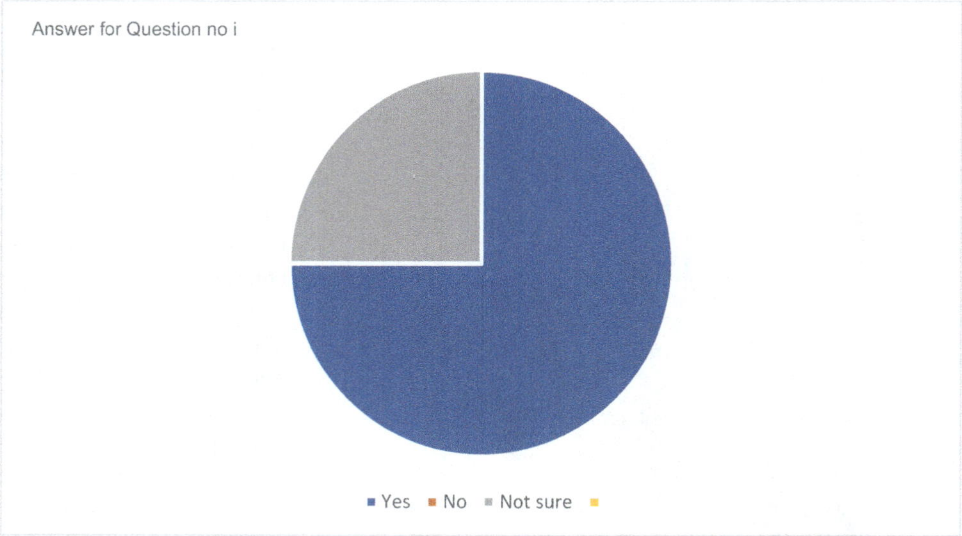

Fig. 7 62.5% of the respondents answer yes, the remaining 37.5% answered no

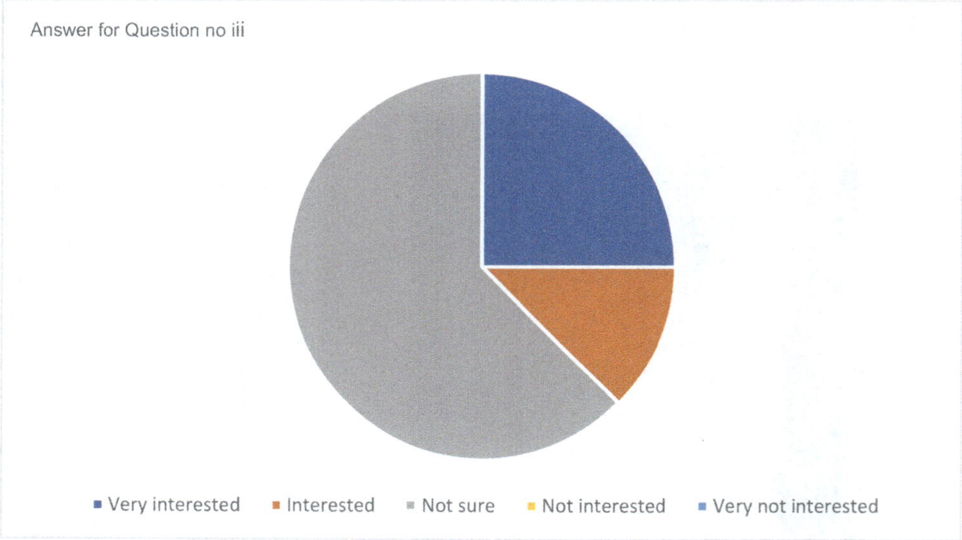

Fig. 8 62.5% are unsure whether they are interested in venturing into fashion/batik business after their practical at Rumah Gahara. 25% are very interested, and the remaining 12.5% are interested

might opt for an easier job with a stable or fix salary. The survey revealed that, most of them opt for a stable monthly income job (62.5%). 12.5% admitted that venturing into real business after graduating needs funding and vast experience that they might not have responded that they get a steady monthly income in other jobs, whilst equally 12.5% replied that they do not have. Others responded that the answers suggested are not related. This is also an expected answer as the earlier survey mentioned that most graduates prefer job security related to a stable income. Therefore, 75% of the respondents answered that their current job is not related to fashion/batik but with fixed monthly income. Only 12.5% said it is somehow related to the batik/fashion business, as stated in Fig. 9. The students may also opt for a job with lesser skill and knowledge as long as it pays their bills since they need to be self-sustained right after graduation.

Nevertheless, 75% of the respondents still felt that batik has a huge potential for the fashion business. The remaining 12.5% believed that there is demand for it but it is not so huge. The rest believed that the batik/fashion business is profitable. The simple survey agreed that batik entrepreneurs normally run as a family related business. This may be because they were exposed to batik making since their early childhood, making the business less scary for them to venture. This is also admitted by Nik Faiz Nik Amin whose grandfather had run a batik workshop. He got the exposure and interest in batik by learning from his family (Amin, 2020). Therefore, it is confirmed that those who were born in the batik-related business are more likely to venture into similar businesses than those who were not.

Finally, H3 stated that producing 'fashionpreneur' will not be easy without fundamental support such as funding

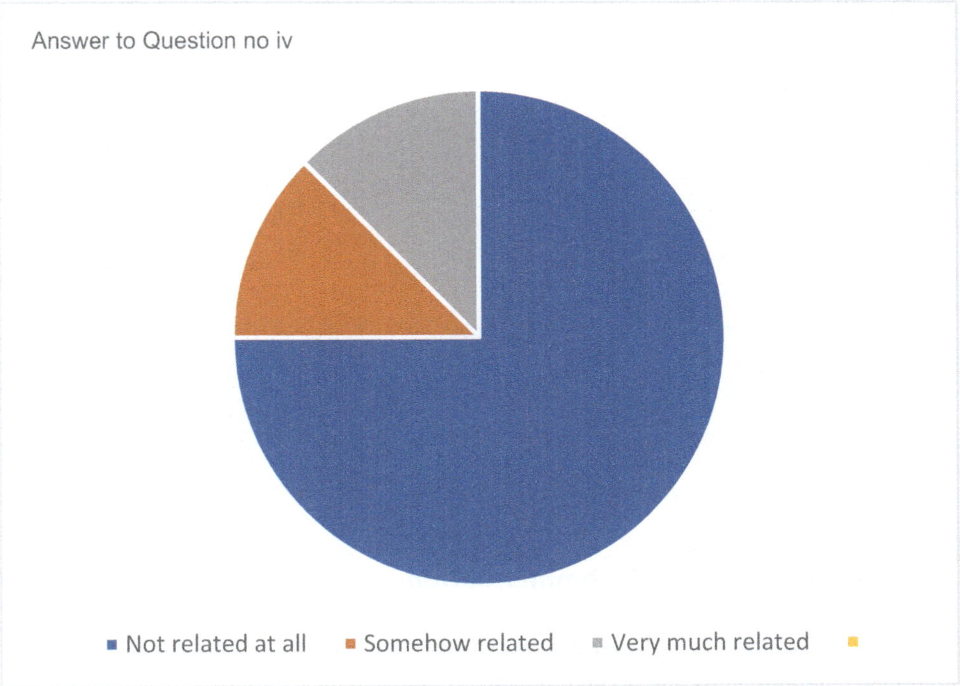

Fig. 9 75% of the respondents answered that their current job is not related at all with fashion/batik. The remaining 12.5% said it is somehow related and the other 12.5% said that it is very much related

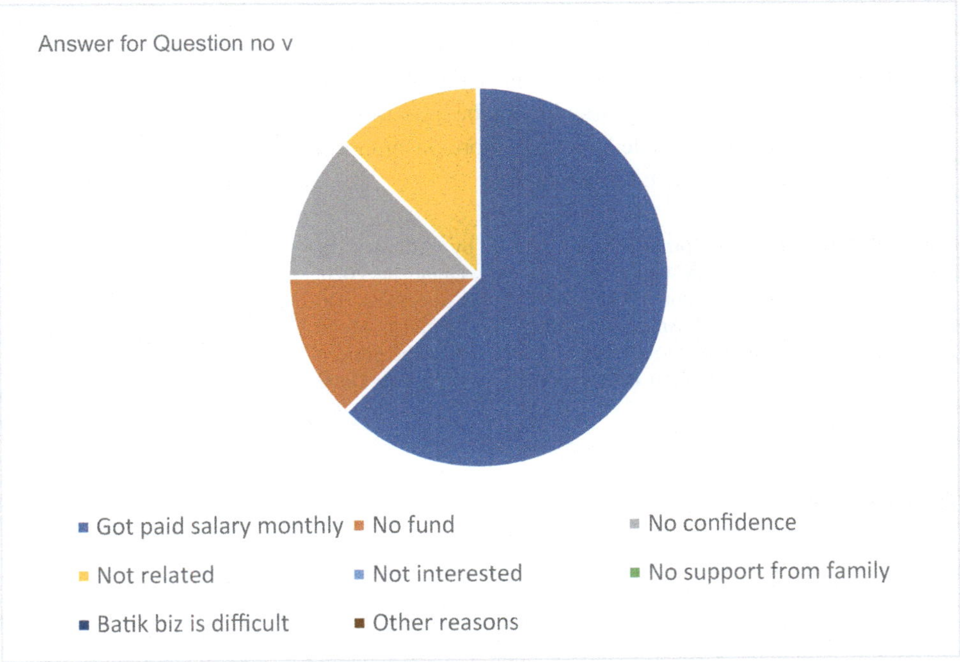

Fig. 10 The majority of them (62.5%) responded that they get a steady monthly income in other jobs, while 12.5% replied that they do not have enough experience and confidence to pursue fashion/batik business. Others responded that the answers suggested are not related

for a start-up business in fashion or batik. This may be due to the fact that all of them came from a B40 family group, in which the funding to start-up a business is limited. This may lead to the fact that most graduates did not work in their field of study as it will add on to their current students' loan. Starting a job and get paid steadily are more important to them than taking a risk on a business with no certainty. It also confirms that producing fashionpreneur will not be easy without fundamental support such as funding for a start-up business in fashion or batik companies.

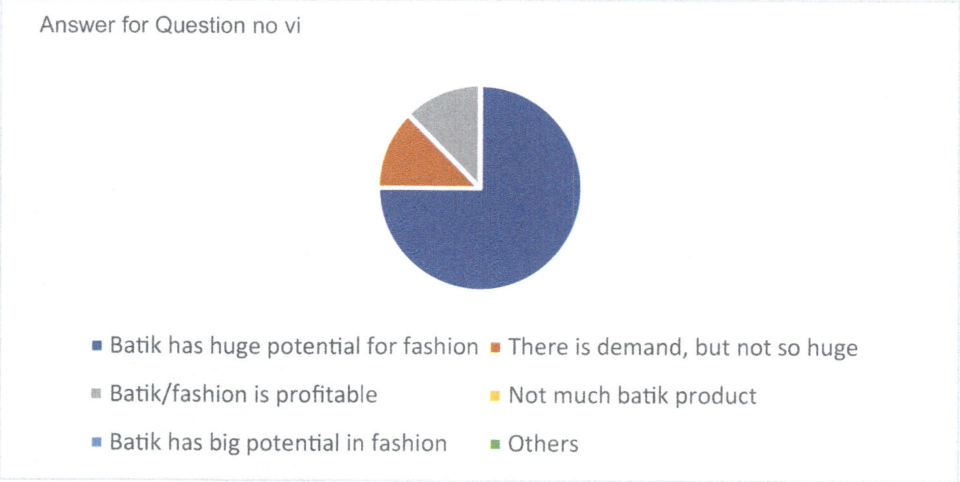

Fig. 11 75% of the respondents feel that batik has huge potential for the fashion business. 12.5% believe that there is demand but not so huge. The remaining noted that the batik/fashion business is profitable

5 Conclusion and Recommendation

Even though this funded project failed to produce any fashionpreneur, similar study should be replicated to understand whether or not taking someone from similar family background will produce better batik entrepreneurs. With the right candidate of respondents, the result may confirm the prior theories that family plays an important role to sustain a batik business. Since this study is only being responded by eight students, the finding should not be generalised. Further investigation is needed so that more fashionpreneurs in batik can be produced in the future.

Acknowledgements The project was funded by Ministry of Education (MOE) Malaysia under High Impact Grant in 2020. The head of this project would like to thank her members, Dr. Norhasliyana Hazlin Zainal Amri, Nurulahda Sulaiman, Dr. Mohd Zaimmudin Mohd Zain and Dr. Arif Datoem for assisting the project. We also like to thank Haida binti Hassan and Dr Suhaida binti Omar for editing this article.

References

Achmad Sobirin, N. F. (2013). Suksesi Pada Perusahaan Keluarga. *Journal Manajemen dan Bisnes Indonesia*, 14–27.

Amin, N. F. (2020). *Gahara*. Retrieved from https://www.gahara.my/: https://www.gahara.my/about-gahara/

Budhiyanto, L. K. (2019). *Family firm sustainability: An ethnographic study of change and adaptation in the traditional batik industry in Lasem, Indonesia*. Oxford Brookes University.

Desy Nurcahyanti, A. S. (2020). Peran Kearifan Lokal Masyarakat Jawa Untuk Melestarikan Batik Tradisi di Girilayu, Karanganyar, Indonesia. *Mudra*, 145–153.

Faradiba Liana Naser, H. K. (2020). Cultural sustainability: A case study of the dying art on Batik. In *Proceedings of the 1st international conference on interdisciplinary arts and humanities (ICONARTIES 2019)* (pp. 78–81). SCITEPRESS.

Ge, B., & Campopiano, G. (2021). Knowledge management in family business succession: Current trends and future directions. *Journal of Knowledge Management, 26*(2), 326–349.https://doi.org/10.1108/JKM-09-2020-0701

HarperCollins. (12 August, 2022). Retrieved from Collins Dictionary: https://www.collinsdictionary.com/dictionary/english/job-security

Hassan, H. (2018). 4th IR and challenges for the creative and cottage industries. In *International conference on design, engineering and computer sciences* (p. 453). IOP Publishing Ltd.

Jr, R. A. (2012). *Government and start-ups*. Wiley Book.

Le Breton-Miller et al. (2004). Toward an Integrative Model of Effective FOB Succession *Entrepreneurship Theory and Practice, 28*(4), 305–328. https://doi.org/10.1111/j.1540-6520.2004.00047.x

Noor Haslina Mohamad Akhir, N. W. (2015). Traditional craftsmanship: The origin, culture, and challenges of Batik industry in Malaysia. In H. B. Roaimah Omar (Ed.), *Islamic perspectives relating to business, arts, culture and communication* (pp. 229–237). Springer.

Noor, H. M., & Baharin, N. H. (2019). *Batik Malaysia terancam*. Berita Harian.

Razali, H. M. (2021a). Current challenges of the batik industry in Malaysia and proposed solutions. In *8th International conference on advanced material engineering & technology (ICAMET2020)*. AIP Publishing.

Razali, H. M., & Ibrahim, M. (2021b). Current challenges of the batik industry in Malaysia and proposed solutions. In *ICAMET 2020. AIP conference proceedings* (Vol. 2347, Issue 1).

Shivam, A. G. (2017). Funding to growing start-ups. *Research Journal of Social Science*, 22–31.

UMK, H. (2022). *SKPG (Saringan Kedua) 2022/23*. UMK.

Woodfield, P., & Husted, K. (2017). Intergenerational knowledge sharing in family firms: Case-based evidence from the New Zealand wine industry *Journal of Family Business Strategy 8*(1), 57–69. https://doi.org/10.1016/j.jfbs.2017.01.001

Yoon Jin Ma, K. H. (2013). Job expectations, job preference, and learning expectations of apparel merchandising and. In *ITAA proceedings* (pp. 35–36). International Textile and Apparel Association Inc.

Mascot and Brand Sustainability in Pandemic Era: Systematic Literature Review

Hedi Amelia Bella Cintya

Abstract

Nowadays, Mascot is widely used for various purposes, such as brand identity and promotion. Japan can be considered the most prosperous country in creating many iconic mascots that have proven to be popular and influential in place branding. The utilization of mascots for place branding is regarded as a common practice in Japan, to the point that its prefectures and provincial policies have their own mascots. Furthermore, several cities in Indonesia, such as Surabaya, Balikpapan, and Malang, have also created city mascots that act as city icons and representatives. In addition to place branding, mascots can be used for communication or events such as the Asian Games 2018, in which the mascot acts as its event branding and represents the eastern, central, and western regions of Indonesia. Are mascots still relevant in this pandemic era? The significance of the mascot's impact on brand sustainability in this pandemic era deserves to be questioned due to the drastic changes that the pandemic has brought to all aspects of life. With the "new normal," most things have shifted online instead of offline to avoid any physical interaction that can trigger the spread of the virus. Moreover, user-friendly websites and applications are now more in demand than ever, and some are using mascots as intermediaries between the website and the user. Therefore, up-to-date research is needed to identify the connection between mascots and brand sustainability. The proposed study aims to summarize the literature on the development of the use of mascots for the past three years during the covid-19 pandemic era from 2020 to 2022. This research utilized the Systematic Literature Review method by assembling two research questions (RQ), (1) Are there any shift in the use and benefits of mascots to brand sustainability during the pandemic era? (2) How has mascot research developed to date? Journals were collected based on journals indexed by Google Scholar, and the keywords used were mascot, brand, and pandemic. The final result of the study answered the two RQs.

Keywords

Mascots · Branding · Sustainability · Literature review

1 Introduction

When someone is selling lemonade, what is the difference between that exact lemonade and the other lemonade sold by the person next door? Branding emerges as the solution to differentiate a product, event, or activity from others. Proper branding can even help sustain a brand during times of crisis (Ash, 2020), which is particularly crucial now. A global-scale pandemic, like the one we are experiencing today, is often considered a "once in a lifetime" event; therefore, to move forward, humanity must adapt to the current situation. Several years after the pandemic hit, multiple parties worldwide, including organizations and companies, have made various innovations to adapt to the situation. To survive the pandemic, organizations and companies have made significant breakthroughs: some have reduced their operational cost, some have temporarily closed their businesses (Antonova & Honcharova, 2021), and others have strengthened their brand identity through various branding strategies. One of the well-known branding strategies is the use of mascots as brand representatives. Research conducted by Esfahani et al. (2022) demonstrates that mascots have a significant effect on increasing the brand's awareness, brand recall, and customer's intention to purchase the brand's product. Compared to a logo, a mascot

H. A. B. Cintya (✉)
Product Design and Management, Creative Industry Faculty, University of Surabaya, Surabaya, Indonesia
e-mail: hediabc@staff.ubaya.ac.id

M. Nawaz Tunio et al. (eds.), *Sustainability in Creative Industries*,
Advances in Science, Technology & Innovation, https://doi.org/10.1007/978-3-031-48453-7_5

offers numerous advantages due to its physical form, which resembles a human and creates familiarity (Ko et al., 2022). On certain occasions, such as the Olympic Games, the mascot becomes a critical component of the overall brand (Palladino et al., 2022), playing a vital role in humanizing modern corporations (Singh, 2021).

2 Research Planning

The changes in conditions before and during the pandemic have also compelled organizations and companies to change their mascot as part of their branding strategies to align with current conditions. This change raises a few questions:

- First, are there any shifts in the use and benefits of mascots to brand sustainability during the pandemic era?
- Second, how has mascot research developed to date?

Understanding the connection between brand sustainability and mascot during the pandemic era will not only give insight for future research but also serve as a strong example of a successful branding strategy that strengthened brand sustainability during the pandemic era. Given that humankind has lived for thousands of years, it is undeniable that another pandemic will occur in the distant future.

The research was conducted to find the answer to the research questions above. The conducted research was done using the Systematic Literature Review Method as follows:

- Database for Research: This research mainly used Google Scholar as its main source. Through GoogleScholar, the search might broaden to another digital library from various publishers such as ResearchGate, SpringerLink, Elsevier, and TandF.
- Keyword for Research: This research focused on five main keywords: Brand, Mascot, Pandemic, covid-19, and Virtual and a combination of the five keywords, (e.g., "Virtual Mascot", "Brand Pandemic", etc.).
- Source of Research: The source that was used for the conducted research was as follows: books, journals, conference proceedings, and theses. The conducted research also used a Website as a supplementary source.

This paper consists of five main sections, in addition to the two opening sections, Introduction and Research Planning. The first section explores the development of mascots unaffected by the pandemic era. Besides research from various researchers, mascots from famous events such as the Olympic Games is also discussed in this section. Next, the second section discusses how the pandemic has compelled brand owners to adapt to the situation by changing strategies to maintain brand sustainability. The third section focuses on the main topics: the changes in mascots during the pandemic era. This section primarily discusses research conducted worldwide on strategies to sustain brand viability. The fourth section is a discussion section providing an analysis of sections one, two, and three. The last section is the conclusion, where the research question is answered.

3 Mascots Without the Pandemic

Even when we set aside the pandemic context, the development of mascots has been undertaken multiple times with different objectives and representations. Research conducted by Kaufmann (2021) attempts to analyze an existing mascot from a football club to determine its suitability for the market using SWOT (Strengths, Weaknesses, Opportunities, and Threats) analysis. In addition to providing recommendations for future work, the research also highlights the less advanced usage of mascots for sports clubs in the European region compared to the United States. It is widely agreed upon that the use of mascots in the United States is highly advanced. Mascots have become a common feature in sports clubs across the United States.

United States College Sports Club employ various methods to create their own mascots that are able to represent their sports club. Whether they take the form of human-like mascots or even real individuals, such as the Leprechaun used by the Notre Dame Fighting Irish football team based in Indiana, United States (Kaufmann, 2020), mascots play a crucial role. In the United States, the creation of mascots can involve collaboration from multiple disciplines. For instance, a collaboration between disciplines resulted in the creation of a robotic goat-like mascot by a student from Worcester Polytechnic Institute (Chang et al., 2020), an institution located in Worcester, United States. This strengthens the previous research's assertion that the utilization of mascots in the United States is undeniably advanced. Different from how the United States employs mascots to represent a particular organization, company, or sports club, Japan frequently uses mascot for tourism purposes. Many mascots, usually known as "*Gotouchi kyara*", are created to represent each of the prefectures or even cities in Japan. Not only are these mascots used as a city branding tool to promote local products and tourism spots, but these mascots are also used as a means of communication between local residents (Lisa, 2014). An example of a well-known *Gotouchi kyara* mascot in Japan can be seen in Figs. 1, 2 and 3.

From left to right, Barii-san serves as the mascot for Imabari City in Ehime Prefecture. Its physical form resembles a chicken, as Imabari is renowned for its chicken-based cuisine, such as Yakitori, meaning grilled chicken. In Fig. 2, the melon bear represents two distinctive aspects of

Fig. 1 Barii-san

Fig. 3 Kumamon

Fig. 2 Melon bear

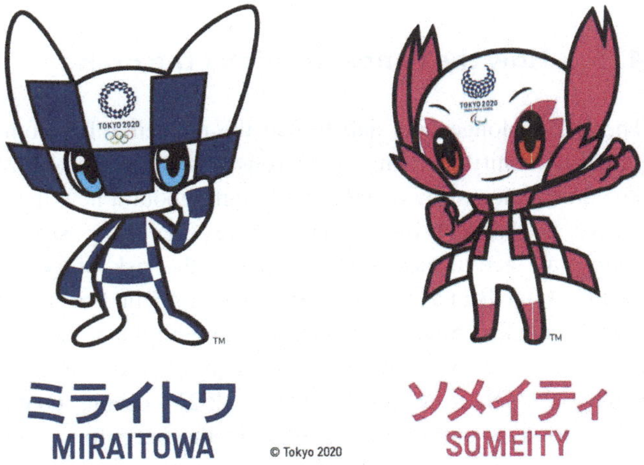

Fig. 4 Tokyo Olympics Games 2020

Yubari City in Hokkaido Prefecture, namely their melons and bears. The final picture, Fig. 3, showcases Kumamon, a widely recognized mascot that represents Kumamoto Prefecture and was voted the best mascot in Japan in 2011 based on public voting (Village House, 2020). Besides mascots for their prefectures, recently, Japan also used mascots for the biggest sports game in the world, Tokyo Olympics Games 2020, as shown in Fig. 4.

Despite the Tokyo Olympics Games taking place in 2021 during the pandemic, the creation of the mascots had been completed before Japan was affected by the pandemic. In Fig. 4 (left), we have Miraitowa, derived from the combination of two words: "mirai," meaning future, and "towa," meaning eternity. The meaning behind the name is to promote a future filled with everlasting hope in the hearts of people worldwide. Alongside Miraitowa, we have Someity, the mascot for the Tokyo Paralympic Games 2020. Unlike Miraitowa, Someity predominantly utilizes the color of cherry blossoms, a flower unique to Japan. Someity's name is derived from "someiyoshino," a widely recognized variety of cherry blossoms, and it sounds similar to "So Mighty," symbolizing the strength of each paralympic participant. Although the mascots have different meanings and representations, both have a friendly and harmless physical appearance (Chiabaut, 2021). This is due to the involvement of thousands of children who were asked to vote for their favorite mascot characters for the Olympic Games. This engagement of children aims to reinforce the concept of yuru kyara mascots, which are imperfect, childlike, and often resemble animals, resulting in cute or kawaii characters (Barrows, 2014). However, this hypothesis still requires further research to determine whether there is a connection between children's participation and the mascots' friendly and harmless appearance.

In a book written by Anderson et al. (2020), the rising popularity and acceptance of hologram culture are discussed. One notable example is Hatsune Miku, a virtual idol who made her debut on August 31, 2007, and became one of the most famous virtual idols worldwide (Karki, 2021). Alongside virtual idols, the emergence of virtual influencers began in 2016. Virtual influencers are computer-generated mascots who perform similar roles to human influencers by promoting products (Audrezet & Koles, 2023). Some renowned virtual influencers include Rozy from South Korea, Zinn from Japan, and Lil Miquela from the United States of America (Conti et al., 2022).

4 How the Pandemic Affect the Brand

There is no longer any debate that the pandemic has compelled humanity to change numerous aspects of their lives. The emergence of the pandemic brought about a new trend known as Work From Home, where everything shifted online, as face-to-face meetings posed the risk of covid-19 infection. Brand owners were also compelled to adjust their strategies. Several studies have highlighted the fact that the pandemic's restrictions have altered customer behavior (Meheli & Swaminathan, 2021) during contagious diseases outbreaks, resulting in the need for extensive re-branding for products that use anthropomorphism branding strategy due to the social withdrawal experienced by the customer (Ding & Xu, 2022). Businesses that could leverage the internet were advised to do so. For instance, with restrictions on football stadiums, football teams were on the verge of losing their fan base. To address this issue, research conducted by Szczepkowski (2021) recommended that football teams utilize social media as a means to stay connected with their fans and maintain brand awareness. In other industries, businesses that relied on face-to-face meetings had to adapt by implementing stricter social distancing protocols for their daily operations (Carr, 2021). Numerous entertainment events, such as the Olympic Games, were postponed for safety reasons. Another example is Hatsune Miku's debut concert at Coachella in 2020, which was delayed due to the pandemic and resulted in the organizer incurring losses of approximately 26.1 billion USD in the United States alone (Anderson, 2021).

5 Mascots During Pandemic Era

Nevertheless, it is up to humankind to determine their perception of this pandemic. In Japan, the pandemic era has provided an opportunity to revive old folklore through the creation of a mascot known as Amabie: a Japanese female creature from the Edo period. Amabie has a human face, a bird's beak, and her body is covered in scales, resembling a mermaid (as seen in Fig. 5). Interestingly, her facial features resemble the famous Plague Doctor mask from the Black Plague era, while her body resembles an Armadillo, a mammal strongly associated with covid-19 (Merli, 2020) Amabie is considered a spiritual icon associated with pandemics, believed to possess the ability to predict the occurrence of a plague and inform the population, thereby protecting the land of Japan from its effects (Hrvatin, 2022).

In different regions of Japan, mascots that previously represented cities or prefectures, like Kumamon in Fig. 3, underwent re-branding to convey messages to the local population. An example of this is Miyazaki Ken, the mascot of Miyazaki Prefecture, which originally consisted of three animals (Fig. 6), but was re-branded as superheroes (Fig. 7).

This re-branding was not done without purpose. The re-branding of the mascot of Miyazaki was meant to convey a message of courage; it was intended to be physiological healing for everyone during the era of pandemic (Occhi, 2022). Another example is the MAT Onion Mascot, a local mascot created during the MAT program based on the research

Fig. 5 Illustration of Amabie

Fig. 6 Mascot of Miyazaki

Fig. 7 Mascot of Miyazaki after rebranding

Fig. 8 Pamphlet of Hatsune Miku

conducted by Rand and Snyder (2021) with the clear message "Onion sometimes makes you cry, but if you prepare it, it will be delicious". It sends same encouragement message to all participants about seeing the good side of the pandemic.

Amul, one of the top leading food brands in India, uses doodles in their advertisements to convey various messages to customers. In addition to messages of encouragement and campaigns for social distancing, Amul also incorporates satire, comedy, and humanity campaigns while promoting its brand. However, there are differences in the messages conveyed by Amul before and during the lockdown, with a greater focus on social campaigns and messages of encouragement during the lockdown period (Chourisa, 2021). In Indonesia, Tokopedia became one of the companies that changed their branding strategy in order to maintain the brand's sustainability. Using #DiRumahAjaDulu, Tokopedia positioned itself as a company that cares about the health of its customer by asking them to minimize outdoor activities while at the same time promoting the use of their application to do shopping from home (Tirtadarma & Darmo, 2020).

Many organizations worldwide have employed social media campaigns using mascots to deliver safety messages during the pandemic. For instance, the World Health Organization (WHO) collaborated with Karrot Animation, the creator of the well-known mascot Peppa Pig, to produce a short video campaign targeting children during the pandemic (Zebua et al., 2022). Additionally, the WHO partnered with virtual influencer Knox Frost for a covid-19 campaign aimed at an older audience (Conti et al., 2022). The Government of Japan appointed several representatives, including mascots, to promote campaigns and countermeasures against covid-19. Notably, Softbank's robotic mascot Pepper and the virtual idol Hatsune Miku deliver safety and encouragement messages to the people of Japan through videos and digital pamphlets (Figs. 8 and 9 illustrate examples of these pamphlets).

Fig. 9 Pamphlet of pepper

6 Discussion

The development of mascots, in general, is vast and creative. However, through the previous sections, there are a few crucial things that are worth mentioning and discussing.

There are slight differences in how the United States and Japan utilize their mascots. As mentioned, both nations often utilize mascots as their branding strategy, but in the United States, most of the mascots are used as a representation of specific companies or organizations, and most of the time Sports Club. While in Japan, the utilization of mascots is not only limited to organizations and companies, but also to represent each of their cities and prefectures for tourism purpose. In addition, the United States tends to use logo and tagline as the representation of their towns as shown in Fig. 10.

There is nothing wrong with both strategies. Even though the research by Ko et al. (2022) does mention the vast benefits of using mascots rather than logos as branding strategies, there is still not enough evidence to point out the conclusion of whether one of the marketing strategies is superior to the other. These, however, could be exciting research subjects for future research. As mentioned above,

Fig. 10 List of state's logo in the U.S.

besides analyzing the benefits and drawbacks of the usage of the logo by the many states in the United States versus the use of Mascots by the prefecture in Japan, the recent global-scale event, World Olympic Games Mascot, also raised a new question. The research conducted by Chiabaut (2021) mentions child participation during the vote. However, is there any connection between the harmless-looking mascots, Miratoiwa and Someity, and the children's participation? Perhaps their harmless-looking style was coming from the "*Kawaii*" style, which is quite popular in Japan, and therefore, the children's participatory have little to zero contribution.

After the pandemic hit, many brand owners were having financial struggles (Anderson, 2021) and were forced to change their strategies to maintain the sustainability of their brands. Due to social distancing and staying indoors for a long time, most of humankind experiences social withdrawal, which changes how they perceive a brand (Ding & Xu, 2022). Besides the social withdrawal, the Internet as advertisement media also gave considerable changes in branding. Social media such as Facebook, TikTok, Twitter, and even Tokopedia have become strategic branding media since the internet can be accessed by everyone, even when a lockdown occurs (Szczepkowski, 2021). However, their effectiveness as advertisement media during the pandemic era requires further research. The ease of use of the internet didn't just change the branding strategy. Because of the pandemic, the term "Work From Home" became popular, and offices and even schools started to undergo these practices. Virtual meeting software such as Zoom, Google Meets, and Microsoft Teams instantly become mandatory software for everyone, regardless of age (Dubey & Tripathi, 2020).

There are enormous differences between a few mascots that have been created or re-created during the pandemic era and before the pandemic area. Most of the mascots created before the pandemic era only represented the company, organization, or event's identity (Lisa, 2014). Mascots that were created during the pandemic era represented solidarity and unity toward the greater good to fight the pandemic with less portion for their own brand's identity, for example, in the case of Miyazaki Ken (Occhi, 2022), where they used to represent Miyazaki prefecture's unique point but they were changed into superhero attire to send encouragement message. The same idea can also be seen in Tokopedia (Tirtadarma & Darmo, 2020) and Amul's Branding Strategies (Chourisa, 2021), where they put their brand's identity before the social message.

Different from prefecture mascots or Amul, where the brand holders or creators themselves changed certain aspects of their mascots to deliver a message or a covid-19 campaign, social media-based campaigns involving virtual idols and virtual influencers are often a result of cooperation between different organizations. For example, the collaboration between WHO and Peppa Pig (Zebua et al., 2022) and virtual influencer, Knox Frost or Hatsune Miku with the Government of Japan. In these cases, the mascots did not undergo drastic changes in their original form. Instead, the virtual idols and influencers conveyed their message through verbal or written speech using videos and image as the main medium. This is likely because humans perceive virtual idols and influencers more as celebrity figures (Anderson et al., 2020) rather than mascots that represent a brand and can be easily changed.

Although there are differences in the methods of delivering the message, both physical and digital mascots share the same spirit. The spirit of "We are in this together" was strong during the pandemic era and had an impact on mascots created during this period, whether in physical or digital form.

7 Conclusion

Based on the conducted research, the study was able to answer both of the before-mentioned research questions as for the research question were:

- Are there any shift in the use and benefits of mascots to brand sustainability during the pandemic era?

Clearly, there has been a shift in the use of mascots and overall branding strategies between the pre-pandemic and pandemic eras. For instance, Amul's doodles, which previously focused on satire, comedy, or the latest trends, shifted toward social campaigns during the pandemic. Similarly, Hatsune Miku and Knox Frost adopted the same strategy. Mascots, in general, moved away from representing individual companies, groups, organizations, or events, and instead conveyed social, encouragement, and safety messages to customers. This shift, although potentially weakening their brand identity, has proven beneficial for brand sustainability. Becoming a sustainable brand involves integrating environmental, social, and economic issues into business operations, and prioritizing human health and a sustainable future aligns with this objective.

- How is the development of mascot research to date?

The research conducted indicates that mascots have been widely used globally. While there are ongoing discussions about the utilization of mascots in the United States and Japan, the development of mascots shows significant innovation and diverse ideas. Mascots are not limited to individuals inside hot costumes waving to passersby; mechanized mascots with electronic components have also emerged. Moreover, with the rise of hologram culture, mascots can even exist as digital entities without any physical presence. The existence of virtual idols and virtual influencers serves as evidence that mascots can be fully operated from a computer without direct human interaction.

Acknowledgements The financial support for this study was provided by a grant from the Product Design and Management, Creative Industry Faculty, University of Surabaya. The author also wishes to thank the Team from Universitas Ciputra and IEREK for their suggestions on preparing the manuscript.

References

Anderson, N. (2021). Hatsune Miku, virtual idols and transforming the popular music experience. *Musicology ECA,* 39–57.

Anderson, W. H., Arceno, J. P., Archer, E., Aryeh, D., Bush, P., Dedeoglu, C., Torseth, R., et al. (2020). Technology and theology. In W. H. Anderson (Ed.), *Holograms and idols: The image of god and artifical transcendence in the cultural phenomenon of the Japanese vocaloid Hatsune Miku* (pp. 263–280). Vernon Press.

Antonova, L., & Honcharova, Y. (2021). The impact of the pandemic on small business. In *Modern scientific and technical research in the context of the language space. СУЧАСНІ НАУКОВО-ТЕХНІЧНІ ДОСЛІДЖЕННЯ У КОНТЕКСТІ МОВНОГО ПРОСТОРУ* (pp. 113–115).

Ash, P. R. (2020). *Unexpected events & brand loyalty*. Victoria University of Wellington.

Barrows, L. (2014). *Planning With Character: Gotouchi Kyara and-Place Branding in Japan* (Thesis). Faculty of Architecture and Planning, Columbia University, New York, USA.

Audrezet, A., & Koles, B. (2023). Virtual Influencer as a brand avatar in interactive marketing. In *The Palgrave handbook of interactive marketing* (pp. 353–376).

Cabinet Secretariat of Japan. (n.d.). *Covid-19 information and resource*. Retrieved from Covid-19 Information and Resource: https://corona.go.jp/en/

Carr, S. E. (2021). *In the library, with a mask, and some plexi-glass: Reopening a small academic library during COVID-19*. College & Undergraduate Libraries.

Chang, J., Dietz, C., Marwha, T., & Roth, G. (2020). *WPI Mascot Robot*. Worcester Polytechnic Institute.

Chiabaut, I. (2021). Tokyo 2020 official mascots and their contribution to the promotion of Japan's culture and legacy. *DIAGORAS,* 89–100.

Chourisa, M. (2021). A study on Amul doodles before and during the lockdown. *Journal of Contemporary Issues in Business and Government*, 2691–2708.

Conti, M., Gathani, J., & Tricomi, P. P. (2022, August). Virtual Influencers in online social media. *IEEE Communication Magazine*, 1–7.

Ding, Y., & Xu, S. (2022). *Detrimental impact of contagious disease cues on consumer preference for anthropomorphic products*. Marketing letter.

Dubey, A. D., & Tripathi, S. (2020). Analysing the sentiments towards work-from-home experience during COVID-19 Pandemic. *Journal of Innovation Management*, 13–20.

Esfahani, D. N., Soltanhoseini, M., Rahbari, S., & Shams, H. (2022). Presenting the model of the impact of brand mascot on brand awareness, brand recall, and intention to purchase among sporting goods consumers. *Sport Management Journal*, 250–276.

Hrvatin, K. (2022). Japanese artists' responses to COVID-19. In *Crisis as danger and hope—Covid-19 in Asia* (pp. 183–209).

Karki, K. (2021). Vocaloid liveness? Hatsune Miku and the live production of Japanese virtual idol concerts. In K. Karki (Ed.), *Researching live music*. Focal Press.

Kaufmann, L. A. (2021). *Opportunities in activating and monetizing mascots in sports as a marketing tool*. Management from the Nova School of Business and Economics.

Kaufmann, T. (2020, October 31). *The 50 best Mascots in college football*. Retrieved from Bleacher Report: https://bleacherreport.com/articles/505613-the-50-best-mascots-in-college-football

Ko, Y. J., Asada, A., Jang, W., Kim, D., & Chang, Y. (2022). Do humanized team mascots attract new fans? Application and extension of the anthropomorphism theory. *Sport management review*.

Lisa, B. (2014). *Planning with character: Gotouchi Kyara and place branding in Japan*. Columbia University Libraries.

Meheli, B., & Swaminathan, V. (2021). Consuming in a crisis: Pandemic consumption across consumer segments and implications for brands. *Journal of Product and Brand Management*.

Merli, C. (2020). A chimeric being from Kyushu, JapanAmabie's revival during Covid-19. *Anthropology Today,* 6–10.

Occhi, D. J. (2022). Idolization of Miyazaki Ken local Mascots and Himukaizer local heroes: The animate spirits of Miyazaki, Japan. In *Idology in transcultural perspective* (pp. 159–185).

Palladino, A., Lee, M., & Zhou, X. (2022). An analysis of olympic mascot design toward attitude and purchase intention. *Human Kinetic Journal*, 190–196.

Rand, P., & Snyder, C. (2021). Bridge over troubled water: A teacher education program's emergent methods for constructing an online community of practice during a global pandemic. *Journal of Higher Education Theory and Practice*.

Singh, V. (2021). *The impact of brand elements on the purchase behaviour of University of KwaZulu-Natal*. School of Management, Information Technology and Governance.

Szczepkowski, J. (2021). Building and maintaining the relationship with football fans during COVID-19. *Journal of Physical Education and Sport*, 1270–1275.

Tirtadarma, E., & Darmo, B. (2020). Analysis of Tokopedia Advertising#DirumahAjaDulu and its correlativity to Tokopedia branding position in pandemic era. In *Advances in social science, education and humanities research* (pp. 8–13).

Village House. (25 September, 2020). *Mascots in Japan*. Retrieved from Village Blog. https://blog.villagehouse.jp/en/japanese-apartment-life/mascots-in-japan/

Zebua, W. D., Rahman, A., Utari, S. A., & Robbani, A. I. (2022). Persuasive communication in the Covid 19 campaign for children. *Journal Komunikasi*, 97–119.

Innovative Approaches to Teaching and Pedagogy in Creative Design Education

Explorative Learning Space for Developing Motoric Skills in the Early Childhood Stage

Emanuel A. Wicaksono and Sharon Aurielle

Abstract

The digital world has dominated and taken over the modern generation, including children. Activities of socializing that used to be done in the past such as playing in the park and reading books in the library are getting replaced by the fun activities that gadgets and social media offer. Children's motor development is equally as important as other aspects of development because a child's inability to carry out physical activities will impact their life negatively onward (Mursid in Pengembangan Pembelajaran PAUD, 2015). Therefore, one solution that could be done is to integrate both activities of playing and learning in one space to encourage the spirit of learning in children. In response to this issue, we aim to create a design that incorporates modular shapes of wood for the construction that forms a space that can accommodate both playing and learning activities. These modules are arranged in a way that provides space for flexible movements of learning and supporting objects such as a bookshelf for storage. Following the Vitruvius design principles, this design has the three important elements of firmitas, utilitas, and venustas. The structure is rigid and strong enough for users to stand on which is supported by the wood material that is used. It is also functional and provides space for children to both play and learn. All the while maintaining an esthetic that could attract children's attention and invite them to use the space. This exploration results in a multi-function space designated for children to support the growth and development of their motoric skills.

E. A. Wicaksono (✉) · S. Aurielle
Universitas Pelita Harapan, Tangerang, Indonesia
e-mail: Emanuel.wicaksono@uph.edu

S. Aurielle
e-mail: 01022200030@student.uph.edu

Keywords

Modular · Children · Playground · Motoric skills

1 Introduction

In this digital era where the world is tremendously dominated by information technology, it is almost impossible to live without gadgets. This also applies to children who are increasingly being introduced to gadgets from an early age. The use of technology is inevitable, especially in today's day and age where it rapidly grows and interferes with our life in easily all aspects. It has also become an effective method for parents to deal with their children when they have a tantrum. In 2011, 38% of children aged 5 used gadgets which then increases in the year 2013 to 72% and 80% in the year 2015 (Sujianti, 2018). With the COVID-19 pandemic that struck the world in an instant, various life activities shifted to the digital world consequently to the increase of gadget use in children. With the early exposure to technology, children are being trained to later be able to adapt and respond well to a technology-dominated world. However, among these positive impacts, the excessive use of gadgets may also hinder children's development, especially physically as they are not moving as much. Psychologically it may also risk children's mental state and cause them to imitate inappropriate behaviors.

According to Kemenko PMK RI (Coordinating Ministry for Human Development and Cultural Affairs of the Republic of Indonesia), currently, 84.4 million Indonesian citizens are children aged less than 18 years old (Supanji, 2021). These children are expected to become Indonesia's Golden Generation in 2045. As Indonesia celebrates its 100th anniversary of Independence in 2045, Indonesia faces many challenges in achieving its independence goals as stated in the State Guidelines Article 4, Indonesia has assets and opportunities supported by the social dynamics, demography bonus, plurality, and natural resources that

M. Nawaz Tunio et al. (eds.), *Sustainability in Creative Industries*,
Advances in Science, Technology & Innovation, https://doi.org/10.1007/978-3-031-48453-7_6

have potential in becoming constructive social energy as a strength (Marwati, 2018).

This study aims to further learn and explore the appropriate learning space for children in the early childhood stage to support their explorative nature and the development of their cognitive, physical, and social skills. It also expects to prove the hypothesis that a certain architectural design can facilitate and support the development of children's motoric skills. The specific research questions to be answered are as follows:

- How to stimulate children's motoric development in the early childhood stage?
- What kind of space is suitable to accommodate the maximum motoric development of a child?
- What are the parameters required in child-friendly design that can support and accommodate motoric development?
- How is that type of space implemented in an architectural form?

2 Theoretical Background

Physically, children experience extraordinary development in their early childhood stage (0–8 years old), starting from the growth of brain cells and other organs to the development of gross motor skills such as walking, running, jumping, climbing and others, also fine motor skills which is the ability to coordinate hand and eye movements such as grasping, reaching, writing, and so on (Sit, 2015). Apart from these, psychologically speaking, children also undergo remarkable developments during this stage when they start to interact with their parents.

Optimal motoric development during the early childhood stage does not only contribute to good physical health, but several other benefits such as:

- Good mental health and self-image from the satisfaction that they can perform various physical activities which will also give them the motivation to take part in more activities.
- Benefits in the emotional catharsis where children can release the restrained energy they hold and free the body from stress tension, anxiety, and hopelessness.
- The independence to do more activities on their own boosts confidence and gives happiness.
- Self-entertainment from carrying out various activities.
- The ability to socialize affects the children's acceptance in a community and more opportunities to develop their motoric skills.
- The concept of self-control from the motoric skills creates a feeling of security psychologically that impacts the behavior of a child.

Motoric development can be boosted for optimal results if the growth environment supports them to move freely and satisfy their curiosity with self-earned knowledge. Outdoor activities are suitable to support and stimulate muscle growth (Tipton & Ferrando, 2008). Other stimuli that support the motoric development of a child are the basic practice of writing and drawing, performing sports like gymnastics and the use of other sports equipment, and performing active movements like jumping, climbing, and running.

2.1 Architectural Implementation

In the effort to shape the future by supporting children's early-age development, these theories can be applied in architectural spaces that can accommodate suitable activities to stimulate children's development in various aspects mentioned. Independence and the opportunity of exploration shall be facilitated in this case involves visual, linguistic, and motor skills in early life to create sturdy neural networks that form the brain's architecture and sustain all ensuing learning. Children develop solid brain connections as they explore their curiosity and try new things, which will be important for their learning and conduct throughout their lives. New abilities gained while doing so also increase their sense of competence and confidence (Gavin, 2021).

In constructing an architectural space for children, some considerations shall be made, namely, safety, entertainment, liberation, adaptability, or openness and understanding (Dobbins, 2018). Among all the criteria, safety is of the utmost importance. A child-friendly space shall foster children's inherent creativity and freedom to play. There are various ways to accomplish this but it essentially comes down to structured play and abstracted play which leads to the next point of liberation. As mentioned in the theories of motoric skill development, the concept of play may depend on allowing kids to interact with or navigate the architecture in a way that is specific to their situation, allowing them to be independent and explorative through the design. As individual beings, no two interactions with children are ever the same, and there may be wide variations in how children use space daily. The ability to change and adapt to any circumstance is a must for spaces. Children are not designed to be confined to the parameters of our adult assumptions of space, thus they must also be open and have access to nature.

2.2 Children's Play Area Design

In a formal children's outdoor playing space (COPS), there are some categories to be considered. Accessibility is a crucial factor to take into account when designing outdoor play areas for kids. If there are not any appropriate purpose-built facilities, kids will use inappropriate locations that can be dangerous or disrupt others. The amount of time it takes a child to leisurely walk from home to the play area without having to cross potentially dangerous physical barriers is the best indicator of accessibility. Along with accessibility, supervision is also an essential consideration for a play area because children of that age group still need maximum supervision to ensure their safety and avoid any unwanted danger.

On the other side, although supervision is very important in a children-focused facility, it is required to be mindful of the urges to assist and resist the impulse to take over control of the activity. Excessive supervision and control in children's learning process hinder their sense of confidence and independence and consequently deters the explorative sense of learning. Educating the child about the task they will be performing is a part of the growth process of their discovery learning to better prepare them to do it independently. Natural curiosity is the driving force behind exploratory learning, and this natural curiosity will result in an established explorer mindset, which is essential for making decisions and solving problems in daily life (Verma, 2019).

According to Şensoy and İnceoğlu (2015), the locations where kids can spend time in today's society are inadequate, poor-graded, and constrained. Children became cut off from nature as a result, their lives became unqualified, and heavy urbanization constrained children's actions and interfered with their social development. As a result, the following considerations must be made while designing a high-quality playground:

- Collaboration by involving the users who are children in the design thought process by doing site safaris, design workshops with models and drawings, and focus groups. Also by observing kids to figure out what they enjoy doing.
- Security and safety must be matched to create surroundings that are engaging and demanding for kids to play in and grow in. The idea of a healthy and safe environment need not conflict with goals for a child's development.
- Social interaction must be facilitated to develop and strengthen their social connections with one another and with the adults nearby in settings like seats and gathering places.
- Children utilize play to push their boundaries and deal with difficult circumstances. Therefore, playground designs must allow acceptable risks that positively impact children's physical and personal development.
- Various equipment shall be available as an opportunity to provide children with different exciting experiences.
- The space also should be maintained with a good standard of maintenance to maintain longevity and ensure the design of the play area is appropriate.
- Hygiene should also be kept and monitored whether or not attendants carry out their duties in keeping a clean environment.
- Informativeness should be present by objects that stimulate brain development. Providing objects that contain new information and visuals to stimulate the brain is necessary and more effective in the child's development, especially when infused into the play area.

Another important element in a child's play area is the surface. Equipment with a fall height of more than 600 mm shall be placed on an impact-absorbing surface. There are two varieties of these surfaces, namely, loose fill and synthetic. Tree bark, sand, lightweight ceramics, and wood chips are examples of materials in this category. Unless daily checks are conducted to guarantee that surfacing levels are maintained and procedures to remove any harmful debris, such as glass, can be carried out, loose-fill surfaces should not be laid beneath moving objects. Meanwhile, synthetic surfaces are mostly made of tile, carpet, or wet-pour materials. Although they might be more expensive, generally they are more durable and require less maintenance than loose-fill surfaces.

3 Methodology

The phenomenon issued in this study is the fact that Indonesian children carry a huge mission as the future "2045 Golden Generation." As an impact of the technology-dominated world we live in now, it is inevitable that even children are being exposed to technology early on in their lives. Although there are various benefits, this may also be a danger from the various aspects of cognitive, physical, and even social skills. In constructing a program that is intended for children, it is essential to formulate an exploratory learning system to improve knowledge and experience in understanding how to relate ideas to context.

This study is conducted with methodological stages as follows:

- Identify the phenomena and problem.
- Literature review and regulations.
- Determination of the design guidelines.
- Design of the project.
- Evaluation of the final design.

Some boundaries and goals are set in this project to optimally deliver a proper function that could bring concrete impact to the development of children's motoric skills. This project aims to create a space that carries both functions of playing and learning.

4 Design Guidelines

From the literature review and the thorough theoretical study, the project design guideline is understood to have characteristics as follows:

- It shall develop gross motor skills such as walking, running, jumping, climbing, and others.
- It shall develop fine motor skills which is the ability to coordinate hand and eye movements such as grasping, reaching, writing, and others.
- It shall stimulate interaction.
- It shall facilitate a growth environment that supports children to move freely and satisfy their curiosity with self-earned knowledge.
- It shall facilitate outdoor activities that are suitable to support and stimulate muscle growth through performing active movements like jumping, climbing, and running.
- It shall provide space for the practice of reading, writing, and drawing.

The guidelines mentioned above are the parameters set to realize a child-friendly design that can support and accommodate motoric development. The fulfillment of this parameter determines whether the design is a success or not. These parameters are set to ensure the optimization of the research in creating an explorative learning space for the early childhood stage that combines both activities of learning and playing into one space so that the children would not feel intimidated but instead invited to play as they acquire knowledge and information for themselves.

5 Project Design

This study took the location of the Pagedangan District Office, Tangerang as the object of study. Pagedangan District is located in Tangerang Regency and administratively consists of 1 sub-district and 10 villages. The total population in Pagedangan District is 120,967 with a land area of 47,782 ha. The site is actively used by the citizens of Pagedangan District with the surrounding environment is relatively filled with busy community life. The location of the Pagedangan District Office is quite strategic because the surrounding area has been developed into the Bumi Serpong Damai area. The documentation in Fig. 1 shows the existing condition of the district office. This site is strategic in terms of vehicle density and pollution as it is located considerably far from the main road and vehicle density which makes it a safer and less polluted area to build a child-friendly area in.

The design in Fig. 2 shows the design development of the district office. The area is developed as a children's play area which also functions as a reading park that is placed close to the district service office. With the presence of this playground nearby the office, the children can have their freedom and explore themselves in the multi-functional playground while still being under their parent's supervision and within a close radius of their parents.

In the design thought process, observation is done beforehand to learn and understand the activities done by children of the early childhood stage to know their interests and what activities they enjoy doing. As a result, some activities that trigger the curiosity and urge to play are found such as climbing, crawling, hanging, running, jumping, and others. These movements are then incorporated into the space by the use of various play equipment that can support their physical growth and motoric development and also freely express their human nature of being free. The dimensions of the design are also adjusted for it to have acceptable risks that are necessary for their physical growth and development to push their boundaries and deal with difficult circumstances, thereby giving them the opportunity to practice problem-solving and spark courage and self-confidence.

The basic form of the playground is a wooden structure that can be assembled and modified to serve a variety of purposes according to the activity that is injected into it. The choice of wood was made since it is a renewable resource and has particular color, texture, and pattern qualities that enhance the esthetic warm impression. The type of wood that is suggested is wood with high exterior durabilities, such as teak, ironwood, merbau, damar laut, or bengkirai, with a coating layer to maintain the wood's durability. As an outdoor facility, this structure needs to be strong enough to withstand external forces such as strong wind, rain, and heat. The use of wood as the basic material is considered strong enough to bear such forces. The use of material, finishing and design arrangement in this project also requires a low effort of maintenance which makes it easy to clean and maintain. The modular design also makes it possible to fix partially without having to replace the whole structure.

The joinery system in this project imitates the Japanese Chidori joinery system. This all-natural joinery technique that is done by the interlocking of timber joints is proven to have higher strength and be more flexible. In this technique, each wooden stick has been uniquely carved, and the unique notches enable them to interlock in one place to

Fig. 1 Padegangan District Office existing condition (personal documentation)

Fig. 2 The proposed design development of the Pagedangan District Office

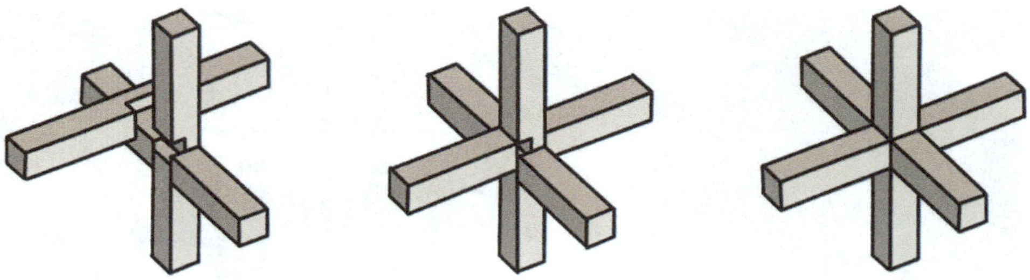

Fig. 3 From Chidori Wood Joint, 3D model by Danilo Hideki

perfectly fit in the gap and therefore produce a clean and neat design. This system is perfect for the design which is an exposed wooden structure. With the neat interlocking of timber joints produced by the Chidori method as shown in Fig. 3, there will be no apparent excess material that can be seen as it is purely made with the existing wood. Not using screws and bolts as the main joinery system in this design helps provide a safer space for children to play in.

This Chidori system is made with a module that adjusts the child's anthropometry, ranging from 70 to 130 cm in the average height of children aged 2–8 years old and 9–35 kg in the average weight, to the size of the wood. In this way, children can reach each module easily and safely. This modular size is an advantage to facilitate the construction process. Children of this age are known to be in the active development stage. The readiness and abilities of a child cannot be compared to the other as they have their own processes of development to go through, affected by various factors such as the environment they live in, nutrition, activities that they do, and genetic conditions. As an essential factor to be considered in a child-friendly space, safety shall be maintained and injury shall be prevented no matter how small.

The module in Fig. 4 shows the dimension, such as the height of the objects installed are also adjusted to make sure it is still a safe space while maintaining its explorative quality. This creates acceptable risks that are enough to stimulate the children's curiosity and confidence, but not endanger them.

The exploded axonometry in Fig. 5 shows the elements in the proposed design. In this design, rubber flooring is installed for security as it cushions and absorbs impact when a child falls, thereby minimizing injury. Rubber flooring is the best and most reliable material to use under playground equipment. Rubber is considered to be the most shock-absorbent and durable material when it comes to flooring, making it ideal for playgrounds where the flooring must be shock-absorbent to prevent injury to children in the event of a fall. This is categorized as the impact-absorbing surface that is synthetic to enable the children to freely play without their guardians having as much trouble avoiding

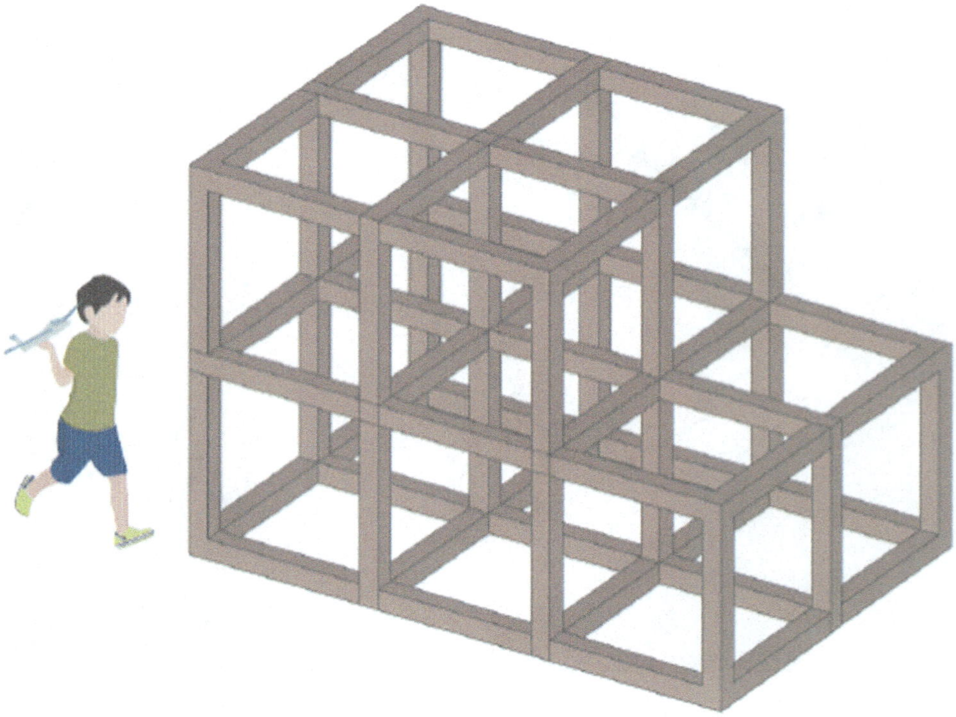

Fig. 4 Module dimension

possible injuries. With the various possible activities and movements to be performed, it is mandatory to utilize such surfaces as a quality assurance standard.

The pictures in Figs. 6 and 7 show the exterior and interior view of the project design. In the effort to combine the functions of learning and playing in one space, the design is created in a way that embodies both tasks as one. The modular shapes of wood as the construction structure are arranged to facilitate spaces for children to perform various activities such as crawling, climbing, hanging, swinging, reading, running, jumping, and others. This provides the children an optimal space for explorative learning as they are able to perform all the movements of kids their age which as mentioned in the theories above are essential in stimulating a child's motoric skills. For children in the early childhood stage, this space is inviting and allows them to answer their curiosity with the endless possibilities that this structure has to offer. With the endless limit of activity and obstacle, as well as challenges to face in this structure, the explorative learning aspect of the space is established and enables children to develop their sense of confidence, independence, and competence.

The illustration in Fig. 8 shows the elements in the playground. In the development of physical and motoric skills, this space provides the opportunity for children to perform various movements they could imagine. The design

Fig. 5 Exploded axonometry

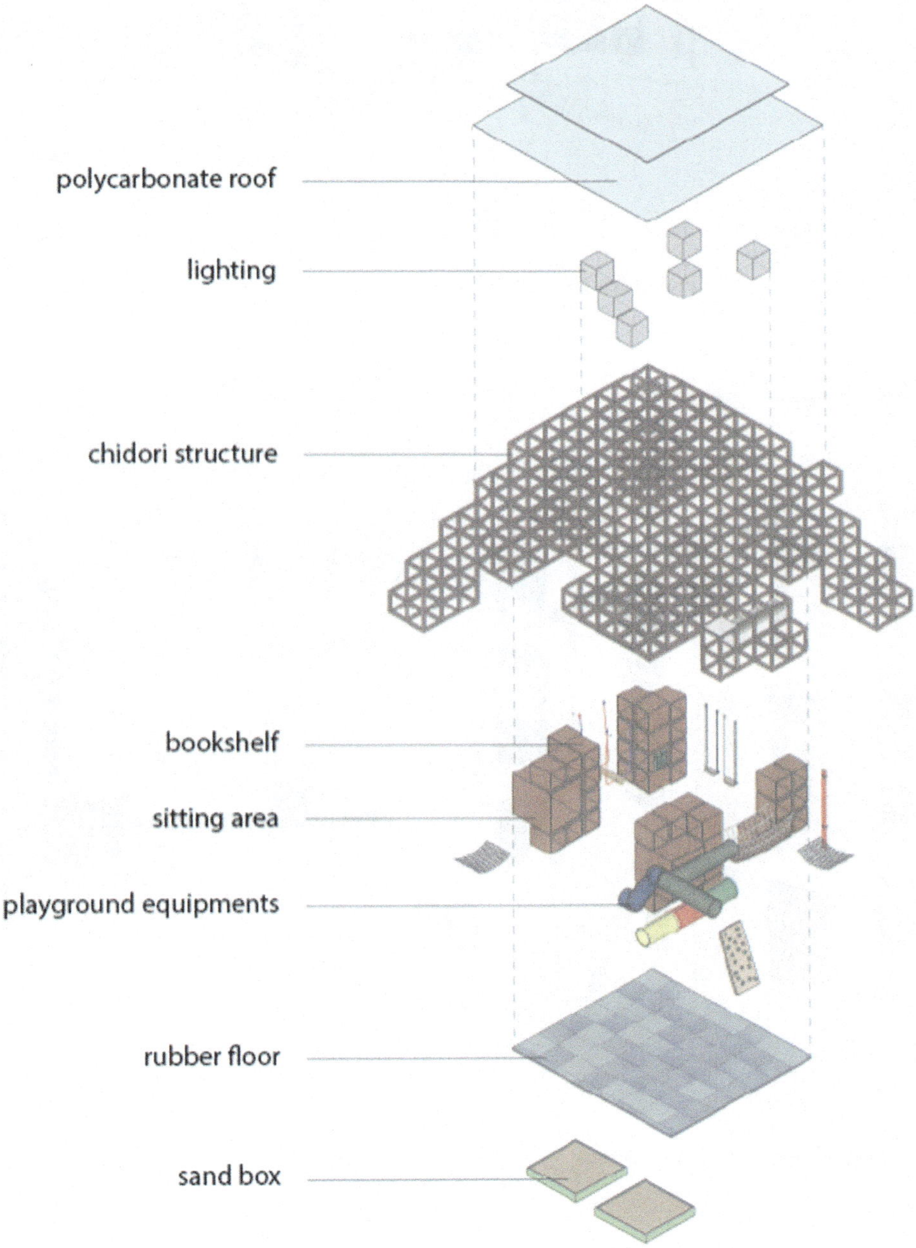

Fig. 6 Exterior view of the project design

Fig. 7 Interior view of the project design

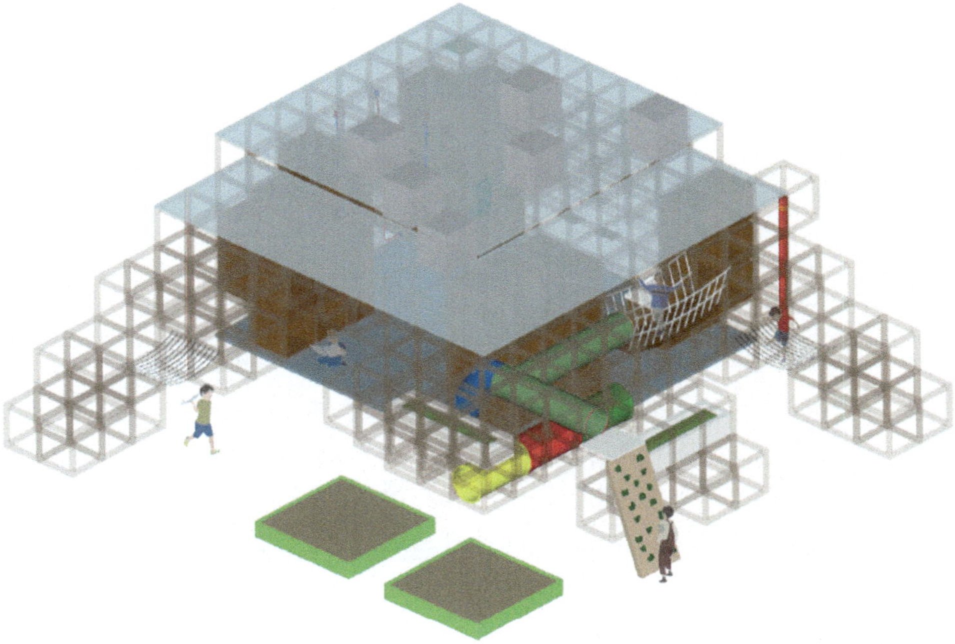

Fig. 8 Elements in the playground

is adaptable and can accommodate various activities that are to be infused into the area. The objects placed such as slides, crawl spaces, climbing stones, and others let children freely move and use their muscles. Moreover, the empty space in the middle of the structure also provides space for children to move even more freely. This stimulates children's muscle ability which develops motoric skills, both gross and fine motor development. By moving their body and putting their muscle to use, their motor skills will excel and later contributes to more benefits such as mental health, self-image, and emotional state that impacts other aspects of development too.

Other than that, the books provided in the installed bookshelf are the embodiment of informativeness in the play area as a concrete use of the playground in the learning aspect of the space. The incorporation of shapes, space, color, and texture in the play area stimulates the brain development of a child. The books here give an informal approach to the learning aspect itself and aim to attract more interest in reading, considering that Indonesia has an alarmingly low rate of interest in reading. Encouraging children to start reading and spark their interest from such an early age is one of the ways to hopefully increase their interest and educate Indonesian citizens to have a better literacy level that is essential in constructing a better future by escalating their critical thinking abilities.

The social and emotional development of a child can also be stimulated in this space. The illustration in Fig. 9 shows the social interaction area. Since this is a communal space that many children of various ages can utilize, it is almost natural that the children in this space get to know other people and start to socialize. Being in this space allows them to speak to new people of their age group and perform activities together. The design was developed with a sociopetal design approach to encourage interaction and the occurrence of interpersonal relationships (face-to-face) in public spaces. The sociopetal design is done by arranging the layout of the space and furniture allowing the children to gather around and interact with each other.

The design layout in Fig. 10 shows the facilities available in the proposed playground design. The various possibility of activities to be conducted in this space provided by the different elements and play equipments mentioned above facilitates children with the tools to develop their motor skills. Both gross and fine motor skills can be developed. Gross motor skills can be developed through the equipments and open space to freely explore their body movements and capabilities. Fine motor skills are developed through the experience they face and activities they conduct. The opportunities to interact with different elements and go through the experiences in this space constructed the explorative aspect in this space.

6 Design Evaluation and Conclusion

The results in Table 1 show the design parameters and contents. Based on the parameters set in the design guidelines, this project has successfully delivered a space that can accommodate and supports the explorative learning space that is aimed to support the development of children's cognitive, physical and social skills in the early childhood stage. The wooden structure that is used gives high structural strength that can withstand various forces. This checks

Fig. 9 Social interaction area

Fig. 10 Project design layout

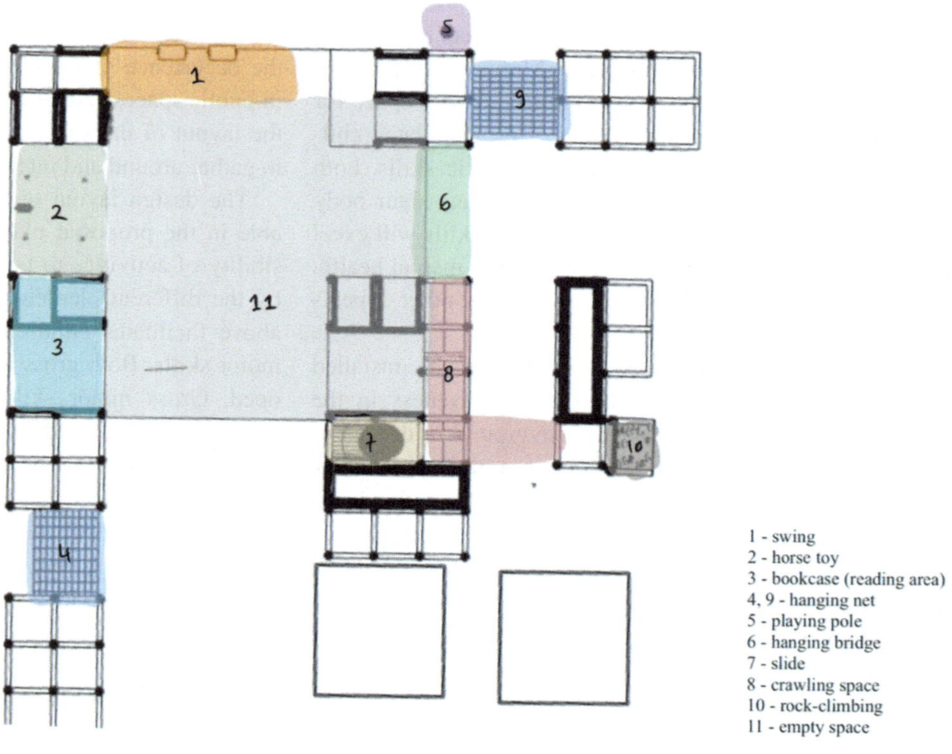

the firmitas principle of Vitruvius in the design created. The multi-function use of space with different objects that supports children to move around such as climbing, crawling, hanging, swinging, and even reading checks the utilitas principle as it can be used by children, especially in the early childhood stage. It is also designed beautifully with the right placement of objects and entryways which not only facilitates optimal use of space but also creates an inviting facade to attract children's attention. The design is also measured suitably to provide spaces that suit the dimensions of the child. The Chidori joinery system used supports in creating a clean and neat look while still performing its function to

Table 1 Design parameters and contents

Accessibility	– Directly in front of the district office that is open for public services and allows the citizen's presence
Supervision	– Enable maximum supervision from the location – Open structure that let guardians see and supervise their kids
Collaboration	- Observation of children's activities beforehand - Thorough research of movements to maximize motoric development
Location	– Far from vehicle access and the road – Strategic location, side by side with citizen housing and daily activities
Security	– Defined field, separated from other activities, especially the road that might endanger – Enable guardian supervision
Social interaction	– Gathering places – Enough space for collective use
Acceptable risk	– Safe equipment, aided with impact-absorbing surfaces
Diversity of equipment	– Different equipment for different motoric experiences – Suitable for children's use
Maintenance	– Periodic checking – Using materials that need low maintenance
Hygiene	– Periodic cleaning
Informativeness	– Books in the play area increase children's interest in reading – Play equipment installed triggers children to explore themselves and cures their curiosity

hold the structure and become safer for children as it does not use screws and bolts that may endanger the children if not installed properly. From these results, this design is considered to be a success and suits the initial intention of serving as both a learning and playing space for children in the early childhood stage. This proves that the hypotheses that a certain design of architectural space can indeed facilitate children to develop their motor skills and stimulate their senses both physically and psychologically through the aspects of explorative learning that are achieved in this space.

Acknowledgements We would like to express our gratitude to each and every individual and organization who have contributed in the publishing of this journal. In particular, to the Pagedangan District Office for giving us the opportunity to conduct our research and exploration on the related project. We also acknowledge the Architecture Department at Universitas Pelita Harapan for giving us the resources and support we required to complete this project. Finally, we would like to express our gratitude to our family and friends for their love and support during the research process as we could not have finished our study without their encouragement and assistance.

References

Dobbins, T. (2018). *Shaping the future: What to consider when designing for children*. ArchDaily. https://www.archdaily.com/901151/shaping-the-future-what-to-consider-when-designing-for-children

Gavin, M. L. (Ed.). (2021). *Safe exploring for toddlers (for parents)*. KidsHealth. Retrieved April 29, 2023, from https://kidshealth.org/en/parents/exploring.html#:~:text=Exploring%20also%20gives%20toddlers%20a,%221%20can%20do%20it!%22

Marwati. (2018). *To Achieve Golden Indonesia 2045, Professors Have Key Role*. Universitas Gadjah Mada. https://ugm.ac.id/en/news/16969-to-achieve-golden-indonesia-2045-professors-have-key-role

Mursid. (2015). *Pengembangan Pembelajaran PAUD*. Remaja Rosdakarya.

Şensoy, G., & İnceoğlu, M. (2015). Design guide for playgrounds. *Journal of Civil Engineering and Architecture*, *9*(11). https://doi.org/10.17265/1934-7359/2015.11.013

Sit, M. (2015). *Psikologi Perkembangan Anak Usia Dini. 1*.

Sujianti, S. (2018). Hubungan Lama dan Frekuensi Penggunaan Gadget dengan Perkembangan Sosial Anak Pra Sekolah di TK Islam Al Irsyad 01 Cilacap. *Jurnal Kebidanan*, *8*(1), 54–65. https://doi.org/10.31983/jkb.v8i1.3735

Supanji, T. H. (2021). Menuju Indonesia Emas 2045, Pemerintah Siapkan Generasi Muda Unggul dan Berdaya Saing. *Kementerian Koordinator Bidang Pembangunan Manusia Dan Kebudayaan*. https://kemenkopmk.go.id/menuju-indonesia-emas-2045-pemerintah-siapkan-generasi-muda-unggul-dan-berdaya-saing

Tipton, K. D., & Ferrando, A. A. (2008). Improving muscle mass: Response of muscle metabolism to exercise, nutrition and anabolic agents. *Essays in Biochemistry, 44*, 85–98. https://doi.org/10.1042/bse0440085

Verma, G. (2019). *Exploration-based learning*. LinkedIn. Retrieved April 29, 2023, from https://www.linkedin.com/pulse/exploration-based-learning-geeta-verma/

Utilizing Virtual Reality to Support Teaching the Design Principles of the Life-Safety System

Mutiara Cininta and K. Khaerunnisa

Abstract

The design studio is a fundamental component of architectural education, providing students with a platform to apply their theoretical knowledge to practical studio projects. Designing fire safety is mainly required to make significant architectural spatial adjustments to meet the safety standards. The current learning method of passive fire protection system design is primarily theoretical. Recent advancements in immersive technologies, such as virtual reality (VR), have made it more accessible and adaptable for integration into the architectural learning process. This research explored the potential of using VR to enhance the understanding of the core design principles of passive fire protection systems among architecture students. Using Sketch-Up and Enscape Walkthrough Render, passive fire protection design was identified with second-year architecture students with a case study of low-rise buildings with pre-test and post-test methods to identify trends in the changing levels of understanding. The results showed that VR complemented the traditional use of engineered drawings and aided in comprehending life-safety system design principles. It was perceived as an exciting visualization medium and positively contributed to the ease of learning. However, the implementation of VR in architectural education presented challenges, such as dizziness for first-time users and the limited availability of compatible devices.

M. Cininta (✉) · K. Khaerunnisa
Department of Architecture, Universitas Atma Jaya Yogyakarta, Yogyakarta, Indonesia
e-mail: mutiara.cininta@uajy.ac.id

K. Khaerunnisa
e-mail: khaerunnisa@uajy.ac.id

Keywords

Virtual reality · Architecture education · Building safety · Emergency exit

1 Introduction

Disasters such as fire events could cause extensive damage to communities, including the aftermath of their recovery (Lee & Lee, 2019). These incidents endanger lives and pose environmental, physical, and economic risks (Kodur et al., 2020). The current implementation of fire safety in buildings is provided by the recommendation of building codes (Kodur et al., 2020). Even with numerous safety precautions, statutory requirements during the design stage, and management laws, fire remains one of the greatest threats to data (Abdul Samad, 2021). Indirect factors can impede the effectiveness of building fire protection systems and the evacuation process (Kodur et al., 2020). In order to execute an evacuation process safely and effectively before a severe building obstruction occurs, comprehensive knowledge and skills are required. This includes familiarity with the space, recognizing exits, and understanding evacuation procedures during emergencies (Kwegyir-Afful et al., 2022; Singh & Saberi, 2020). Architects, therefore, need to have a holistic understanding of fire protection system design techniques. By acquiring this knowledge, they can contribute to designing safer buildings and mitigating the risks associated with fire incidents.

Building safety techniques, including fire protection systems, are integral to architectural practices and have been implemented as fundamental competencies and ethical codes in architecture institutes worldwide. This includes the 2020 Code of Ethics and Professional Conduct, Canon I, ES 1.5. and Rule 1.403 (American Institute of Architects, 2020) emphasizes their importance. The Royal Institute of British Architects (2021) stated that health and life safety was one of the three mandatory competencies, while the

M. Nawaz Tunio et al. (eds.), *Sustainability in Creative Industries*,
Advances in Science, Technology & Innovation, https://doi.org/10.1007/978-3-031-48453-7_7

Indonesian Board of Architects (2017) outlined 13 architectural competencies, with the first item on building safety. The Indonesian architectural education curriculum was formulated in collaboration with the Indonesian Board of Architects (IAI) to ensure a comprehensive learning experience aligned with professional architectural standards (Mutaqi, 2018). Traditionally, fire protection system plans are often developed after the building configuration is completed. The passive fire protection systems depend on the layout configurations, number of stairs, and other relevant aspects of the building design. The incorporation of passive fire protection systems from the onset enables architects to enhance the overall safety of the building.

The graph provided mapped the relationship between 87 selected publications from The Lens database (https://www.lens.org/) that focused on the intersection of virtual reality and education in Architecture from 2017 to 2022. Notable shifts in research trends can be observed by using VOSviewer to analyze keyword occurrences alongside referring to the publication year. The topics in this context have shifted from theory and problem to design studio and interaction, as represented by the blue and green sections on the diagram. The current research focuses on using immersive learning technologies (ILT) to enhance perceptual and spatial skills. There is an expanding body of research that explores the integration of virtual reality (VR) with other immersive technologies such as augmented reality (AR), mixed reality (MR), and extended reality (XR). This growing interest in ILT underscores the broader scope of research and its potential impact on architectural education and practice.

On the other hand, it could be observed that the phrase "architecture student" is not directly linked to the fire safety education system in previous research, indicating there have not been many investigations in this particular domain. The research on this subject primarily dates back several years as shown in Fig. 1, where the fire safety education system is represented by the purple shade, indicating its relevance in 2018. Future research to explore the use of VR in teaching fire safety to architecture students is required.

VR has proven to be a valuable tool in improving user performance during hazardous situations in a multilingual environment. Afzal and Shafiq (2021) conducted research in the United Arab Emirates (UAE) where a simulation was implemented using a four-dimensional (4D) Building Information Modeling (BIM) and VR to train the construction crews. The result showed that a combination of 4D BIM-enabled VR significantly improved their recognition of hazards and understanding of safety protocols compared to traditional written and verbal instructions. Additionally, it enabled them to respond to risks and mitigate hazards proactively. VR simulation has shown significant advantages over traditional methods and has the potential for widespread application across various fields. Three research reported positive outcomes from VR training. For instance, Lovreglio et al. (2021) compared non-interactive video and VR fire extinguisher training involving random volunteers at a university in Auckland. Zhao et al. (2021) compared the passing examination rate of medical students trained using VR versus traditional methods to evaluate the effectiveness of VR-based training in medical education. An assessment of situational safety awareness, hazard mitigation training, and evacuation drills in fire emergency scenarios was conducted by Kwegyir-Afful et al. (2022) for both experienced and inexperienced workers in the engineering field.

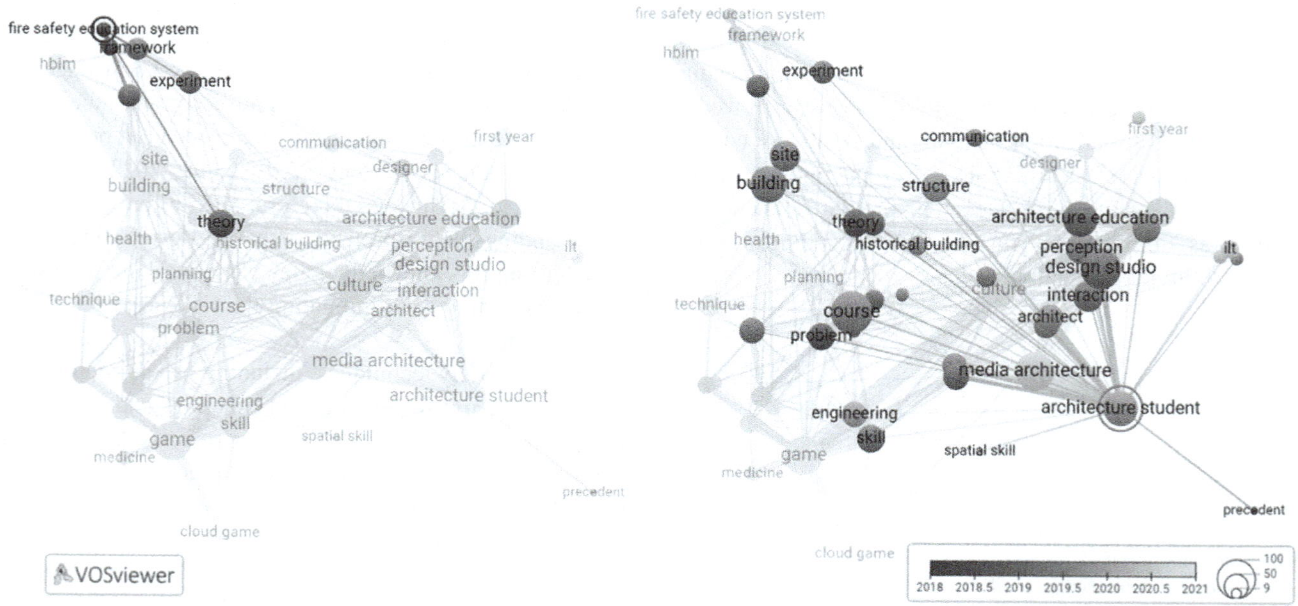

Fig. 1 Keyword occurrences mapping on VOSviewer. *Source* Authors

Extensive research showed the effectiveness of using VR in architectural contexts, particularly in problem-solving activities. Experimental research was conducted in a primary design education setting by dividing participants into VR and paper-based groups (Özgen et al., 2021). The results showed distinct differences between the two groups, with the VR-based group displaying exceptional enjoyment, expressing their intention to use the technology, and acknowledging its effectiveness in problem-solving. Considering these remarkable findings, exploring the potential of VR in teaching building evacuation plans becomes an indispensable opportunity. This research explores opportunities in using VR to increase understanding of the core design principles of passive fire protection systems for architecture students.

2 Utilizing Virtual Reality in Education

VR is a technology that creates artificial 3D domains resembling real-world settings (Farshid et al., 2018; Win et al., 2018). Although there are many types, it is generally classified into three types, namely non-immersive, semi-immersive, and fully immersive VR (Win et al., 2018). With the help of a VR head-mounted display (HMD), users can experience fully immersive 3D environments, including walkthroughs and 360 stills (Evans, 2019; Farshid et al., 2018; Gebczyńska-Janowicz, 2020; Hui et al., 2020). Some popular, affordable standalone HMDs, are Oculus Rift, Samsung Gear VR, and HTC Vive, which can be set up with wired or wireless configurations (Gebczyńska-Janowicz, 2020). In order to maximize the potential of a VR system, additional components such as the VR engine, software and databases, input/output devices, users, and tasks are required (Win et al., 2018). The architectural field has long been interested in utilizing immersive technologies, including VR, for visualization purposes (Hui et al., 2020). Many architects integrate VR into the design process because of its compatibility with various BIM, CAD, and 3D modeling software (Gebczyńska-Janowicz, 2020). CAD software assists in creating detailed drawings, while VR enables users to virtually explore the design (Farshid et al., 2018). VR enables seamless integration of multiple software tools, simplifying the technical aspects of the design process. It facilitates real-time interactions and collaborations between users and designers (Hui et al., 2020; Özgen et al., 2021). The Covid-19 pandemic has significantly accelerated the adoption of digital teaching methods, including within architectural education (Hui et al., 2020). VR, with its immersive and walkthrough features, provides students with a better understanding of architectural concepts. One of the integrated software commonly used by students is Enscape, which is equipped with a VR integration facility (Hui et al., 2020). Therefore, implementing VR in architectural curricula, specifically when teaching about fire protection systems, holds great promise.

3 Building Fire Protection Standards

The components of building performance are classified into four, namely passive fire protection (PFP), active fire protection (AFP), occupants fire response (OFR), and fire fighters response (FFR) (Wang et al., 2022). Fire safety facilities in buildings are generally classified as active (e.g., sprinklers, fire detectors, fire extinguishers, etc.) and passive fire protection systems (structural and non-structural fire rating, material, etc.) (Kodur et al., 2020). Active fire protection systems address predictable and unpredictable situations (de Wilde, 2019). Passive fire protection systems do not need human or automatic intervention to protect the building, but it relies on structural components and layout configurations (Kodur et al., 2020). The ability to safely evacuate a building during a fire incident depends not only on individual characteristics and the use of the building regarding its occupancy classification but also on factors such as the layout, floor plan, and interior design. Fire safety measures, such as escape routes and emergency exits, significantly influence the likelihood of surviving a fire. Additionally, the availability and accessibility of these safety features could be impacted by the location of the fire (Abdullahi et al., 2022). Alterations to a building could also impact its fire performance (Wang et al., 2022).

Management prevention is crucial to fire life safety (Wardani et al., 2018). The purpose of the life-safety system is to save the occupant, specifically in the event of a fire, through various prevention methods (Shapiro & Maras, 2021). The critical measures to improve the security level of this system involve planning, management, and execution (Wardani et al., 2018). Life-safety design aims to prevent accidents and injuries during emergency evacuations by providing effective evacuation measures that allow occupants to exit the building without obstruction safely. Life-safety design aims to avoid mishaps or injuries during emergency evacuations by providing effective evacuation measures that allow occupants to exit the building without obstruction safely. In countries such as Indonesia, with high population density, life-safety systems have become increasingly important. For instance, in DKI Jakarta, the most crowded province in the country, of the 1.505 fire incidents in 2020, residential, public, and commercial buildings were the primary objects that burned the most, with 461 and 269 events, respectively (DKI Jakarta Provincial Fire Management and Rescue Service, 2020). Therefore, this present research would mainly be focused on the aspect of life-safety systems.

The technical requirements for fire protection systems in buildings and the environment in Indonesia are governed through *Regulation of the Minister of Public Works and Public Housing concerning Technical Requirements for Fire Protection Systems in Buildings and the Environment* issued by the Minister of Public Works and Public Housing number 26/PRT/M/2008 (Ministry of Public Works and Public Housing, 2008). This document specified the

Table 1 Summary of life-safety system technical requirements (Minister of Public Works and Public Housing, 2008)

Features	Article number	Technical requirements
Reliability of exit means	3.6.2.1	The furniture, decorations, or other objects should not be placed in the doorway or obstruct its access
	3.6.2.2	Avoid railings, barriers, or doors that divide an open area into separate sections, obstructing the means of egress
	3.6.2.4	Design and construct doors and main entrances to make them clearly visible from the outside
Doors	3.7.1.1	Design and install doors to swing from any position to the fully open position
	3.7.1.1.4 (3)	Doors and grills must not cause the door to close if the room is filled
	3.7.1.1.5	Horizontal exit doors are not required to open in the direction of the exit
	3.7.1.4	Provide an unobstructed space of at least half the required width of an aisle, corridor, canal, or stair landing during swinging
	3.7.1.4 (2)	Protrusions, when the door is fully opened, should not exceed 18 cm from the required width of aisles, corridors, canals, or stair landings
Exit access	3.9.4	The width of the exit corridors must be sufficient to accommodate the capacity required by all exits
Exit capacity	3.2	Every building should provide sufficient capacity for exits to ensure a safe evacuation process
	3.10.1.1	The total capacity of the egress facility for each floor, balcony, terraced seat, or other occupied area must be sufficient for the occupancy load
Exit corridor	3.10.3.2	Determine the capacity requirements of a corridor based on the floor occupancy load and the number of associated exits
	3.10.3.3.2	When the exit access is the only access to the exit, its capacity should be at least the required capacity of the doorway leading to it
	3.10.3.3.3	When there is more than one exit access to the exit, each exit access must have a width sufficient for the number of people served
Exit clearance	3.10.3.3.1.1	Exit access widths are formed by movable furniture and partitions, which serve no more than six people and have a length of not more than 15 m
	3.10.3.3.1.1 (1)	Width must not be less than 455 mm at points below 965 mm in height and not less than 710 mm above 965 mm in height
	3.10.3.3.1.1 (2)	The width should not be less than 915 mm for new exit access and not less than 710 mm for existing exit access without making permanent wall changes
Number of exits	3.12.1.1	Every balcony, mezzanine, or floor level must have at least two exists
	3.12.1.6	Elevator lobbies must have access to at least one exit without requiring keys, special knowledge tools, or special efforts
Exits arrangement	3.13.1.2	Corridors must provide exit access without obstructing space other than permitted areas, such as corridors and lobbies
	3.13.1.5	Exit access must be arranged so there are no dead ends in the corridor
	3.13.1.1	Exits should be strategically located, and accesses must be easily accessible
	3.13.2.2	Exit access from exit doors must be easily recognizable
Exit release	3.14.1	All exits must terminate directly on a public road or outside the exit discharge
	3.14.2.3	The view of the exit should not be obstructed by any objects
Exit visibility	3.17.1.2	Exits other than the main exit doors on the exterior of the building should be marked with visible signs
	3.17.1.5.2	New sign placement should ensure visibility within a certain distance in the exit access corridor

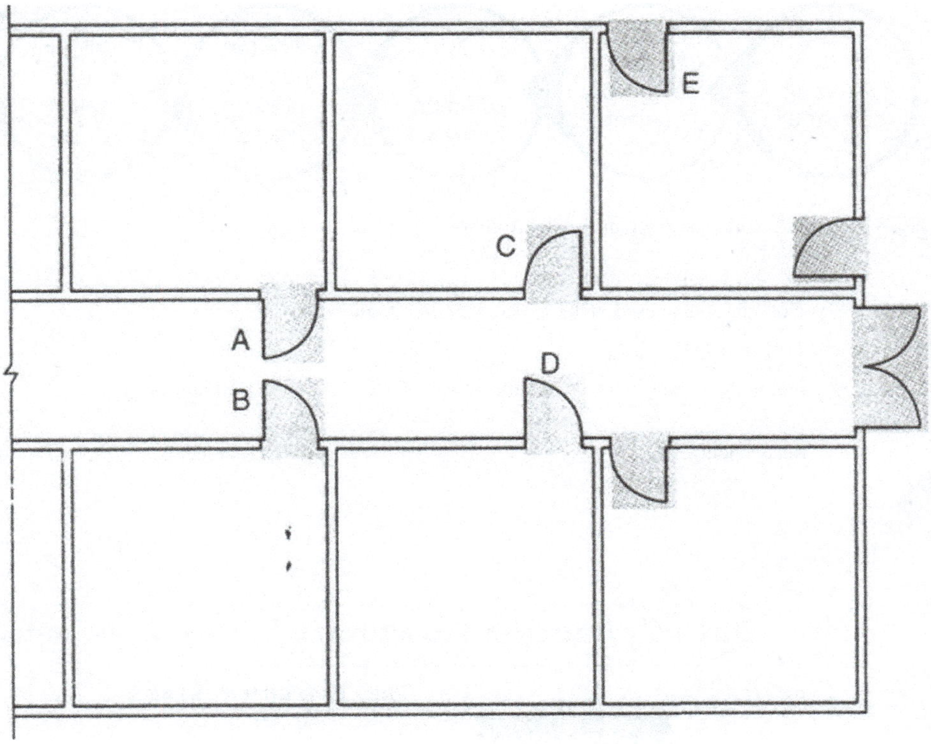

Fig. 2 Door placement and opening direction possibilities (Ministry of Public Works, 2008)

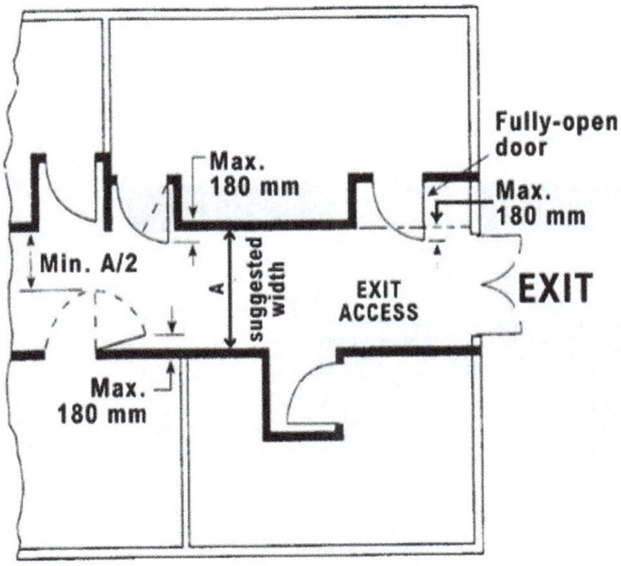

Fig. 3 Suggested doorway and exit routes clearance (Ministry of Public Works, 2008)

technical requirements for life-safety measures through 16 main features (Table 1). Examples of these building life-safety features include exit route clearance (Fig. 3), door placements, and opening direction (Fig. 2). However, this present research would focus specifically on nine criteria that are consistently emphasized among students, namely furniture layout, exit or clear opening of doorways, space configuration, corridor width, travel distance, the number of emergency exits, dead-end corridors, emergency exit points, and visibility.

4 Methodology

4.1 Analytical Method

This research employed a quantitative descriptive approach and an experimental design method. Descriptive research could be performed by merely comparative analysis (Siedlecki, 2020). An experiment was conducted with the same group subjected to different treatments (conventional and VR identification), and the results were compared. This descriptive research aimed to describe the sample and variables without manipulating any of the variables (Siedlecki, 2020).

This present research focused on using VR to enhance comprehension of the core design principles of passive fire protection systems by architecture students. In Indonesia, the current teaching approach to this topic is primarily theoretical. In order to assess students' practical understanding of life-safety principles, the experiment was conducted in a second-year design studio in an undergraduate architecture study program at one of the private universities in the country. Studio projects were selected as examples because it provides a more accurate reflection of the ability of students to implement life-safety principles compared to theoretical

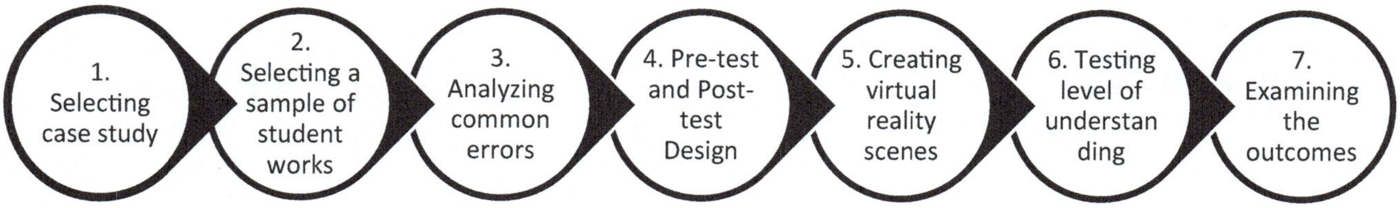

Fig. 4 Research workflow diagram. *Source* Authors

Table 2 Type of data source used in the research. *Source* Authors

No	Type of data source	Number	Notes
1	Fire protection design from architecture studio 2	23 Building designs	Quantitative analysis based on nine basic principles
2	Respondents	Nine students	Identifying mistakes in passive fire protection design using conventional and VR methods

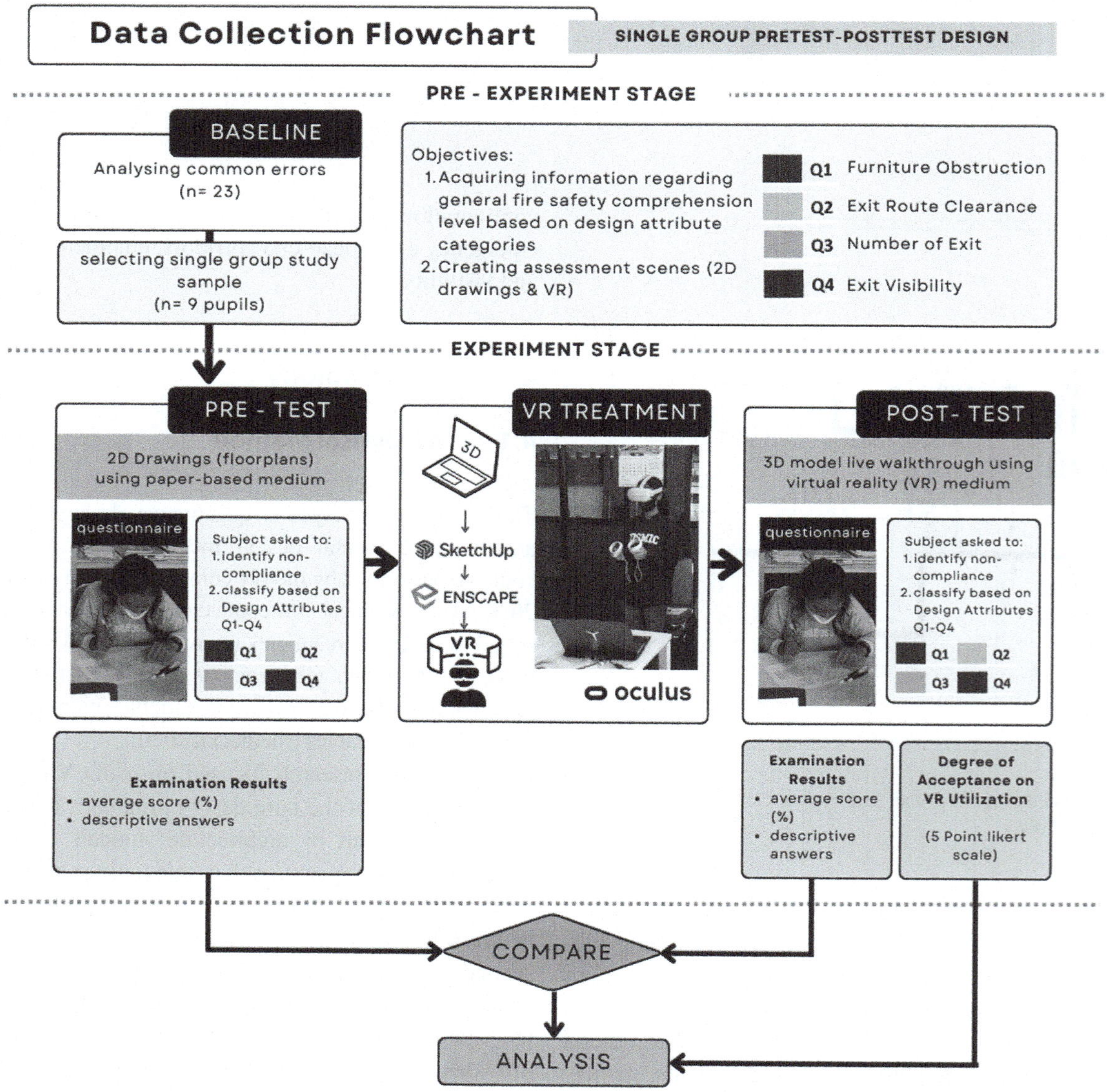

Fig. 5 Data collection flowchart. *Source* Authors

class assignments. The research project focused on evaluating the effectiveness of VR tools in teaching life-safety aspects of building design.

In order to realize this, a hypothetical assessment scenario was created based on a specific location in Indonesia, taking into account the common errors made by architecture students in designing life-safety systems. The design indicators for life safety were incorporated into the scenario to ensure comprehensive coverage, allowing for quantitative measurement of the significance of VR tools implementation. The assessment involved comparing the pre-test and post-test results to evaluate the impact of VR on the understanding and application of the principles by the students as shown in Fig. 5. In addition, the overall research workflow is shown in Fig. 4. Table 2 displays the type of data sources used in the research. A total of 23 students building designs from Architecture Studio 2 were analyzed, focusing on nine basic principles in designing passive fire protection. The respondents created and evaluated building models based on those results, as shown in Table 1. Nine second-year bachelor degree students in the architecture program were selected. They had limited experience in applying fire protection building standards in their studio design project and had not taken an elective course on fire protection design. Regarding the limited number of respondents, this research focused on evaluating the first-time VR user experience that is the target of the application of this learning method.

During the data collection process, the research approach considered several essential phases:

1. Selecting a relevant case study from the construction course on buildings that effectively supports the objectives of the architectural design studio.
2. Selecting a sample of student works from the building construction course. The selection process specifically focuses on projects related to the topic of passive building protection.
3. Analyzing common errors in designing passive fire protection appearing on student works.
4. Composing the pre-test (paper-based medium) questionnaires and post-tests (VR).
5. Creating virtual reality assessment scenes.

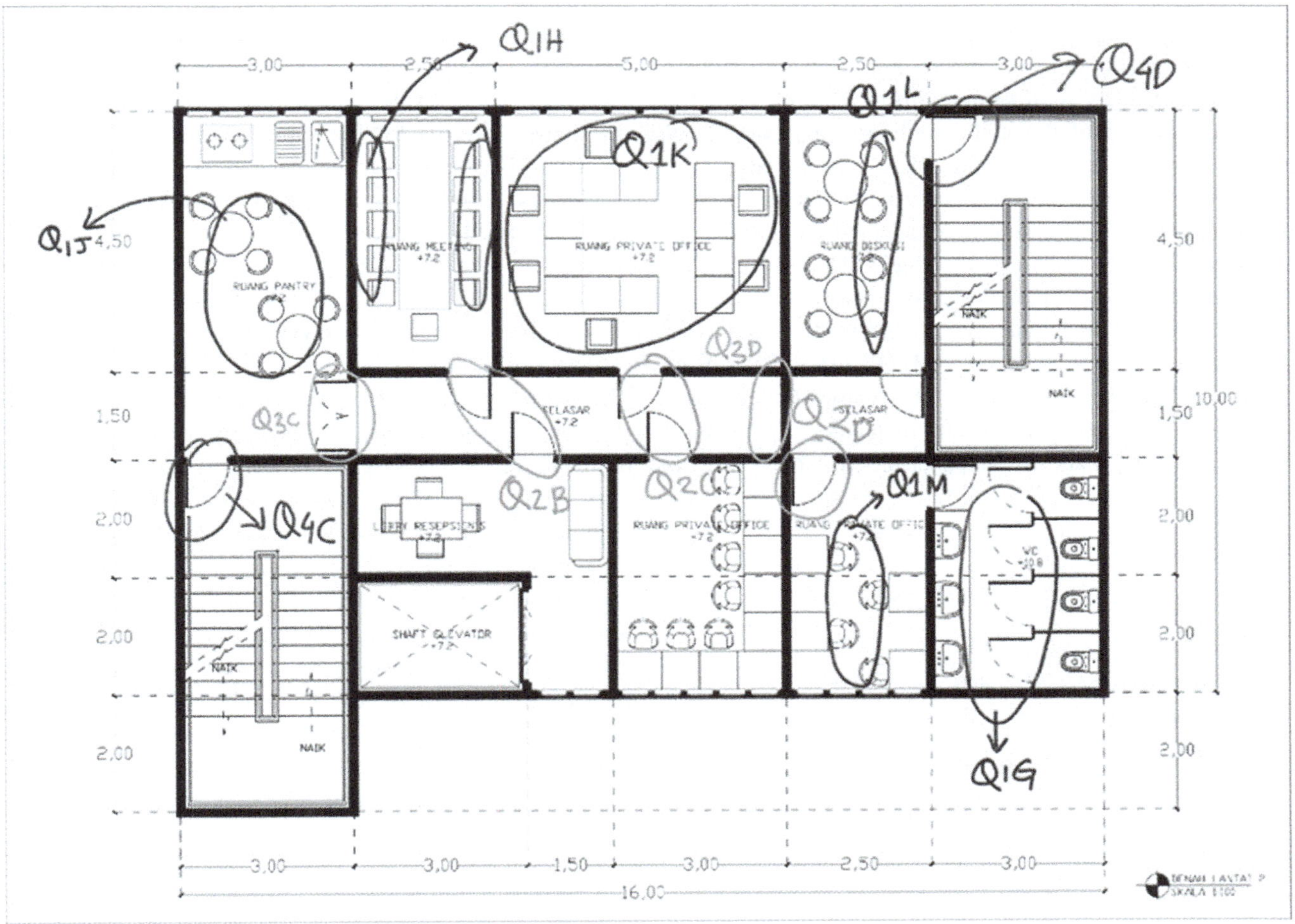

Fig. 6 Sample of paper-based pre-test: second floor examination scenario. *Source* Authors

6. Testing the students' level of understanding of passive building fire protection principles. This is realized by using pre-test and post-test quantitative methods to compare the effectiveness of the VR simulation in identifying common mistakes in passive fire protection design. The respondents are a sample of students enrolled in a building construction course.
7. Examining and discussing the outcomes.

4.2 Questionnaire Design

This research aims to evaluate the effectiveness of the VR method in architectural education compared to traditional 2D drawings. Therefore, the design of virtual space scenarios is vital in preparation for conducting the pre-test and post-test. The questionnaire is divided into three main parts, namely a paper-based pre-test, a VR-based post-test, and a five-point Likert scale to measure the degree of acceptance regarding the implementation of VR in architecture. This approach allows for the measurement of changes in understanding levels and the evaluation of participants' acceptance of VR technology.

In order to accurately assess the effectiveness of the virtual space design, it is important to ensure that it represents the recurring non-compliance issues identified in the findings. The analysis of samples from 23 students led to the determination of the non-compliance and compliance percentages for each indicator. The students' awareness levels were classified into five descriptive categories using the Compliance Rate framework, adapted from Paguigan and Jacinto (2018) as shown in Table 3. These levels were evaluated based on the nine design principles of passive fire protection in buildings and were subsequently categorized into four distinct groups, namely Q1, Q2, Q3, and Q4. Each design attribute was associated with specific sub-attributes, which served as the assessment indicators. The sample of sub-attributes marking on paper-based can be seen in Fig. 6. The awareness levels were quantified based on these indicators to provide a comprehensive evaluation.

Table 3 Categories of students' awareness levels

Scale	Max (%)	Min (%)	Descriptive rating
5	100	84	84% > fully aware
4	83.9	68	68–83.9% aware
3	67.9	52	52–67.9% neither aware nor unaware
2	51.9	36	36–51.9% unaware
1	35.9	20	< 35.8% fully unaware

Adapted from Paguigan and Jacinto (2018)

Specific equipment is required to facilitate the VR setup as shown in Fig. 7. The hardware includes an Oculus Quest 2, Oculus Link Cable, and a laptop. The laptop specification used is Lenovo Legion Pro 5i, with an AMD Ryzen 7 5800H processor, NVIDIA GeoForce RTX 3060 graphics card, 1T SSD storage, and 16 GB RAM. The 3D building model was created using SketchUp Pro 2021, while the walkthrough simulation was rendered with Enscape 3.20. Although the Oculus Quest 2 could be used as a standalone VR headset, for this research, an Oculus Link cable was used to connect the headset to the laptop and the HMD, allowing access to the VR experience. This connection method requires the graphic card of the laptop to meet the minimum specification for optimal VR performance. The setup allows the respondents to stream their point of view in real-time and to use the controllers to identify design in compliance. The mentioned software also provides respondents with the option to choose a walking mode, allowing them to explore the virtual space naturally. The aforementioned software also provides respondents with the option to select a walking mode, allowing for the natural exploration of the virtual space. It is recommended to provide a clearance of 1500 mm to accommodate movements during the VR experience.

5 Result and Discussion

The discussion focused on the preliminary identification of common mistakes made by students in designing exit accesses. A design scenario was developed according to the results that showed attributes with the lowest level of awareness. Subsequently, students' awareness levels taken during the pre-test and post-test were compared based on the four designated categories Q1, Q2, Q3, and Q4. A detailed classification of the four design attributes and their corresponding sub-attributes are shown in Table 4. The potential use of VR technology in improving the understanding of exit access designed among students was evaluated. The degree of acceptance among students regarding the utilization of VR as a medium for designing exit accesses within the context of building design was also examined.

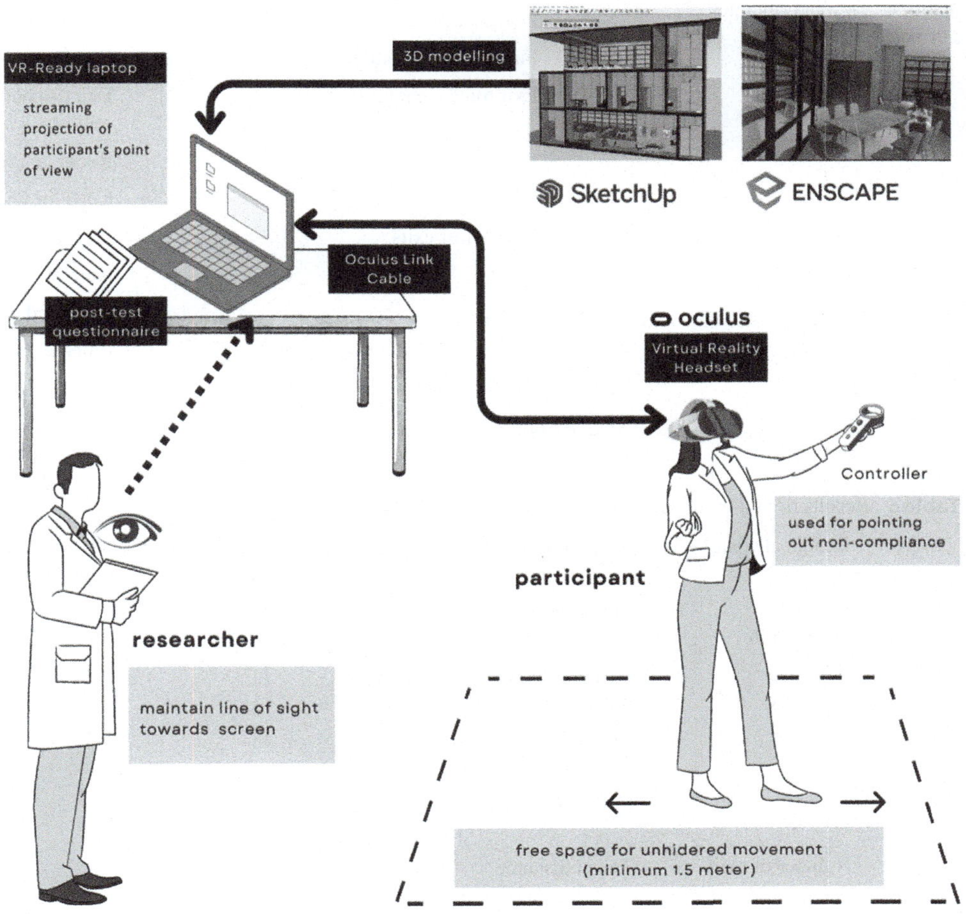

Fig. 7 Schematic diagram of hardware, software, and survey instruments used in the data collection. *Source* Authors

5.1 Common Mistakes in Designing Exit Access

The students' awareness levels across nine design indicators, which were then grouped into four main attribute categories, namely Q1, Q2, Q3, and Q4, were examined. Further details are shown in Table 5. The findings (Fig. 8) indicated that students generally had a good understanding of emergency exit location (70%) and emergency exit visibility (74%). The lowest level of understanding was observed in the layout of the furniture (Q1) and the obstruction of the exit route (Q2, Indicator 1), both of which demonstrated a compliance ratio of 26%, equivalent to the fully not aware category. On average, students possessed an intermediate level of understanding of exit route clearance (Q2). This was primarily attributed to challenges related to room configuration (Indicator 3, 57%) and narrow corridor width (Indicator 4, 65%). The level of understanding on the fulfilment of the standard number of exits remained at an intermediate level, with discrepancies found in the non-fulfilment of travel distance, number of stairs, and dead-end corridor access, by 43%, 39%, and 35%, respectively. The virtual reality scenario emphasized on the potential benefits and challenges of VR in addressing the low awareness of furniture layout compliance and exit route clearance.

5.2 Potential Contribution of VR in Designing Exit Access

The decline in Q2 observed among students tested with the paper-based method was attributed to the absence of certain features provided by the VR method (Fig. 9). The VR visualization medium offered a human-eye view perspective, which enabled users to interact with and perceive the elements that defined the spatial environment. Respondents closely examined the furniture arrangement, assessed the suitability of its sizes for the corresponding space, and analyzed circulation patterns when the furniture was rearranged. In contrast to the floor plan drawings, this medium offered detailed pictures of spatial configuration and overall dimensions. Conventional media provided a broader overview, while VR offered a more immersive experience in the spatial environment. Lovreglio et al. (2021) stated that using

Table 4 Design principles attributes

No	Design attributes	Sub-attributes	Assessment indicator
Q1	Furniture obstruction	1	Furniture obstruction
Q2	Exit route clearance	2	Exit discharge opening direction/blocked exit routes
		3	Room configuration
		4	Corridor width
Q3	Number of exits	5	Travel distance
		6	Number of stairs
		7	Dead-end corridor access (not exceeding 9–12 m)
		8	Emergency exit location
Q4	Exit visibility	9	Emergency exit visibility

Source Authors

Table 5 Results of non-compliance identification from examples of student works

No	Design attributes	Assessment indicator (sub-attributes)		Non-compliance ratio (%)	Compliance ratio (%)	Awareness criteria
Q1	Furniture obstruction	1	Furniture obstruction	74	26	< 35.8 fully unaware
Q2	Exit route clearance	2	Exit discharge opening direction/ blocked exit routes	74	26	<35.8 fully unaware
		3	Room configuration	43	57	52–67.9 neither aware nor unaware
		4	Corridor width	35	65	52–67.9 neither aware nor unaware
Q3	Number of exits	5	Travel distance	43	57	52–67.9 neither aware nor unaware
		6	Number of stairs	39	61	52–67.9 neither aware nor unaware
		7	Dead-end corridor access (not exceeding 9–12 m)	35	65	52–67.9 neither aware nor unaware
		8	Emergency exit location	30	70	68–83.9 aware
Q4	Exit visibility	9	Emergency exit visibility	26	74	68–83.9 aware

Source Authors

VR in fire extinguisher training resulted in better knowledge acquisition compared to video-based training groups alone. Feng et al. (2019) reported that VR training exhibited similar behavioral outcomes to real-world training scenarios. Therefore, VR and 2D drawings served as a valuable complement to traditional engineered drawings in facilitating an understanding of life-safety system design principles. Both mediums possessed their respective advantages and used in conjunction for optimal learning outcomes.

5.3 Respondents Degree of Acceptance

The results showed a widespread acceptance of VR as a valuable tool for learning life-safety design principles. Most statements received scores above 4 points as shown in Fig. 10, indicating a positive perception among respondents. In contrast, statement number 3, which pertained to user satisfaction, was slightly below 4 points, precisely 3.6. Regarding functionality, most respondents strongly agreed that VR offered a realistic point of view and enhanced their understanding of basic design principles. The highest average score was attributed to the statement that VR utilization eased the identification of design errors. This interpretation was based on the scoring level 3 (normal) as the neutral attitude. The finding was supported by previous research by Afzal and Shafiq (2021), Kwegyir-Afful et al. (2022), Lovreglio et al. (2021), Zhao et al. (2021) and the recommendation by Bashabsheh regarding the applicability of VR in various architectural education courses. Bashabsheh et al. (2019) also stated the effectiveness and impact of VR in learning building construction. These findings indicated

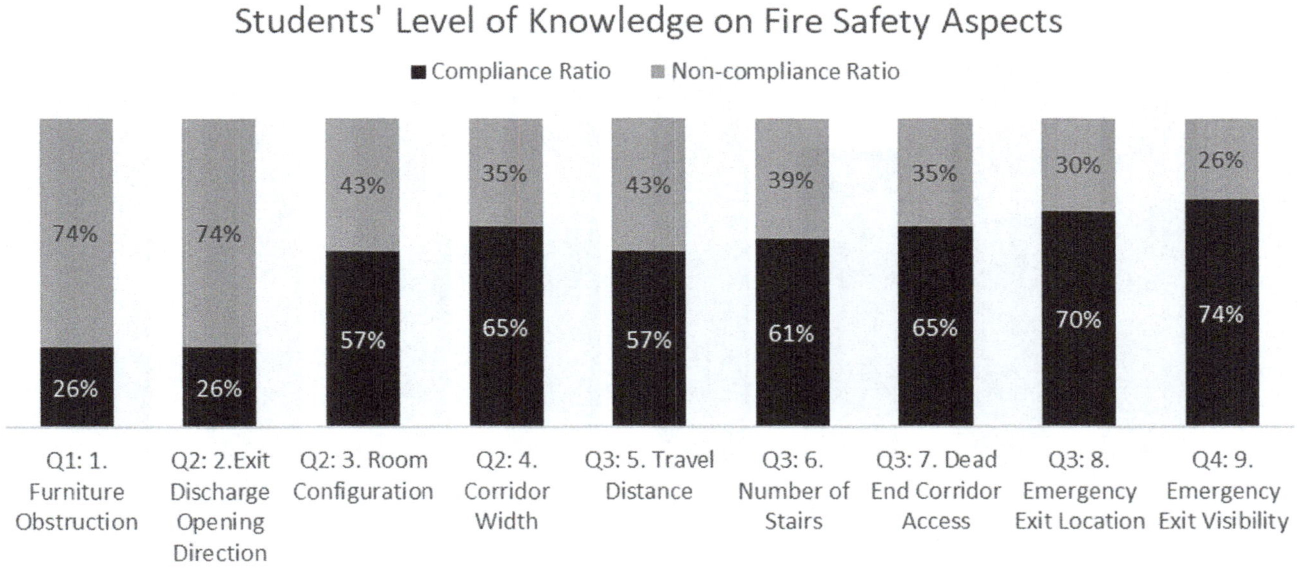

Fig. 8 Students' level of knowledge on fire safety aspects. *Source* Authors

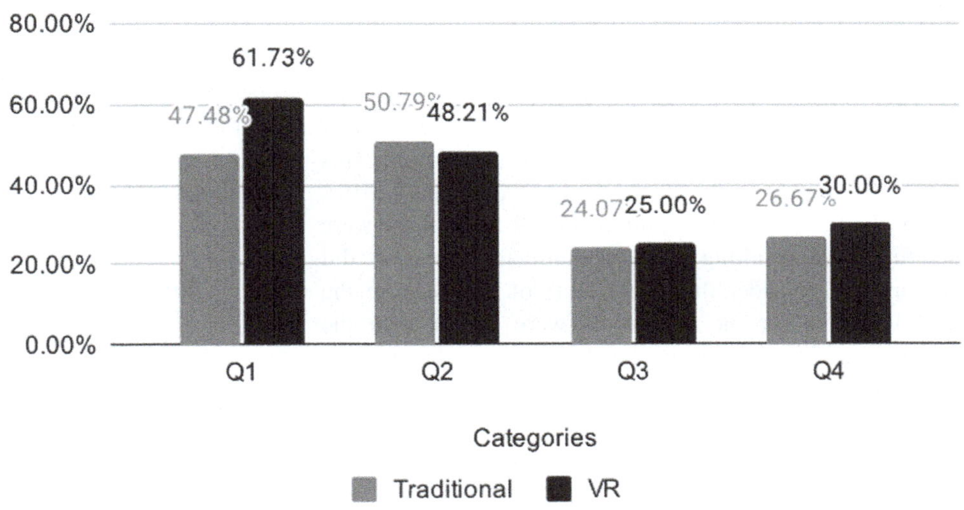

Fig. 9 Comparison between the student awareness level of the traditional and VR methods. *Source* Authors

that VR improved 3D environment recognition across various settings and supports the architectural learning process by enabling the identification of design errors that cannot be captured on 2D drawings. The immersive experience in VR offered a more convenient way of identifying non-compliance in designs. In terms of motivation, respondents expressed a strong interest in incorporating VR into their studio works (4.2 points) and learning the basic principles of passive fire protection (4.1 points). This indicated that VR was perceived as an exciting medium for visualization and potentially enhanced the learning experience.

Although the overall perception of VR was positive, disagreements were expressed in statements 1, 3, 5, and 8. These findings showed that few respondents had a lower level of acceptance in these aspects. The reasons for this lower acceptance were attributed to the unfamiliarity of first-time users with the VR experience, which led to dizziness and navigation issues. This finding is consistent with the previous research by Evans (2019), which also highlighted the occurrence of dizziness and discomfort after using VR, particularly among first-time users. Additionally, the experiment revealed that conducting a VR walkthrough simulation for each person took approximately 45 min, and this could be perceived as a limitation in terms of time. Maheshwari and Maheshwari (2020) identified various limitations of VR usage, including time, the associated costs, content development for learning purposes, and potential health effects. Despite these limitations, Maheshwari and Maheshwari (2020) reported that the enhancement of technology in the future had extensive possibilities in tackling these issues.

Experienced users positively perceived the VR interface, because it was easily understood. However, as shown

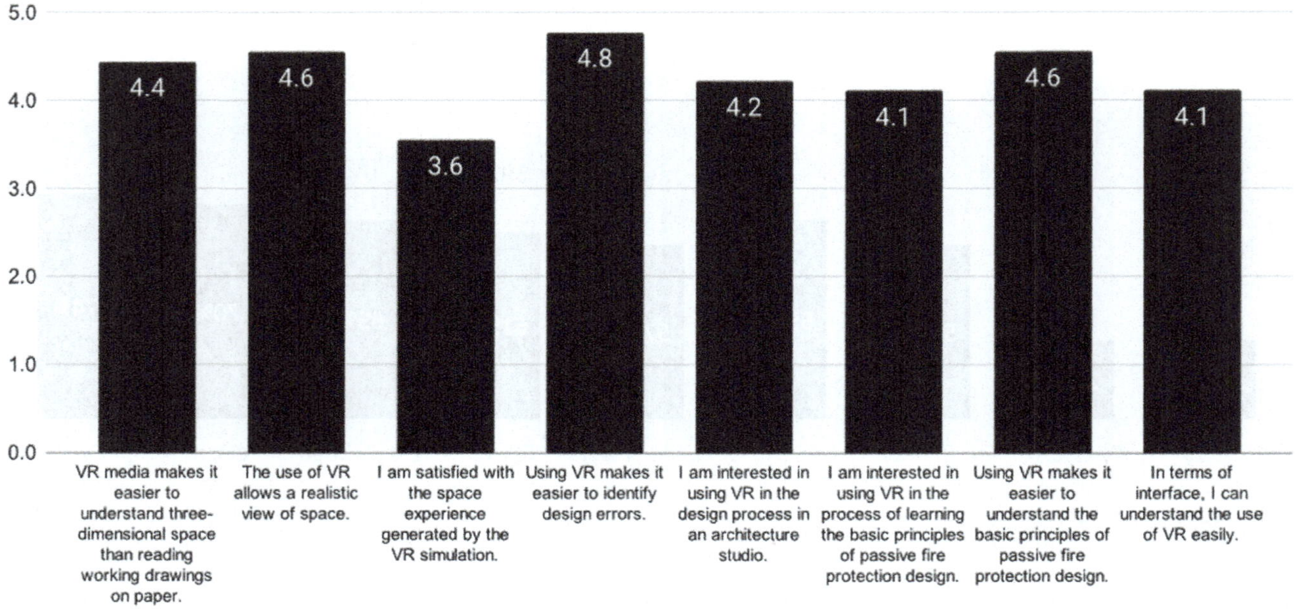

Fig. 10 User perception of utilizing VR. *Source* Authors

in Fig. 11, there was a significant gap between the perceptions of users at Statements 1 and 3, with the satisfaction level of Statement 3 below the neutral threshold. This dissatisfaction aligned with the findings of Evans (2019), where the users high expectations of VR technology were not fully met, resulting in disappointment. The users did not initially consider the side effects of VR because the overall scores for the statements were higher than the threshold of 3 points, except for the satisfaction in the VR space by experienced users. In other words, users were satisfied using VR to support life-safety systems design principles.

6 Conclusions

In conclusion, based on the analysis of the respondents' level of understanding, VR offered an immersive point of view that enhanced their spatial experience. Most of the second-year students perceived the use of VR in the architecture studio learning process as helpful. VR was perceived to significantly help students understand whether furniture configurations obstruct the exit access, specifically, the visibility of exit routes from different points in the room. Additionally, it proved valuable in identifying the optimal number of exits.

VR for learning also had some challenges, specifically regarding technology usage. The average respondents expressed dissatisfaction with the resulting virtual environment due to feelings of dizziness, while some early adopters struggled to navigate the VR space. The availability of tools that met the required specifications also affected the use of VR in the learning process, particularly in the school of architecture.

6.1 Future Research Recommendation

This research was constrained by the limited number of respondents, as conducting VR walkthroughs for each person required a considerable amount of time. In order to enhance the precision of data analysis, future research should aim to include a larger number of respondents. Additionally, this research was restricted by the availability of only two pairs of VR headsets. The experiment took longer than expected, as some participants needed a short break due to motion sickness. Considering a three-story building scenario, each respondent typically needed a 30- to 45-min walkthrough simulation in the VR environment. Most first-time users would occasionally need a break in between floor levels as they experience dizziness. Aside from that, some of them also faced difficulties in navigating the VR interface.

Future research holds the potential to explore the utilization of web-based or social VR platforms to enhance distance learning. It is anticipated that VR technology will be more affordable and integrated with the digital world. Moreover, such VR mediums can offer opportunities to explore and incorporate various learning subjects beyond architecture and spatial design.

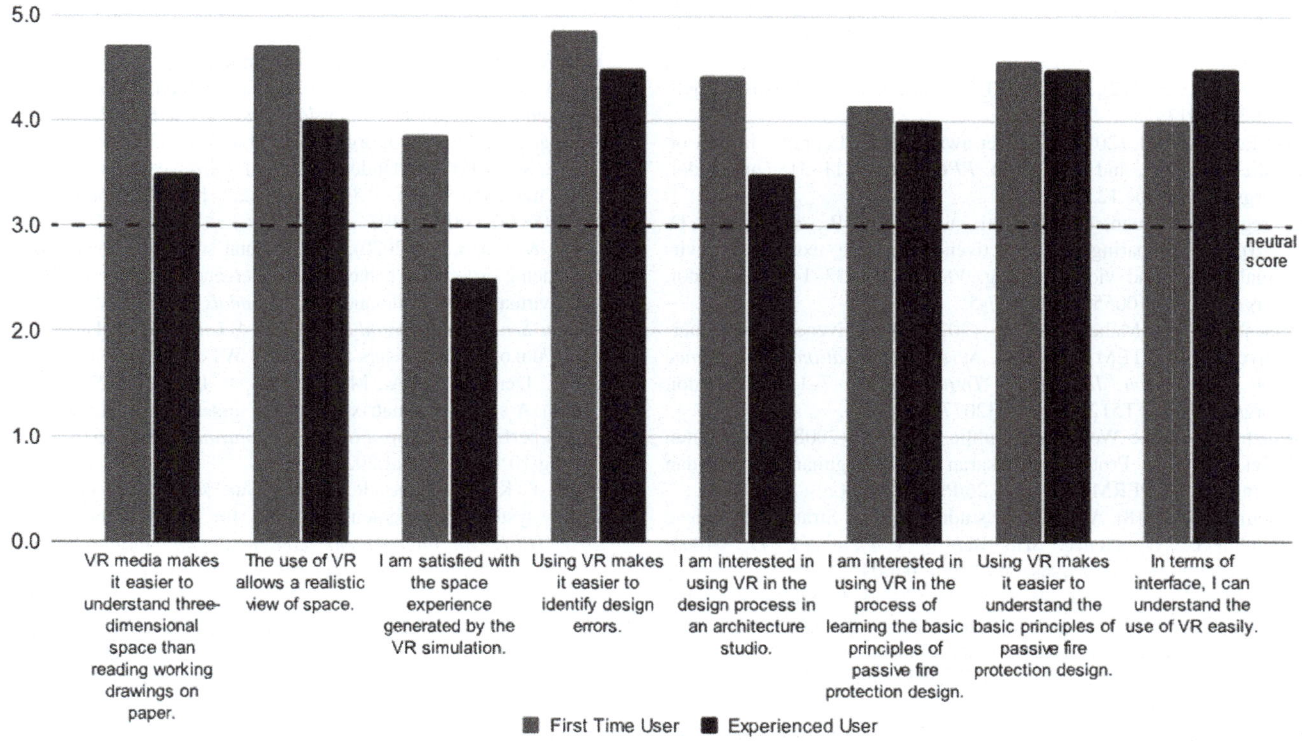

Fig. 11 First-timer user and experienced user's perception of utilizing VR. *Source* Authors

Acknowledgements We thank the Department of Architecture at Universitas Atma Jaya Yogyakarta for providing the resources and equipment needed to complete this research and all the participants in this study for their time and willingness to share their experiences.

References

Abdul Samad, M. H. (2021). Fire safety regulations: A comparative study of Malaysia and England and Wales. 2021 CABE Malaysia Chapter. https://www.buildingengineer.org.uk/intelligence/fire-safety-regulations-comparative-study-malaysia-and-england-and-wales

Abdullahi, Z. Y., Saad, A. M., Aliyu, A. M., & Isah, A. (2022). Building fire safety should be taken into account. In *1st international conference on innovative academic studies (September)* (pp. 608–613). https://www.researchgate.net/publication/363813106_Building_fire_safety_should_be_taken_into_account

Afzal, M., & Shafiq, M. T. (2021). Evaluating 4d-bim and vr for effective safety communication and training: A case study of multilingual construction job-site crew. *Buildings, 11*(8). https://doi.org/10.3390/buildings11080319

American Institute of Architects. (2020). *2020 code of ethics and professional conduct*. Apress. https://content.aia.org/sites/default/files/2020-08/2020_Code_of_Ethics.pdf

Bashabsheh, A. K., Alzoubi, H. H., & Ali, M. Z. (2019). The application of virtual reality technology in architectural pedagogy for building constructions. *Alexandria Engineering Journal, 58*(2), 713–723. https://doi.org/10.1016/j.aej.2019.06.002

de Wilde, P. (2019). Ten questions concerning building performance analysis. *Build, 153*, 110–117. https://doi.org/10.1016/j.buildenv.2019.02.019

DKI Jakarta Provincial Fire Management and Rescue Service. (2020). *Fire incidents in Dki Jakarta in 2020*. Statistik.Jakarta.Go.Id. https://statistik.jakarta.go.id/kejadian-kebakaran-di-dki-jakarta-tahun-2020/

Evans, L. (2019). Barriers to VR use in HE. In *Proceedings of the virtual and augmented reality to enhance learning and teaching in higher education conference* (Vol. 2018, pp. 3–13). https://doi.org/10.1255/vrar2018.ch2

Farshid, M., Paschen, J., Eriksson, T., & Kietzmann, J. (2018). Go boldly!: Explore augmented reality (AR), virtual reality (VR), and mixed reality (MR) for business. *Business Horizons, 61*(5), 657–663. https://doi.org/10.1016/j.bushor.2018.05.009

Feng, Y., Duives, D., Daamen, W., & Hoogendoorn, S. (2019). Pedestrian exit choice behaviour during an evacuation—A comparison study between field and VR experiment. In *Transportation research board 98th annual meeting transporation research board*. https://www.lens.org/lens/scholar/article/010-194-715-527-772/main

Gebczyńska-Janowicz, A. (2020). Virtual reality technology in architectural education. *World Transactions on Engineering and Technology Education, 18*(1), 24–28. https://mostwiedzy.pl/pl/publication/virtual-reality-technology-in-architectural-education,151347-1

Hui, V., Estrina, T., Huang, A., & Agma, S. (2020). Virtual reality as a response to emergent challenges in architectural education. In *Abstracts & proceedings of ADVED 2020—6th international conference on advances in education, 5–6 October 2020* (pp. 207–216). https://doi.org/10.47696/adved.2020119

Indonesian Board of Architects. (2017). *13 Butir Kompetensi—Sertifikat Keahlian (SKA) Arsitek IAI*. https://iai.or.id/sertifikasi/13kompetensi

Kodur, V., Kumar, P., & Rafi, M. M. (2020). Fire hazard in buildings: Review, assessment and strategies for improving fire safety. *PSU Research Review, 4*(1), 1–23. https://doi.org/10.1108/PRR-12-2018-0033

Kwegyir-Afful, E., Hassan, T. O., & Kantola, J. I. (2022). Simulation-based assessments of fire emergency preparedness and response in virtual reality. *International Journal of Occupational Safety and Ergonomics, 28*(2), 1316–1330. https://doi.org/10.1080/10803548.2021.1891395

Lee, E., & Lee, H. (2019). Disaster awareness and coping: Impact on stress, anxiety, and depression. *PPC, 55*(2), 311–318. https://doi.org/10.1111/ppc.12351

Lovreglio, R., Duan, X., Rahouti, A., Phipps, R., & Nilsson, D. (2021). Comparing the effectiveness of fire extinguisher virtual reality and video training. *VR, 25*(1), 133–145. https://doi.org/10.1007/s10055-020-00447-5

Maheshwari, I., Maheshwari, P. (2020). Effectiveness of immersive VR in STEM education. *Seventh International Conference on Information Technology Trends (ITT)*, 7–12. https://doi.org/10.1109/ITT51279.2020.9320779

Ministry of Public Works and Public Housing. (2008). Persyaratan Teknis Sistem Proteksi Kebakaran pada Bangunan Gedung dan Lingkungan. PERMEN PU No.26/PRT/M/2008.

Mutaqi, A. S. (2018). Architecture studio learning: Strategy to achieve architects competence. *SHS Web of Conferences, 41*, 04004. https://doi.org/10.1051/shsconf/20184104004

Özgen, D. S., Afacan, Y., & Sürer, E. (2021). Usability of virtual reality for basic design education: A comparative study with paper-based design. *International Journal of Technology and Design Education, 31*(2), 357–377. https://doi.org/10.1007/s10798-019-09554-0

Paguigan, G. J., & Jacinto, D. C. (2018). Occupants conservation attitudes on energy consumption: The case in Isabela State University in Cabagan, Isabela. In *4th international research conference on higher education* (Vol. 3, issue 6, pp 270–282). https://doi.org/10.18502/kss.v3i6.2386

Royal Institute of British Architects. (2021). *Mandatory competences.* https://www.architecture.com/knowledge-and-resources/resources-landing-page/mandatory-competences

Shapiro, L. R., & Maras, M.-H. (2021). Encyclopedia of security and emergency management. *Encyclopedia of Security and Emergency Management.* https://doi.org/10.1007/978-3-319-70488-3

Siedlecki, S. L. (2020). Understanding descriptive research designs and methods. *CNS, 34*(1), 8–12. https://doi.org/10.1097/NUR.0000000000000493

Singh, S., & Saberi, M. (2020). Experimental study to compare factors influencing exit choice behaviour in emergency evacuation situations using virtual reality techniques. *ISPRS Annals of the Photogrammetry, Remote Sensing and Spatial Information Sciences*, *6*(3/W1), 115–122. https://doi.org/10.5194/isprs-annals-VI-3-W1-2020-115-2020

Wang, H., Dembsey, N. A., Meacham, B. J., Liu, S., & Simeoni, A. (2022). A sensitivity matrix method to understand the building fire egress performance gap. *Fire Safety Journal, 127*, 103516. https://doi.org/10.1016/j.firesaf.2021.103516

Wardani, T. K., Nurcahyo, R., & Dachyar, M. (2018). Jakarta fire safety system management practices for high-rise building. In *2018 IEEE 5th international conference on engineering technologies and applied sciences (ICETAS)* (pp. 1–4). https://doi.org/10.1109/ICETAS.2018.8629119

Win, L. L., Aziz, F. A., Hairuddin, A. A., Abdullah, L. N., Jen, Y. H., Saito, H., Seyajah, N. B. (2018). Virtual reality application in engineering assembly education: A review. *Journal of Engineering and Science Research*, *2*(6), 17–23. https://doi.org/10.26666/rmp.jesr.2018.6.4

Zhao, G., Fan, M., Yuan, Y., Zhao, F., & Huang, H. (2021). The comparison of teaching efficiency between virtual reality and traditional education in medical education: A systematic review and meta-analysis. *Annals of translational medicine*, *9*(3), 252–252. https://doi.org/10.21037/atm-20-2785

Virtual Personal Branding Education Workshop for GenIUS School Students Using Framework for Innovation Participatory Design Method

Alfiansyah Zulkarnain, Ellis Melini, Brian Alvin Hananto, Kartika Magdalena Suwanto, Caroline Heliawanto, Stefanny Kusuma, and Daffa Andrian

Abstract

Generasi Indonesia Jaya Untuk Semua (GenIUS) School is a private school located in Binong, Karawaci, Indonesia. It is a scholarship-based school for disadvantaged region students from East Indonesia (Sekolah GenIUS in Latar Belakang Sekolah GenIUS, 2021a; Sekolah GenIUS in Program Sekolah GenIUS, 2021b). As part of a collaboration between GenIUS School and UPH's Visual Communication Design, the university lecturers and students were to conduct a personal branding education workshop for select GenIUS students. The design problem was how to design and execute the workshop through participatory design approach in a virtual setting as the covid-19 pandemic enforced social distancing between the two parties. The Framework for Innovation's evolved Double Diamond design methodology was chosen to develop the workshop, as its participatory model is clear and comprehensive enough for the students to discover the problem and find the best medium to deliver the workshop in innovative ways (Design Council in Framework for Innovation: design council's evolved double diamond, 2019). The research method was quantitative–qualitative. Data were gathered and analyzed through questionnaire with students, interviews with experts, and literature (discover and define step). After that, design solutions were developed through iteration (develop step) before the workshop was conducted as the design solution (deliver step). The digital workshop itself was delivered through three sequences of mini-workshops that were led by separate teams of students: introduction to personal branding workshop, personal branding in social media workshop, and designing a logo for personal branding workshop. Each group developed the best solution for their workshop and came up with innovative digital platforms to deliver the workshops. The result was a comprehensive digital personal branding education workshop conducted in four weeks that solved the geographical problem through integrated digital platforms based on participatory research.

Keywords

Virtual workshop · Branding · Participatory design · Framework for innovation · Social design

1 Introduction

GenIUS (Generasi Indonesia Jaya Untuk Semua) School is a private school established in 2013. The School's location is in Karawaci, Tangerang City, Banten Province, Indonesia. It aims to develop East Indonesia by focusing on accepting and providing a platform for Eastern Indonesian students to get the best education (Background, n.d.). GenIUS School specializes in math and research-based science curriculum to produce mature and qualified graduates (Programs, n.d.). To achieve that, the school also developed several academic

A. Zulkarnain (✉) · E. Melini · B. A. Hananto · K. M. Suwanto · C. Heliawanto · S. Kusuma · D. Andrian
Visual Communication Design Department, Universitas Pelita Harapan, Tangerang, Indonesia
e-mail: alfiansyah.zulkarnain@uph.edu

E. Melini
e-mail: ellis.melini@uph.edu

B. A. Hananto
e-mail: brian.hananto@uph.edu

K. M. Suwanto
e-mail: kartika.suwanto@uph.edu

C. Heliawanto
e-mail: 01023190036@student.uph.edu

S. Kusuma
e-mail: 01023190006@student.uph.edu

D. Andrian
e-mail: 01023190006@student.uph.edu

M. Nawaz Tunio et al. (eds.), *Sustainability in Creative Industries*,
Advances in Science, Technology & Innovation, https://doi.org/10.1007/978-3-031-48453-7_8

and non-academic programs with one of them called the Digital Literation program. It is an hourly, three times a week program that aims to equip students with the understanding of digital literacy and know-how in navigating and presenting themselves with the help of digital tools.

To run the program, GenIUS School cooperated with Universitas Pelita Harapan's Visual Communication Design Department. From January 2022 to April 2022, a team of the design department's lecturers and third year university students organized a workshop as part of the digital literation program. For the university team, the activity itself was part of the Project-Based Learning (PBL) for Design for Society course. In this course, students will learn how to utilize design as a response and solution to problems in society.

After initial research through interviews with the school's stakeholders (select principals, teachers, and students) to identify the problem, it was found that there was a need to equip the GenIUS school students, especially the junior high students from Grade 11 and 12, with digital branding and brand design knowledge. The university team then planned to develop and conduct a branding workshop for them using participatory approach.

The main issue in conducting this workshop would be the covid-19 virus pandemic still raging in Indonesia. With the social distancing policy by the government still strictly enforced, it nullified direct contact between the university team and the school students and forced interactions only in a virtual setting. It would then hinders the research, development, and implementation stages of the workshop. There would be a need to adapt and innovate if the university team uses the participatory approach in designing the workshop. The design objective was to design and conduct the workshop through participatory design approach in a virtual setting as the covid-19 pandemic created social distancing between the two parties.

Studies regarding operating virtual design workshop in covid-19 pandemic is limited as it is still an ongoing event, but there are some findings on online learning in pandemic situation that could provide initial insights. Allo (2020) stated that online learning could be very helpful in pandemic situation. This could be the ground where the university team and school stakeholders decide on online setting as the way to carry on the workshop. Furthermore, He also pointed out that teaching material and assignment should be accompanied by clear explanation first. The workshops may need an introductionary lecture to be effective in online and virtual setting. In developing online learning, flexibility, accessibility, and transportability are important elements (Key, 2022). There is also a threat of online learning producing a lower performance than a face-to-face situation because of the unfamiliarity of the new medium (Contreras et al., 2021). The participants needs to have previous knowledge in using certain online applications to make sure the event goes smoothly (Mohalik & Poddar, 2020). Mohalik and Sahoo (2020) concluded that there is a need for digital knowledge and skills training for both students and teachers to prepare them in online or virtual learning environment. The availability of digital device and internet connection as infrastructure is crucial in facilitating online learning (Nagar, 2020). These aspects would helps the university team in designing the workshop for the school students.

2 Theoritical Framework

2.1 Social Design

Social visual communication design can be defined as a form of design that has social properties in it (Hananto, 2020). In his further explanation, these social qualities can be found in either its message (content), social or cultural background (context), or the methodology of design (form). In this project, we can first identify the context of creating and conducting a nonprofit workshop project for elementary/middle school students as a social quality, as the primary purpose of this project is to equip them with digital literacy skills. In creating the workshop, the university team will use the social design framework as its design methodology. This formal quality also made the project a social visual communication design project. Brand and branding in social media are the content to be delivered to the students. As they are rooted in business and marketing discipline, it does not have social qualities. We may conclude that this design project qualifies as a social design project as it has social properties in its formal and contextual aspects.

2.2 Participatory Design

Participatory Design can be defined as a form of collaborative design of an information technology that involves the people using it. Participation of the user and stakeholder is essential in the co-design process (Simonsen & Robertson, 2013). The participatory approach puts the user as the expert on the subject matter. In this project, the GenIUS school students act as an influencer in how the workshop will be conducted. Besides that, the content of the workshop was also driven by them. The school teachers and staff involved as stakeholders.

3 Methodology

Framework for Innovation is a design process model developed by Design Council in 2019 (Design Council, 2019) from the UK. It is an evolution of the Double Diamond model they formed in 2005. It represents a process of exploring an issue more widely or deeply and then taking focused action, as the model is based on the common design methodology of divergent and convergent thinking, and also its nonlinear process.

There are four nonlinear phases in Framework for Innovation's Double Diamond model as stated by Design Council: Discover, Define, Develop, and Deliver (Fig. 1). Discover and Define phases (first diamond) helps to not only to identify but also to understand the problem. In these phases, the activity is in engagement through speaking and spending time with the users and stakeholders. In Discover phase, interviews and discussions are conducted to gather as many insights as possible on the issue. In Define phase, insights collected from the previous step are then used to determine the challenge differently. In Develop and Deliver phases (second diamond), an innovative solution is developed from the clearly defined problem. Develop phase focuses on design exploration. Meanwhile, Deliver phase involves refinement. In the last phase, design deployment and testing are done on a small scale, and then identify its effectivity before doing iteration to improve it.

This model was created to solve social problems in both the public and private sectors through innovation, by delivering services that can achieve significant and long-lasting effects to the user. To achieve that purpose, the Framework for Innovation's Double Diamond design methodology was chosen to develop this workshop. Its participatory model is clear and comprehensive enough for the students to discover the problem and find the best medium to deliver the workshop innovatively.

This project used a quantitative–qualitative approach in its research and development of the workshop. Data were gathered through questionnaires to students, interviews with experts, and literature (discover and define phases). After that, design solutions were developed through iteration (develop phase) before the workshop was conducted as the design solution (deliver phase). The timeline for the project can be seen in Fig. 2.

According to the Design Council, the four core principles of the Framework for Innovation model are: put people first, communicate visually and inclusively, collaborate and co-create, and iteration (2019). These principles will drive the design of the workshop. The workshop will be designed based on the user's needs, which means gathering insights on the school students are imperative. The workshop will then be delivered through various visual forms and interactivity, such as presentation slides, personal interactive booklet, photography, illustration, social media feeds design, and logo design. The collaborative aspect will be done with the stakeholder and the school students to create practical workshops. This aspect put the workshop design into a participatory design project. Iteration will be done through various phases. Back-and-forth discussion and approval between university student teams and the lecturer team, between the university team and the school staff, and between the university teams and the school students teams

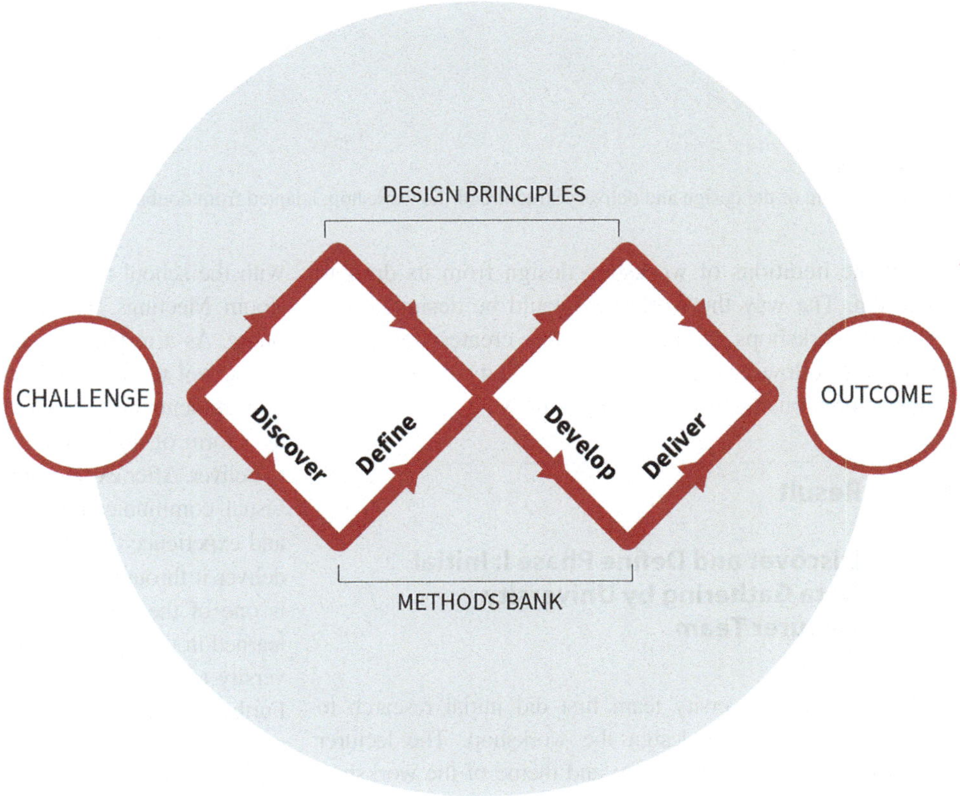

Fig. 1 Framework for innovation design process model by British Council

Design Phase	Jan-22			Feb-22				Mar-22				Apr-22
	Week 2	Week 3	Week 4	Week 1	Week 2	Week 3	Week 4	Week 1	Week 2	Week 3	Week 4	Week 1
Discover	Lecturer Team	Lecturer Team	Lecturer Team			Student Teams	Student Teams	Student Teams				
Define				Lecturer Team	Lecturer Team	Lecturer Team	Student Teams	Student Teams	Student Teams			
Develop								1st Student Team	2nd Student Team	3rd Student Team		
Deliver									1st Student Team	2nd Student Team	3rd Student Team	
Evaluation/ Closing												Lecturer Team

Fig. 2 Timeline of the branding workshop for GenIUS School, from design process to implementation

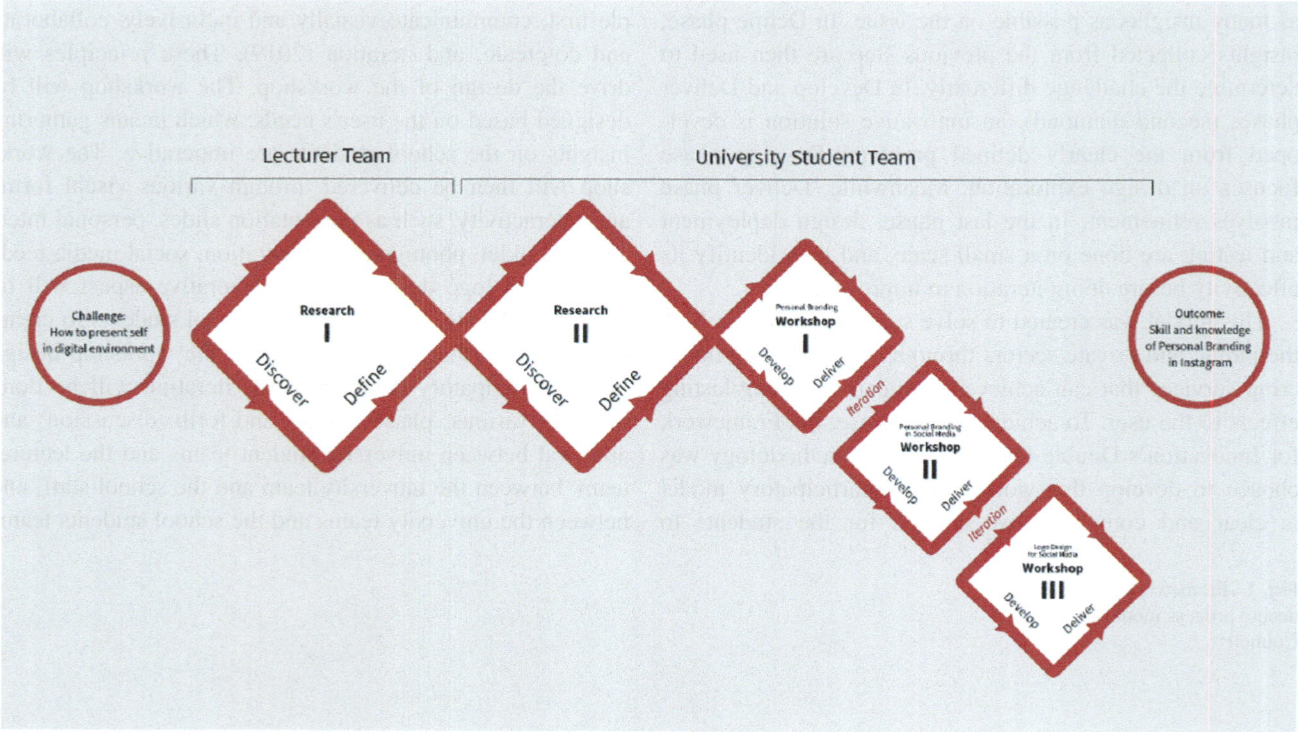

Fig. 3 Diagram of the design and delivery structure of the workshop, adapted from double diamond model

will create iterations of workshop design from its draft to execution. The way the workshop would be designed as a series of workshops in succession also creates opportunities for the following teams to observe, analyze, and iterate from the previous team's workshop (Fig. 3).

4 Result

4.1 Discover and Define Phase I: Initial Data Gathering by University Lecturer Team

In this step, the university team first did initial research to gather primary data to design the workshop. The lecturer team discussed the possible form and theme of the workshop with the school teachers and staff as the stakeholder through Zoom Meetings. The timeframe was also addressed in this phase. As a result, we gathered important informations. First, the school teachers and staff requested that knowledge about how students present and conduct themselves in social media as a form of digital literacy is what the lecturer team needed to deliver. After exploring and assessing its correlation with the visual communication design discipline, and the knowledge and experience that the lecturer team had, then we decided to deliver it through the context of brand identity. Brand identity is one of the studio courses the university student team had learned in their previous semester, so it is deemed that the university team was equipped to conduct a workshop about this. Further discussion and approval of the general theme/content of the workshop with the school staff were conducted before moving to the next step.

Second, based on the constraint of the timeframe given, the workshop needs to be delivered three days a week between February 16th to April 4th, 2022, from 3 to 4 PM for each workshop session (after the students finish their day school activities). During that timeframe, there was a week off because the students had a mid-term exam. The information and constraint regarding the timeframe available would drive the workshop's design into a series of workshops. It was decided with the stakeholder that the university team will deliver three continuous workshops: introduction to personal branding, personal branding in social media, and how to design a logo as a visual identity in social media. These workshop series will provide a gradual understanding of branding in social media, from the introduction to concepts of personal branding, how it works in particular social media feeds, and how to represent personal branding in social media in the form of logo design.

Third, it was decided that the workshops would be conducted through Zoom Meetings based on the Pandemic situation at the moment (early January). The Banten Province covid-19 positive case number hiked three times compared to a week before in December, making it the third highest in Indonesia at the moment. 46.37% of the case comes from Tangerang (Satuan Tugas Penanganan covid-19, 2022). Because of this, the university students involved are not allowed by their parents to do any on-site campus activity. Another context was the dorm-based system in GenIUS School, where students lives full time in the school area and had not been allowed to go outside of it. This fact limited the option for the university team to conduct the workshop on-site as it will increase the risk of covid-19 spread. After assessing the risk and opportunities available through multiple discussions with the stakeholder, conducting the workshop utilizing the Zoom Meeting platform is deemed the best option available.

Fourth, there are going to be 38 students as users. Those numbers consist of junior high students from Grade 11 and 12. The students are participants of the Digital Literation program, making this workshop part of it. The lecturer team also gathered some insight regarding the students from Zoom Meetings discussion with the school teachers and staff. All students are from Eastern Indonesia, so there will be a cultural and social barriers. Although all of the parties involved speak Bahasa Indonesia, there are different dialects and lingo between the university team and the school student, which created some language barriers. The students are also described as curious but shy and could be quite blunt in their communication. They can be active in discussions, but we must build trust and comfort with them first. The communication noise formed by the digital platform (Zoom Meeting) will add to the challenge. These insights will be a consideration when designing the communication approach in the workshop.

4.2 Discover and Define Phase II: User Insights by University Students Team

Based on the information from the initial research, the lecturer team divided the university students into three groups, with around five to six university students on each team. The teams then researched into the school students as the user and developed the content for this Project-Based Learning program. The first team was tasked to run a workshop about personal branding in general. The second team was tasked to continue the program by conducting a workshop about personal branding on social media. The third and final team continue deeper into the visual communication design discipline, executing a workshop about logo making as personal branding. Based on the schedule given by the school, the university team then sends each group to conduct lead workshop on three-days-a-week timeframe. As the university student teams were preparing the workshops in the Develop phase, the lecturer team presented lectures about the introduction to brand and personal branding to the participants to help kickstart the workshop series. These lectures aimed to provide a foundational understanding of the subject.

The university student teams designed the content of the workshop by gathering data through relevant literature and interviewing experts. The first team interviewed Mr. Joshua Sudihman, a design agency owner with a decade-long experience in the design industry. The second team interviewed Mrs. Lia Herna, an experienced design lecturer qualified in branding and UI/UX design. The third team was focused only on literature as their primary reference, as they were already equipped with logo design skills and knowledge as design students. Further data was gathered through digital questionnaires to understand more about the users, gauging their initial understanding of their personality, personal branding, social media, design literacy, and pop culture references.

From the analysis of the data gathered by the university student teams, some insights were found. First, most of the school students were able to define their personality through the Myers-Briggs Type Indicator test. The test was conducted through the 16personalities.com website. 54% of them positively described themselves (e.g., nice, smart, attentive), 31% in a negative way (hotheaded, fussy, or lazy), and 15% were neutral (good in sport, tall, good looking). There were also 16.28% of the school students who had contrasting opinions about themselves (for example, choosing both diligence and laziness as their personality). These neutral and contrasting numbers show that some students might not yet understand the concepts of personality and how to define it. Second, the way the school students answer the questionnaire indicates that they are quite sincere in their answers, as the answers tend to be quite plain and blunt. This information echoes the insight gathered by the lecturer team from the school staff. Third, despite being prohibited by the school from accessing social media such as Instagram, they were pretty knowledgeable in social media space. They understand common references to mainstream pop culture such as celebrities and social media influencers, both local and global. Fourth, most of the school students (77%) have a basic understanding of the logo and its purpose. They have certain visual preferences in visual shapes, color, and typeface categories. Around 90% of the school students were able to recognize popular randomized products, companies, and celebrity logos through their color and shape. This means they have a basic understanding of logos, although they were certainly never exposed to brand identity and visual communication design theories and principles before. From the data and insight gathered from literature review, expert interview, and user questionnaire, the university student teams could then develop a workshop that can communicate its content effectively to the participants.

4.3 Develop and Deliver I: Personal Branding Workshop

In Develop phase, the university team designed a three-day workshop based on this structure. A lecture on day one, workshop on day two, and evaluation of workshop result on day three. The design process was done through discussion and iteration with the lecturer team and the school staff. This structure was chosen because there is a need for education to know about the subject matter, then a workshop to understand the subject matter deeper, and evaluation to reflect on the subject matter. This structure would then be applied to the three workshop series by the university student team.

Initially, the first team created a timeline for better planning and time management. After that, the team did brainstorming sessions to pick the topic and to consider what are the most effective media to deliver the workshop. Then it was decided that personal branding as the topic, and the workshop would be done through Zoom Meetings in conjunction with a physical activity booklet. The booklet was an essential element as it created a tangible form of the workshop. It was also used to record the activity result. It was designed through a couple of revisions and input from the school staff and lecturer team. The revisions managed to push for sharper and clearer activities and directions in the booklet.

The team also planned the workshop rundown. They also created a guideline for the teachers as the school student mentor and the Digital Literation program team. The guidelines consist of directions for the teachers in assisting with the workshop. They would help mitigate and solve technical problems from the school side, as the workshops were going to be conducted remotely.

In the Deliver phase, the first team then run the workshop. On the first day (March 8th, 2022), they presented a lecture about the definition, function, and benefit of personal branding. There was Q&A session after the lecture session to have a further discussion and make the school students more active participants of the day. The printed booklets were sent to the school at the end of the day.

On the second day (March 9th), the main workshop was conducted (Fig. 4). Each school student was given a booklet to be filled along the workshop session, with guidance from the university student team with the teacher mentors assisting from the site (Fig. 5). The participants was divided to three Zoom Meetings breakout rooms. There are two university students leading each room to guide the participants. Each breakout room will have a smaller number of participants for a more conducive and effective workshop session. Each room will also be able to have a more intense and personal discussion regarding the participants' difficulties and questions. At the end of the workshop session, the university team came to collect them. After that, the first team recaps the answers into data to be analyzed quantitatively and qualitatively.

On the third day of the workshop (March 10th), the first team reviewed the workshop conducted on the previous day. They presented the result from the booklets and discussed it with the participants. They also asked for open feedback from the school students. It will be useful to gauge how effective the workshop was. It will also be helpful insight for the other university teams in designing the following workshops.

4.4 Develop and Deliver II: Personal Branding on Social Media Workshop

During the workshop session by the first team, the second university team was present to observe and gather insight into the participants' behavior and response. The second team also received the data collected at the end of the first workshop. Those will helps the second team in designing the second workshop and creating iterations of the previous one.

In the Development phase for the second team, they created lecture materials regarding personal branding on social media and how to express themselves adequately, particularly on Instagram. Instagram was chosen as it was one of the most popular social media platforms in Indonesia (We are Social, 2022). To ensure it, the team created a questionnaire for the school students regarding their current understanding of social media. The result indeed showed the school students were most familiar with it (Fig. 6). It also showed some of the role models and influencers they were familiar with.

Fig. 4 Screen capture of the second day of the first workshop

Using the similar structure used in the first workshop of the series, the second team then began to develop the workshop. Based on the data gathered from the Discover phase, the team decided that the topics of the workshop: the correlation between personal branding and social media, how to create personal expression through said media, the right way to express yourself on social media, and some study cases. Considering many topics that needed to be delivered and the time constraint, it was decided that the first two days of the workshop would be about the lecture after much discussion with the lecturer team.

Meanwhile, the university team would deliver the workshop activity and evaluate it in the last day. For the workshop activity itself, the team designed the "Yourself in Images" activity, where its aim was to ask the participants to upload photos or pictures that represent themselves in Instagram templates through the Canva design app. The university designed the Instagram template and design instruction to be as simple and intuitive as possible for the school students to use. This phase also developed through various revisions through discussions and approval from the lecturer team and GenIUS school staff. Like the first team, the second team also designed the event rundown and guidelines for the teacher mentors to ensure a swift workshop.

The Deliver phase saw the second team conduct a three-day workshop from March 22nd to 24th. On the first day, after the introduction of the university team members to the participants, they first reviewed the result of the first workshop. Then, they gave a presentation about the introduction to social media and how to build personal branding on social media. They also explained the forms of personal branding in social media using Instagram. They used some influencers to show the different purposes of personal branding. A Q&A session was conducted after the lecture to continue the discussion in interactive form. On the second day, the second university team explained the advantages and shortcomings of using social media and gave study cases from celebrities and athletes. These influencers were chosen because of the participants' familiarity to the them. It was then followed by a Q&A session and briefing for the workshop activity on the third day.

On the third day, the second team separated the participants into three breakout rooms on the Zoom platform and gave the Canva web app link to the GenIUS school students in each breakout room. The participants were tasked to fill the blank squares with pictures that represent themselves, simulating the Canva template as their own Instagram account (Fig. 7). The university team, assisted by teacher mentors, acts as a guide for the participants in filling in the template. After the activity session, the second team gave a short review of the result before giving conclusions.

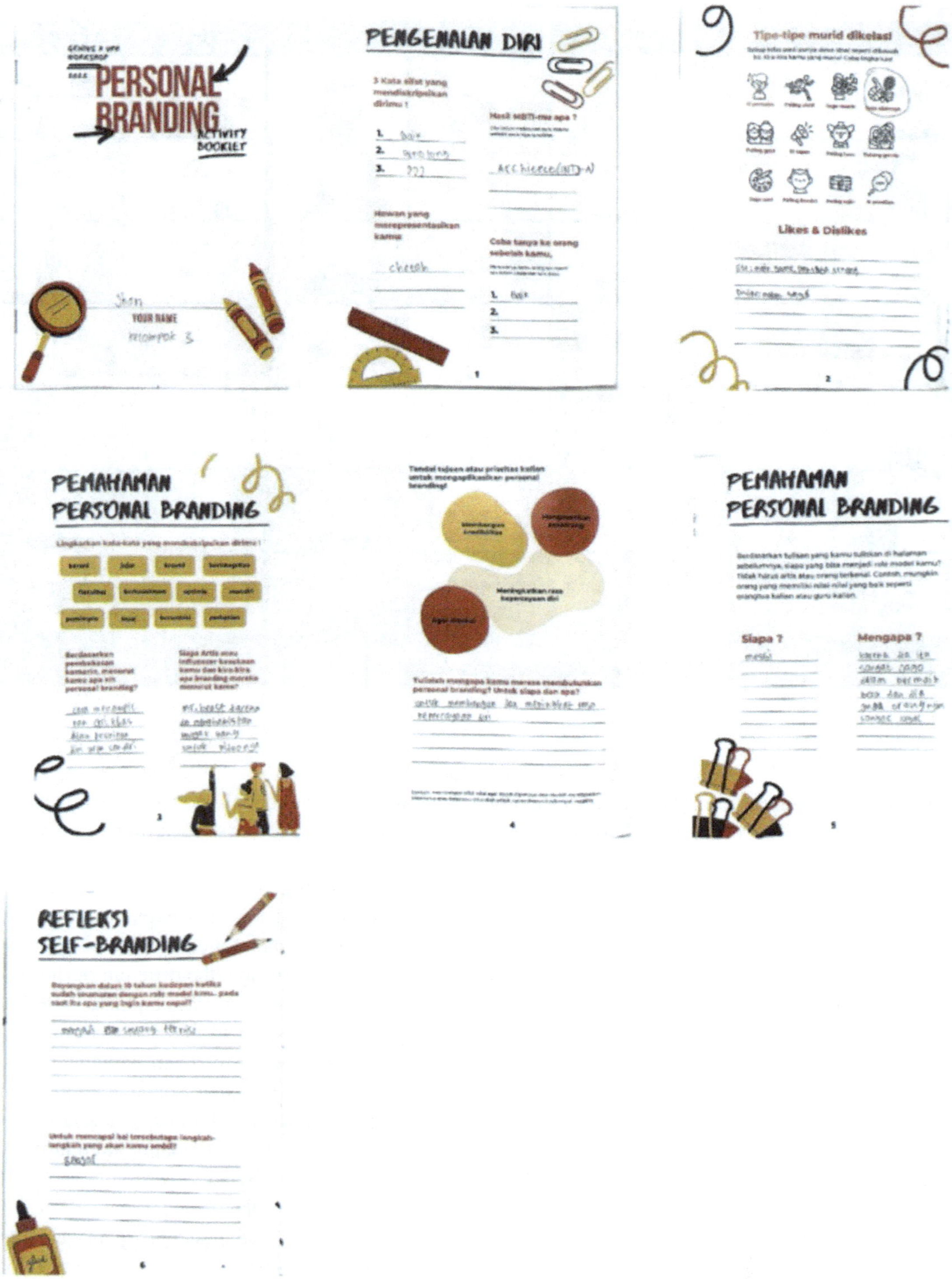

Fig. 5 Digital scan of one of the participant's booklet

4.5 Develop and Deliver III: Logo Design for Social Media Profile Workshop

The third university team was tasked with the same observational position to the first and second workshop. This would help them iterate more on the workshop they would design and deliver. In the Develop phase, the team prepared questionnaires for the participants to gauge their understanding of art and design, such as color, visual shapes, and typography. The questionnaires also asked about their

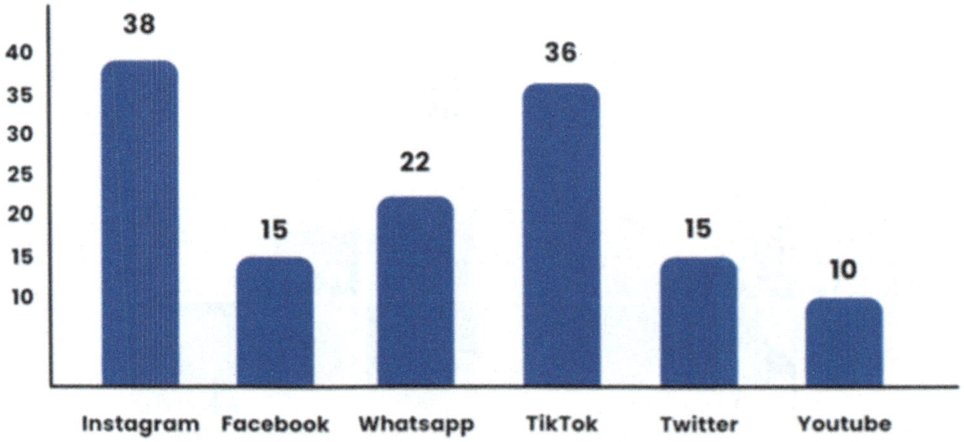

Fig. 6 Data result questionnaire on the social media the participants used the most

preferred logo design, color, and typeface they like. The third team hoped that it would inform them better in developing a more suitable workshop tailored for the GenIUS Junior High students. The third team then tried to create suitable workshop material based on the data gathered on the school students' personalities, preferences, and knowledge of logo design. Knowing that the school students did not have the proper hardware to create logo design professionally, the university team then agreed on free, web-based, and amateur logo design and color picker apps, both by Adobe. The participants would only need to create an account, choose the type of logo and color they preferred from the templates provided, and then the app will automatically produce some logo alternatives for them to decide. Following the proven steps created in previous workshops, event rundown and teacher mentor guidelines were then made.

The Deliver phase of the third workshop was conducted from March 29th to 31st of 2020. On the first day, the third team started with a review of the result of previous workshops for the participants. After that, the team gave a presentation on basic art and design elements, such as the definition and types of logo, shape, color, and typography. On the second day, it was followed by a lecture on how to create a color palette using Adobe Colors and design a simple logo using Adobe Logo Maker. The school students were immediately tasked to create a logo that could represent themself and be used for their Instagram accounts. At this step of the workshop, the university team acted as an adviser and gives technical and artistic guidance to the participants remotely through the Zoom platform, where the teacher mentors assisted on-site. After the participants managed to create the logo design, they were collected by asking them to upload their designs using Google Forms. We can see some of the them in Fig. 8.

On the last day, the third university student team lead a review session. The design results were uploaded to the Google Jamboard website, and then the team gave design critiques to them, pointing out the strength and ways to improve each design. They also asked for feedback and comments from some of the participants.

4.6 Further Findings

The lecturer team was allowed to have a closing session at the end of the workshop series on Tuesday, April 5th, 2022. At the session, besides giving an overall review of the workshop series and conducting a closing ceremony with the university students, GenIUS school staff, and school students, the team gave a feedback form to the students to see whether the workshops were effective through their lens. The result was quite positive (Fig. 9). On a 1–5 scale where 1 is the least satisfied and 5 is the most satisfied, 61.1% of respondents gave 4 as the satisfaction score, and 27.8% gave 5 as the satisfaction score.

On the question about input for further development for the workshop, 88.9% said that they were no additional input, with two respondents (11.1%) abstaining. Based on the answers, most of the respondents were satisfied because of the clarity of the delivery of the workshops. They could understand the lecture and directions on the activities well. The main factors considered that contribute to this success were the many dialogues done and questionnaires given to the participants. It was an effort to understand their nature, behavior, state of knowledge, cultural preferences, and manner, which in the end helped immensely when delivering the workshops. With the cultural and societal barrier between the university team and the school students, gathering data and insights about the participants was imperative in executing a successful workshop.

We can also gather insight about the effectiveness through the question about what kind of knowledge they received after experiencing the workshop series. Mostly answered about logo design knowledge (70.5% of total respondents). Some of them also added personal branding

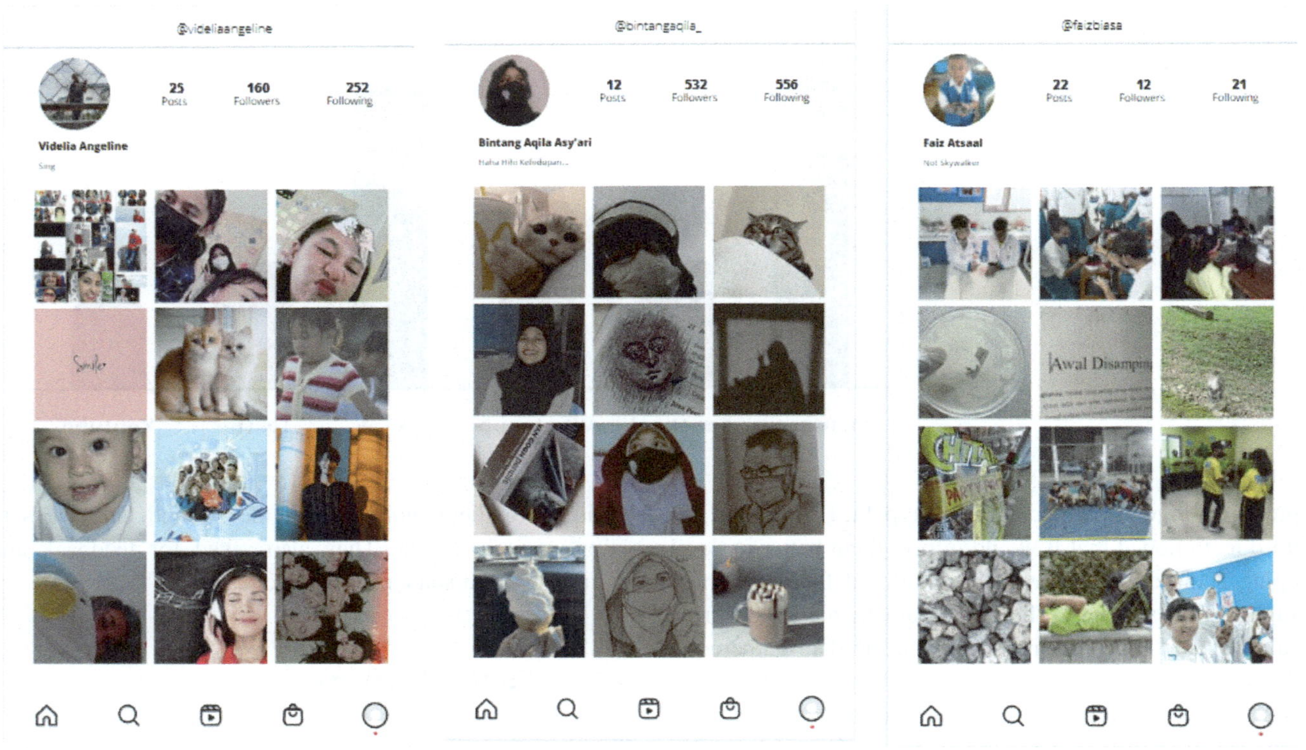

Fig. 7 Screen capture of results of the second workshop

Fig. 8 Some logo design as the results of the third workshop

knowledge to their answer (41.1% of total respondents), and some answered they knew themselves better than before (29.4%). There are 47% of total respondents managed to respond in a more holistic way, able to utilize their understanding of their personality to create a personal brand through social media and use logo design to represent themselves. Meanwhile, one respondent answered they did not know what new knowledge they got from it (5.8%).

On the other hand, some shortcomings were identified through postworkshop evaluation within the university team, and also between the lecturer team and the GenIUS school staff as the stakeholder. First, with communications done remotely through Zoom Meetings, the lack of internet infrastructure in the school created some obstacles. Signal interference because of bad weather and slow internet bandwidth produced communication noise and stutter

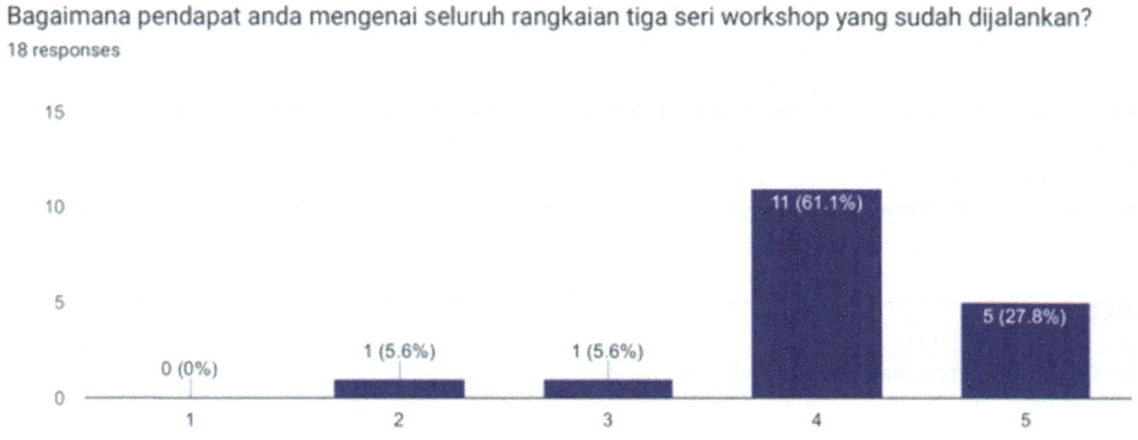

Fig. 9 Data result of the satisfaction rate of the participants

in dialogue in some instances, and also made session delivery take longer than scheduled. Patience on both sides and flexibility in scheduling became the answer to the problem. Improvisation was also done a couple of times in the face of technical and time-limit issues. Limited numbers of computers also created some problems as the ideal condition of the workshop was for each student to have a laptop or computer so they could take part in the workshop individually. This condition forced some activities to be led in groups or alternately. This problem also adds to the time needed to finish the workshop session. Moreover, the stern timeframe of the workshop necessitates the workshop session to be strictly finished on time as they have dorm activities waiting. Meanwhile, the transition between the end of school activity and the start of the workshop session made some sessions start much later than the allotted schedule. Fortunately, the university team identified these problem in the first workshop, and able to mitigate them in its iterations by adjusting the workshop on each session to be shorter.

The implication of covid-19 forced the university team to carry out the workshop remotely through online platforms and some physical forms. The issue pushed innovation in adapting various digital platforms to deliver the workshop series. Zoom Meeting was the primary conference medium in delivering each workshop and carrying on discussion. Google Forms were used to conduct research and collect data. A PowerPoint presentation and physical booklet were useful in delivering the first workshop. In the second workshop, we utilized Canva. It was used again in the third workshop alongside Adobe Logo Maker, Adobe Colors, and Google Jamboard. These various apps and platforms work together to deliver a successful online workshop, and most of them act as substitutions for physical media. The ability to find a suitable app or platform and have the skill not only to operate but to give clear and concise instructions in executing them is an important aspect to note here.

One notable finding is a concern regarding participation when designing and delivering the workshop. The limited time and the remotely nature of the delivery hinders the full participation of the school students as co-designers of the workshop. As a compromise, participation with the school students was done mainly through discussions, questionnaires, and direct observations of their response during the workshops. The school staff's participations were in the form of discussions, providing data, approval and direction of various design proposal. The way the participation was done, not fully adhering to the definition of participatory design defined by Simonsen and Robertson made the proposed participatory approach in the workshop questionable. But these compromises were testimony that the participatory design approach was being exercised the best way the situation and condition allowed. This issue creates a deeper question regarding the definition of participatory design.

5 Conclusion

The personal branding workshop for GenIUS Junior High students was successfully delivered. The proper research (discover), analysis (define), and design (develop) process contributes to it. The implementation of iterations in the process of delivery transforms the workshop into a series of three continuous workshops. The pandemic condition that made the workshop to be remotely executed created technical obstacles that needed solution in the prior phase leading to workshop delivery. The way the participation was done through compromises brings further questions regarding the boundary of the definition of participatory design. Various apps (both digital and physical) were used as tools and platforms in carrying the workshops. In the end, the result could be summarized into a new model as the result of applying the Framework for Innovation model to this workshop project.

This model could be applied in further workshops to find new insights, strengths, and weaknesses for iteration purposes. Additional discussion and literature regarding the participatory aspect of this project could be done more to determine its validity. We could also test alternative tools for delivering similar workshops compare their effectiveness to the one used in this project.

References

Allo, M. D. G. (2020). Is the online learning good in the midst of Covid-19 pandemic? The case of EFL learners. *Journal Sinestesia, 10*(1), 1–10.

Contreras, C. P., Picazo, D., Cordero-Hidalgo, A., & Chaparro-Medina, P. M. (2021). Challenges of virtual education during the COVID-19 pandemic: Experiences of Mexican university professors and students. *International Journal of Learning, Teaching and Educational Research, 20*(3), 188–204. https://doi.org/10.26803/ijlter.20.3.12

Framework for Innovation: Design Council's evolved Double Diamond. (2019). Retrieved August 2, 2022. https://www.design-council.org.uk/our-work/skills-learning/tools-frameworks/framework-for-innovation-design-councils-evolved-double-diamond/

Hananto, B.A. (2020). What is social visual communiation design? Proposals on how to identify and define it. In *Proceedings of international conference of innovation in media and visual design (IMDES 2020)*. ISBN 978-94-6239-284-8/ISSN 2352-5398. https://doi.org/10.2991/assehr.k.201202.049

Key, L. (2022). 'Beyond the crisis': Accepting and adapting to the virtual academic skills workshop. *Journal of Learning Development in Higher Education*, (25). https://doi.org/10.47408/jldhe.vi25.974

Mohalik, R., & Poddar, S. (2020). Effectiveness of webinars and online workshops during the COVID-19 pandemic. *SSRN Electronic Journal*. https://doi.org/10.2139/ssrn.3691590

Mohalik, R., & Sahoo, S. (2020). E-readiness and perception of student teachers' towards online learning in the midst of COVID-19 pandemic. *SSRN Electronic Journal*. https://doi.org/10.2139/ssrn.3666914

Nagar, S. (2020). Assessing students' perception toward e-learning and effectiveness of online sessions amid COVID-19 Lockdown phase in India: An analysis. *Tathapi, 19*, 272–291.

Satuan Tugas Penanganan COVID-19. (2022). *Analisis data COVID-19 Indonesia: Update Per 09 Januari 2022*. Retreived August 2, 2022. https://covid19.go.id/artikel/2022/01/16/analisis-data-covid-19-indonesia-update-9-januari-2022

Sekolah GenIUS. (2021, July 23). *Latar Belakang Sekolah GenIUS*. Retrieved August 2, 2022. https://genius.sch.id/latar-belakang/

Sekolah GenIUS. (2021, July 23). *Program Sekolah GenIUS*. Retrieved August 2, 2022. https://genius.sch.id/program-kurikulum/

Simonsen, J., & Robertson, T. (Eds.). (2013). *Routledge international handbook of participatory design*. Routledge.

We Are Social. (2022). *Digital 2022 Indonesia: The essential guide to the latest connected behaviors*. Retreived August 2, 2022. https://andi.link/wp-content/uploads/2022/02/Digital-2022-Indonesia-February-2022-v01_compressed.pdf

The Impact of a Connectivist Learning Environment on Indonesian Design Students' Learning Experiences Through MOOC

Lisa Indriati and Neo Mai

Abstract

This study sought to investigate the impact of Connectivism on design education; specifically for Project Managerial Class for undergraduate design students in Indonesia. Research has shown that design students need to have critical thinking skills and graduate with useful life skill sets, which can increase their employability, especially in post-pandemic situations. As such, this study used a mixed-method research design to investigate the development of effective learning content guided by Connectivism and assessed using the Revised Community of Inquiry (RCOI) theoretical framework. Forty-nine design students in an Indonesian University took part in this study. A 7-point Likert scale questionnaire was administered to them to gauge their attitudes and perceptions toward four elements of the RCOI, namely Cognitive Presence, Social Presence, Teaching Presence, and Learner Presence. This study also investigated to what extent Connectivism can be applied to the design learning process and how it can be explored to create an engaging learning experience for design students. Findings from the study show that students were very positive toward these four elements of the RCOI within a Connectivist learning environment. These results strongly encourage the use of Connectivism's social characteristics in the design learning process during the rapidly emerging digital era.

Keywords

Connectivism · Digital learning · Learning environment · Learning experience · RCOI

1 Introduction

In 2020, the world has declared an emergency, namely a global pandemic, transforming how individuals undertake activities and learn. The pandemic has impacted many industries, resulting in no industrial output, rising inflation, severe job loss/unemployment, and deteriorating living conditions, with a few millions more people expected to fall into poverty (Grima et al., 2020). However, within two years, the situation gradually improved.

The Indonesian government is trying to accelerate the improvement of conditions, especially in health, economy, and education sectors (Dewi, 2021). Even though the situation has improved at the end of 2021, humans are starting to feel comfortable with the current situation, thus, all activities will not return to the same as before the pandemic.

Many predicted that the situation would not return to normal but become a new normal. Innovation, technological advancement, and widespread acceptance of these disruptions resulted in faster recovery from these situations. As more significant percentages of the population become more dependent on digital networks for their jobs, education, health care, everyday commercial transactions, and critical social interactions, people's ties with technology will strengthen (Anderson et al., 2021).

Amidst all this instability and uncertainty, there are many changes, particularly in the education system. Regardless of the circumstances, students must have critical thinking skills and graduate with useful life skill sets to find work (Croy, 2018). It is imperative to integrate these outcomes and teaching methods to achieve both. Research has

L. Indriati (✉)
Universitas Ciputra Surabaya, Surabaya, Indonesia
e-mail: lisa.indriati@ciputra.ac.id

N. Mai
Multimedia University Malaysia, Cyberjaya, Malaysia
e-mail: neo.mai@mmu.edu.my

M. Nawaz Tunio et al. (eds.), *Sustainability in Creative Industries*,
Advances in Science, Technology & Innovation, https://doi.org/10.1007/978-3-031-48453-7_9

shown that for the time being, the most significant problem in educational institutions is ensuring a smooth educational process without creating excessive disruptions to students and teachers.

A person's behavior changes relatively permanently as a result of experience or training that takes into account both physical and psychological factors, such as going from not knowing something to knowing it, from not having skills to having skills, and so forth (Syarifuddin, 2011). The relationship between learning as a process of self-adjustment and the learning environment, which will result in changes in pupils, might affect how the learning process determines the student's perspective. In addition to helping students build creativity, critical thinking, analytical, and precise skills in identifying and applying the learning materials, as well as their capacity to create new knowledge to understand the material under study better, appropriate learning can also help students improve their capacity for communication and teamwork, and comprehend clinical issues (Zakaria & Awaisu, 2011).

The lecturer must create a learning plan before beginning the learning process since it is a scheduled activity designed to help students learn and develop the desired competencies. Learning planning is known as a list of objectives and plans for educational activities (Maria & Sediyono, 2017). If this transition doesn't take place, learning will stop being continual. A teaching philosophy that establishes a structural connection between teaching, students, and subject matter is required to accomplish a deeper understanding.

To ensure that students develop their understanding, teaching should be collaborative and include identifying student misunderstandings, interventions to resolve them, and establishing an engaging learning environment. The instructor guides the methods by concentrating on the fundamental issues that describe the barriers to students' learning, the subject matter to be learned, and the problems with students' learning. To facilitate understanding, teachers might combine a deeper learning perspective, ongoing skill improvement, and different teaching styles, claims Ramsden (2016). Learning theories can also be applied to guide educators as they adapt to new learning settings. New ideas must be developed if existing theories can no longer fully or only partially explain learning in specific environments.

Massive Open Online Courses (MOOC) popularity has skyrocketed (Impey, 2020). The top three MOOC platforms, Coursera, EdX, and FutureLearn, each witnessed a gain in membership of about 50% and an increase in website traffic of over 300% in just six months, according to Class-Central.com (Shah, 2020a). In total, 950 colleges worldwide have announced 16,300 MOOCs by the end of 2020, roughly four times the total number of MOOCs released by 2015 from just over 500 universities (Shah, 2020b).

MOOCs are at the forefront of education in an increasingly digitized, distant, and globally competitive world. In the realm of MOOCs as well, there is fierce competition. Tens of thousands of students or more may enroll in the best courses. For instance, 1.5 million people had registered for Yale's happiness course, "The Science of Well-Being," by April 15, 2020, little than a month after the first covid-19 shutdown (Ofgang, 2020). However, even on the best platforms, most MOOCs may only attract one or two thousand students annually, creating a large tail of interest given the over 16,300 MOOCs available (Shah, 2021).

Rai and Chunrao (2016) identified five categories of student engagement in MOOCs: Viewers, Solvers, All-rounders, Collectors, and Bystanders; however, despite the availability of high-quality course materials in MOOCs, there are several negative trends emerging among learners including: (1) a growing tendency among learners to collect and store large quantities of materials without utilizing them, (2) inadequate guidance for learners to determine the necessary and useful materials required to complete the course successfully, (3) an increasing preference among learners for audio, video, and PowerPoint materials over traditional text-based materials, and (4) a decline in face-to-face interactions and problem-solving opportunities with fellow learners, with most learners depending on discussion forums to ask and answer questions.

Although some literature discusses the advantages and disadvantages of MOOCs, there has yet to be an analysis of the factors that contribute to the success of MOOCs in design programs, which leads to the question of "What makes a great MOOC for design students? What motivates design students to continue learning? How to make design students engage with the lesson?". This is especially crucial given the environment in higher education today. Covid-19 and rising educational delivery costs exacerbate the decline in college enrolment. Over 11% fewer students have been enrolling in colleges since 2012 (U.S. Coursera, 2020). A successful MOOC can provide the opportunity to test a new curriculum before it is offered on-campus and help students who lack the resources or credentials transfer directly to colleges with higher entrance criteria.

2 Theoretical Framework

2.1 Learning Environments

Behaviorism, cognitivism, and constructivism are the three basic learning theories most widely utilized in creating instructional settings. These theories, however, were developed before technology has influenced learning, and technology has revolutionized how we live, communicate, and learn in the last two decades. The social environments in which learning occurs should be reflected in the learning demands and theories that define learning principles and processes (Vaill, 1996).

The rapid increase in information raises further issues; we must act by taking information from sources other than our prior knowledge. The process of learning is more important than the lessons themselves. How students can obtain knowledge in a world full of networks is worth investigating. Further alteration is no longer feasible due to the significant changes in the underlying conditions; a whole new approach is required. Learning theories are being brought into the digital age through the use of technology and developing connections as learning activities. We no longer have the ability to experience and gain the knowledge we require to act personally. This led to the emergence of a new theory known as connectivism.

According to Connectivism theory, learning is a process that occurs in a group of groups that provide information to each other; thus, the learning experience is an information cycle (Siemens, 2017). The learning process begins when a piece of knowledge is presented by a person connected and participates in a learning community and when others add more related information. For example, a topic is given in class, so students interact and collaborate to explore the topic through the lecturer's learning activities.

In the connectivist model, a learning community is referred to as a node, and nodes are invariably a part of more extensive networks. Nodes are created by linking points that exist on a network. A network is made up of two or more nodes that are connected to share resources. Depending on the information they hold and the volume of users traveling through them, nodes might vary in size and strength (Downes, 2008).

According to connectivism, knowledge can be saved in several digital formats and dispersed across an information network. It is believed that knowledge and learning "depend on the diversity of ideas" (Siemens, 2008, para. 8). Both the cognitive and affective domains are utilized in learning, and both of these domains play a big part in the education process. Furthermore, Siemens believes that learning in the twenty-first century cannot be adequately understood through the principles of behaviorism and cognitive psychology, given its complexity. Instead, it is necessary to depend on a community of individuals and advancing technology to store, retrieve, and utilize knowledge. This viewpoint is illustrated in Fig. 1 (Simui et al., 2018).

It is essential to filter materials at the individual level to determine whether the information is critical to the user because a lot of information is available in the connected network, which changes frequently. In addition to filtering content, the person must also believe that new details can change the way they think, enabling them to make decisions in the future based on the most recent data. Through social networking sites and other collaborative technologies, students can connect through the connectivism teaching technique. Many ideas hold that a person's mind is where all of their learning takes place.

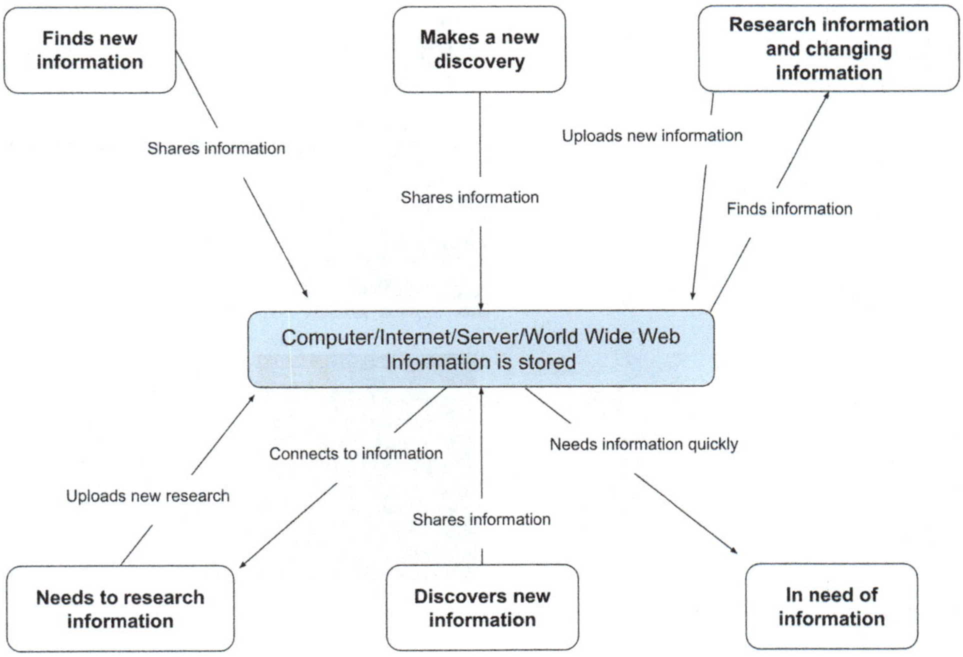

Fig. 1 Diagram of connectivism

2.2 Massive Open Online Courses (MOOC)

In a massive open online course (MOOC), students have open access and can sign up for it for nothing or for a small fee (McAuley et al., 2010). MOOCs differ from regular online courses in that they are accessible to all potential students and usually do not require registration fees if students do not want to receive a certificate.

During the pandemic, the public's interest and awareness of MOOCs has coincidentally grown as the annual number of participants rises (Gaevi et al., 2014; Pappano 2012). It should not be surprising that empirical study into MOOCs has recently increased to examine their acceptance, benefits, designs, implementations, impacts, and outcomes (Deng & Benckendorff, 2017; Veletsianos & Shepherdson, 2016).

2.3 Revised Community of Inquiry (RCOI)

This study investigates online learning performance using the Revised Community of Inquiry (RCOI) theoretical framework (Garrison et al., 1999; Shea & Bidjerano, 2012; Shea et al., 2012; Swan et al., 2012). Because this framework implies that collaborative interactions between engaged students in the learning environment produce knowledge acquisition, it will be possible to determine the effectiveness of online learning using this model (Shea & Bidjerano, 2012).

Cognitive Presence, Social Presence, Teaching Presence, and Learner Presence are the four components of the RCOI which is depicted in Fig. 2. The knowledge that requires both critical and creative thinking is called cognitive presence. Collegial connections are encouraged by social presence. The teacher's presence in a learning setting plays an instructional role, and learner presence includes self-learning and co-regulation. It will be feasible to assess the efficacy of online learning using this paradigm since it assumes collaborative interactions between involved students in the learning environment resulting in the generation of knowledge.

According to Fig. 3, the three components of the Community of Inquiry are interconnected: social presence, cognitive presence, and teaching presence. Learning presence is influenced by both lecturer and societal presence. As shown in Fig. 2, social presence is a mediating component in the blended learning educational process by providing context. In the Revised Community of Inquiry, the lecturer's presence is critical because he/she must guarantee that the learning process is relevant and purposeful. Following the learners' engagement with the activities and practices, the link between cognitive presence and teaching presence

Fig. 2 MOOC design project management

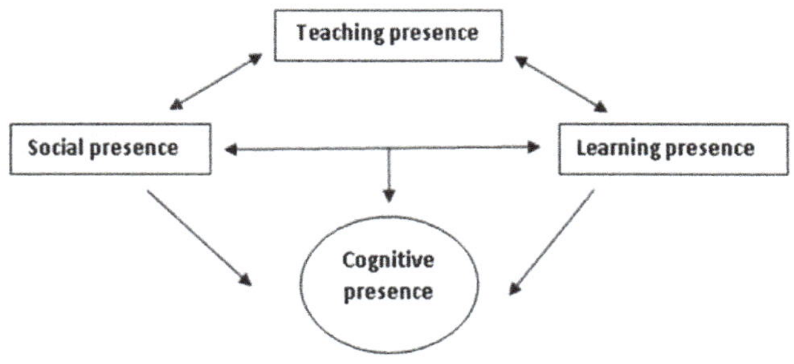

Fig. 3 Revised community of inquiry

becomes obvious. Furthermore, once learner engagement is achieved, the lecturer's presence directly impacts students' learning processes [8].

The knowledge that requires both critical and creative thinking is called cognitive presence. Collegial connections are encouraged by social presence. The teaching presence serves as an instructional role in a learning setting. Among the components of the learner, presence is co-regulation and self-learning.

3 Research Methodology

This research employs a mixed-method approach and consists of two phases. First, the study was conducted during the first iteration of Digital Spark, a MOOC platform developed by a university in Indonesia. The MOOC uses Project Managerial course, and follows the same syllabus as the face-to-face course delivered on campus. The course lasted for sixteen weeks and the learning materials included video lectures, articles, and an e-book. The course had three types of graded assignments: biweekly quizzes, recorded presentations, and a final project. Second, at the end of the course, a survey with 20 questions was distributed to the students, and comments from the students were also collected.

The Project Managerial course for undergraduate design students in Indonesia in the 2021 academic year used a connectivism approach in online classes. The class is divided into four modules, as shown in Fig. 4.

Each module has an output that fits the needs of today's creative industries, and students will collaborate in solving problems. Learning activities are designed to accommodate the principles of connectivism, so that the learning process and assessment media are divided according to each principle, as in Table 1.

4 Results and Discussion

The survey was conducted on forty-nine design students in Indonesia using Google Forms. The questionnaire used a 7-point Likert scale to determine students' perceptions and behavior towards elements of the RCOI, namely Cognitive Presence, Social Presence, Teaching Presence, and Learner Presence. The survey results were then analyzed using SPSS V28. The results are presented in each table showing the means (M), Standard Deviation (SD), and positive responses from students (i.e., value 5 (more or less agree), value 6 (agree), value 7 (strongly agree) on the survey).

Table 2 presents teaching presence and shows that according to students, the presence of lecturers in online learning activities is very high. The lecturer also communicated the learning themes and objectives well, with a positive response of 91.8% (Items 1–2, M = 6.12). Students also felt helped by the presence of lecturers as facilitators of their learning activities, with positive responses of 89.9% (Item 3, M = 5.86). In addition, 85.7% of students felt that the lecturers helped them in an online learning environment because they received directions to find new ideas (Item 4, M = 5.80). Lecturers are also deemed sufficient in providing input according to students' needs (Item 5, M = 5.82).

In the aspect of teaching presence, we see that 2 questions have the highest average score. Namely, the lecturers successively inform the theme of online learning, and the lecturer communicates the online learning objective. In this case, it appears that the theme and the learning objective given by the lecturer can be understood by students. This interpretation shows that the average value of teaching presence tends to be close to the maximum number, which is number 7.

Fig. 4 Project managerial course module

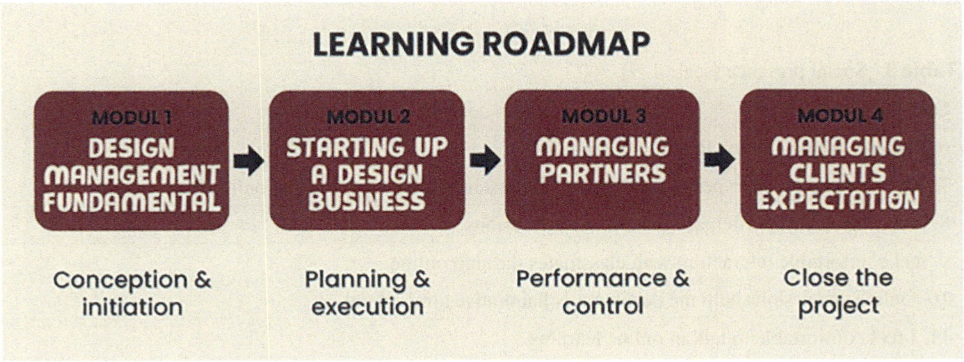

Table 1 Connectivism and learning activities

Connectivism principles	Learning activities
• Learning and knowledge rests in a diversity of opinions	• Small group discussion • Class forum • Collaborative learning
• Learning is a process of connecting specialized nodes or information sources	• Study case • Research and analysis • Observation • Interview
• Learning may reside in non-human appliances	• Learn through learning management system (moodle) • Discovery learning
• Capacity to know more is more critical than what is currently known	• Study case • Research and analysis • Observation • Interview • Validation
• Nurturing and maintaining connections is needed to facilitate continual learning	• Project based learning • Role play and simulation
• Ability to see connections between fields, ideas, and concepts is a core skill	• Small group discussion • Case study • Collaborative learning • Project based learning
• Currency (accurate, up-to-date knowledge) is the intent of all connectivist learning activities	• Study case • Discovery learning
• Decision-making is itself a learning process	• Discovery learning • Self-directed learning

Table 2 Teaching presence (scale 1–7)

Questions	Mean	SD	%
1. Lecturer communicates the theme of online learning	6.12	0.949	91.8
2. Lecturer communicates the online learning objective	6.12	1.073	91.8
3. Lecturer help student engage and participate in the online learning process	5.86	1.155	89.9
4. Lecturer encouraged student to explore new ideas	5.80	1.172	85.7
5. Lecturer gives feedback for student as needed	5.82	1.112	85.7

Overall social presence (Table 3) also shows a fairly positive response; 85.7% of students felt that lecturers could create an online learning community well (Item 6, M = 5.67). 91.8% of students also felt that they were involved in the learning process (Item 7, M = 5.37), and they were also quite comfortable discussing online with positive responses of 75.5% (Item 8, M = 5.35). In addition, they were quite comfortable talking in learning

Table 3 Social presence (scale 1–7)

Questions	Mean	SD	%
6. Lecturer created online learning community	5.67	1.231	85.7
7. Getting to know other people makes me have a sense of belonging to the online learning process	5.37	1.550	91.8
8. feel comfortable participating in online discussions	5.35	1.267	75.5
9. feel comfortable interacting with classmates through online	5.33	1.375	67.3
10. Online discussions help me develop a collaborative atmosphere	5.02	1.407	65.3
11. I feel comfortable to talk in online learning	4.94	1.651	85.7

Table 4 Cognitive presence (scale 1–7)

Questions	Mean	SD	%
12. I am motivated to explore materials related to online learning themes or topics	4.92	1.470	91.8
13. I use various learning resources during the online learning process	5.57	1.225	93.9
14. Online discussion helps me to appreciate different points of view	5.90	0.918	81.6
15. Online learning activities help me to provide solutions	5.53	1.138	89.8
16. I feel helped by the process of reflection and discussion of online learning materials	5.45	1.292	75.5

Table 5 Learning presence (scale 1–7)

Questions	Mean	SD	%
17. I am motivated to explore materials related to online learning themes or topics	5.90	0.941	91.8
18. I use various learning resources during the online learning process	5.10	1.342	69.4
19. Online discussion helps me to appreciate different points of view	4.73	1.381	53.1

Table 6 Learning satisfaction (scale 1–7)

Question	Mean	SD	%
20. Overall, I am satisfied with online learning	5.35	1.200	77.6

activities online with a positive response of 85.7% (Item 11, M = 4.94). However, only 67.3% of students felt comfortable interacting online with classmates (Item 9, M = 5.33), so the collaborative atmosphere was not maximally formed, with a positive response of only 65.3% (Item 10, M = 5.02).

Social Presence plays a role in encouraging relationships between participants in the online learning process (collaborative relationships with colleagues). From 6 questions, we found that lecturers were able to form a community atmosphere in online learning. Students can recognize their classmates well, shown by their ability to have different impressions of each of their friends. Furthermore, students can get to know their colleagues, who give them a sense of belonging to the online learning process. They also feel that their peers can accept them; this is indicated by their feeling that their views or opinions can be obtained by their colleagues. Besides, they also feel comfortable participating in online discussions.

In regards to cognitive presence (Table 4), 91.8% of students were motivated to look for supporting materials according to the learning topic (Item 12, M = 4.92), and 93.9% stated that they used various sources in the online learning process (Item 13, M = 5.57). Online discussion activities also helped 81.6% of students to get various points of view in viewing the problem (Item 14, M = 5.90), and 81.6% of students agreed that the resulting solutions were more diverse (Item 15, M = 5.53). In general, 75.5% of students think that the process of reflection and discussion for online learning materials helped them to understand the topic better (Item 16, M = 5.45).

Cognitive Presence is the knowledge that involves students' critical and creative thinking skills. From the cognitive presence aspect, it can be seen that students have the ability to gather new information to help them answer questions in online learning. Students also benefit from online discussions as they can help them appreciate different points of view. Students are also proven to be able to carry out conversations with their colleagues in fulfilling their assignments. Furthermore, online learning students can also use various learning resources. Online learning can also help students through the process of reflection and discussion about learning materials.

From the students' perspectives, the learning presence in Table 5 shows that 91.8% of students interacted with classmates to help each other complete online assignments (Item 17, M = 5.90). 69.4% of students stated that they better understood the concepts taught in class through online learning (Item 18, M = 5.10). The thing that is still lacking is the learning experience, especially in this class, which is still considered not good, with positive responses of only 53.1% (Item 19, M = 4.73).

Learning Presence shows students' ability to learn independently. Students think that they can easily access online learning materials, and they can also maintain the quality of assignments and set standards for online learning assignments. An exciting and important finding is that students can interact with their classmates to do online assignments.

Overall, in Table 6, 77.6% of students are quite satisfied with online learning activities (Item 20, M = 5.35). The survey is supported by the students' comments about the class, such as "*the material and delivery of assignments and the lecture timeline are good.*". In addition, students also said that: "*lecturers provided a lot of inspiration and lessons that are easy to understand. The lecturers also gave fast responses*". Another student also mentioned that: "*the materials in class has been delivered clearly and opportunities to discuss the tasks that are being done were always provided.*"

5 Conclusion

The findings indicate that students' responses are very positive toward the Connectivist learning environment, so it is necessary to explore its use in other design courses. Connectivism is an effective learning approach that enables students to acquire critical thinking skills and the necessary soft skills for the workplace. In particular, there was an improvement and positive responses from students regarding the Revised Community of Inquiry (RCOI) constructs of Cognitive Presence, Social Presence, Teaching Presence, and Learner Presence that were embedded within this Connectivist learning environment. The findings suggest that the Connectivist learning approach, with its emphasis on social and collaborative learning, can provide an effective alternative to traditional learning approaches in design education. It is recommended that educators incorporate Connectivist learning approaches into their curriculums and leverage technology to facilitate communication and collaboration among students. Further research is needed to explore the potential of Connectivist learning approaches in other educational contexts and to identify the most effective use of technology in these environments.

References

Alfathoni, M. A. M., & Manesah, D. (2020). Pengantar Teori Film. *Deepublish*.

Anderson, J., Rainie, L., Vogels, E. A. (2021, April 5). *Experts say the 'new normal' in 2025 will be far more tech-driven, presenting more big challenges*. Pew Research Center: Internet, Science & Tech. Retrieved April 6, 2022, from https://www.pewresearch.org/internet/2021/02/18/experts-say-the-new-normal-in-2025-willbe-far-more-tech-driven-presenting-more-big challenges/

Blakey, N., Guinea, S., & Saghafi, F. (2017). Transforming undergraduate nursing curriculum by aligning models of clinical reasoning through simulation. In R. Walker, & S. Bedford (Eds.), *Research and development in higher education: Vol. 40. Curriculum transformation* (pp. 25–37). Higher Education Research and Development Society of Australasia. http://www.herdsa.org.au/research-and-development-higher-education-vol-40-25

Croy, S. R. (2018). Development of a group work assessment pedagogy using constructive alignment theory. *Nurse Education Today, 61*. https://doi.org/10.1016/j.nedt.2017.11.006

Dewi, A. (2021, May 22). Anggaran.kemenkeu.go.id. Retrieved April 6, 2022, from https://anggaran.kemenkeu.go.id/in/post/penanggulangan-pandemi-covid-19-melalui-program-pengadaan-vaksin-dan-pelaksanaan-vaksinasi-covid-19

Grima, S., Dalli Gonzi, R., & Thalassinos, E. (2020). The impact of COVID-19 on Malta and its economy and sustainable strategies. *Journal of Corporate Governance, Insurance and Risk Management, 7*(1). https://doi.org/10.51410/jcgirm.7.1.5

Ndhlovu, D., Simui, F., Mwewa, G., Chota, A., Kakana, F., Mundende, K., Thompson, L., Mwanza, P., & Namangala, B. (2018). "WhatsApp" as a learner support tool for distance education: Implications for policy and practice at University of Zambia.

Rai, L., & Chunrao, D. (2016). Influencing factors of success and failure in MOOC and general analysis of learner behavior. *International Journal of Information and Education Technology, 6*(4), 262.

Shea, P., & Bidjerano, T. (2012). Learning presence as a moderator in the community of inquiry model. *Computers and Education, 59*, 316–326.

Siemens, G. (2004). A learning theory for the digital age. Retrieved from http://www.elearnspace.org/articles/connectivism.htm

Siemens, G. (2007). Connectivism: Creating a learning ecology in distributed environments. *Didactics of Microlearning. Concepts, Discourses and Examples*, 53–68.

Siemens, G. (2017, January 1). Connectivism. Go to the cover page of foundations of learning and instructional design technology. Retrieved April 13, 2022, from https://lidtfoundations.pressbooks.com/chapter/connectivism-a-learning-theory-for-the-digital-age/

Syarifuddin, A. (2011). Penerapan model pembelajaran cooperative belajar Dan Faktor-Faktor Yang mempengaruhinya. *Jurnal Pendidikan Islam, 16*(1), 36–113.

Vaill, P. B. (1996). *Learning as a way of being* (p. 46). Jossey-Blass Inc.

Zakaria, S. F., & Awaisu, A. (2011). Shared-learning experience during a clinical pharmacy practice experience. *American Journal of Pharmaceutical Education, 75*(4). https://doi.org/10.5688/ajpe75475

Competences, Capabilities, and Skills in Teaching and Learning Fashion Design for Sustainability

Erminia D'Itria and Federica Vacca

Abstract

Today, education continues to be seen as a central part of the solutions for sustainability. There is a tendency in the educational literature to underline the importance of defining the proper competencies to design and teach successfully. These competencies provide a clear and collective agenda for creating different and identifiable profiles of the educational field, the schools, graduates, the professionals, and so forth. The following paper deals with a radical vision of education for sustainability, which informs the need for a concrete transformation of the educational processes and objectives that specifically outline the knowledge, competencies, and skills required for designers who intend to work at various levels in the field of sustainability. In order to define the relationship between learning goals/results and trace the evolution of the curriculum itself, the work systematizes the main taxonomies about processes and objectives related to the field of education (Bloom BS, Krathwohl DR in Taxonomy of educational objectives: The classification of educational goals. Cognitive domain. Longman, London, 2020), education for Sustainability (Wiek et al. in Sustainability Science 6:203–218, 2011), and the theories about the Education for Sustainable development pedagogical levels of transformation (Sterling in JLTHE 2011:17–33, 2010). The result of this process illustrates the definition of Competencies, Capabilities, and Acquired Skills in teaching and learning fashion design for sustainability through a holistic approach based on the sustainability pillars: Environment, Economy, Society, and Culture.

E. D'Itria (✉) · F. Vacca
Design Department, Politecnico Di Milano, Milan, Italy
e-mail: erminia.ditria@polimi.it

F. Vacca
e-mail: federica.vacca@polimi.it

Keywords

Fashion design for sustainability · Education · Paradigm transition · Competences · Capabilities · Skills

1 Introduction

At present, education is indicated as one of the main drivers for activating a Sustainability transformation (Loorbach & Rotmans, 2006; Sterling, 2010). According to the literature, defining the proper competencies is essential for successfully designing and teaching (Baartman et al., 2007; Bowden & Marton, 1998; Burke, 1989; Spady, 1994; Voorhees, 2001) as these competencies inform the framework for defining the necessary profiles of the academic field, the schools, the graduates, the professionals, and so forth.

This work addresses a specific perspective that contributes to building the basis for an effective change in the educational field through the re-elaboration of its processes and objectives that shape the knowledge, competencies, and skills required of future designers who plan to enter the fashion field with a Sustainability ethos.

To identify them, the research employs the definitions proposed by the EQF—European Qualification Framework (https://europa.eu/europass/en/europass-tools/european-qualifications-framework), that classifies (1) knowledge as the outcome of the assimilation of information through learning; (2) competencies as the proven ability to use knowledge, attitudes and personal development; and (3) capabilities and skills as the ability to apply knowledge and use know-how to complete the task and solve problems. As Fornasier (2011) argues, knowledge corresponds to knowledge itself. It refers to all knowledge accumulated by the person throughout life. Skills are related to know-how, which is the ability to perform a task and apply the acquired knowledge. "Attitudes correspond to learned

M. Nawaz Tunio et al. (eds.), *Sustainability in Creative Industries*,
Advances in Science, Technology & Innovation, https://doi.org/10.1007/978-3-031-48453-7_10

predispositions, judgment, values, or individual beliefs that determine the course of action or the behavior" (Fornasier, 2011, p. 158).

The discussion is informed starting from the analysis of the main taxonomies about processes and objectives related to the field of education (Bloom & Krathwohl, 2020), education for Sustainability (Wiek et al., 2011), and the theories about the Education for Sustainable development pedagogical levels of transformation (Sterling, 2010; Williams, 2016).

In order to define the relationship between learning goals/results and trace the evolution of the fashion curriculum itself, the presented theories have been systematized through main sustainability key enhancement concepts and topics (Black, 2008; Fletcher, 2008; Niinimäki, 2010; Orr, 2002; Salcedo, 2014; Von Busch, 2009; Williams et al., 2009) codified through an original perspective derived from the outcomes of the researches on the Fashion Design for Sustainability (FDfS) field carried out, in the last five years, by the Fashion in Process—FiP Research Lab.—within the Design department at Politecnico di Milano, which is involved in studying the sustainable evolution of fashion and creative industries through design research. The result of such a process is the definition of Competencies, Capabilities, and Acquired Skills in teaching and learning Fashion Design for Sustainability. Such a perspective is informed by a holistic approach through which the focus moves from single and short-term thinking to long-term sustainable strategic planning and future-oriented thinking based on the four pillars: Environment, Economy, Society, and Culture (UNESCO, 2010).

The proposed approach appears consistent with the times and demands of the current education system that the European Union explicitly supports through its Erasmus+program. This initiative promotes systemic, inclusive, high-quality education and training to attain the qualifications and skills needed for active participation in a democratic society, genuine intercultural understanding, and transition to the labor market. An example of good practice in this area is the Erasmus+FashionSEEDS project (https://www.fashionseeds.org/), which offers tutors evidence-based knowledge to structure educational materials for a new generation of design graduates. Through the FashionSEEDS platform, tutors are provided with tools and resources to understand what sustainability topics to teach, how to teach them, and how to transform their teaching through the critical concepts of Fashion Design for Sustainability. Also, both authors participated in this project as principal investigators for their HEI. This supported their academic growth, nurtured the development of their critical perspective, and helped them to define their original vision further.

According to those mentioned above, the definition of such a vision has been instrumental in informing competencies by identifying necessary features for tutors in nurturing graduates with more profound knowledge of fashion design for sustainability.

Similarly, the selected perspective advised the capabilities through the definition of needs and objectives, as well as the assessment of tasks and related results that should be performed in order to achieve a holistic student learning experience. These capabilities support tutors in building and evaluating the learning experiences and help students deepen their knowledge of fashion design for sustainability.

Finally, this vision instructed the acquired skills considering both the transformational dimension of teaching/learning and the evolution from the educational entry level to the exit one. The established skills help tutors perform and evaluate their teaching activities and support students in defining and assessing their fashion design for sustainability knowledge.

2 Methodology: Building the CCS Tool Framework

Higher Education Institutions (HEIs) are considered key players in delivering sustainability awareness and integrating a holistic approach. According to Setó-Pamies and Papaoikonomou (2016), the importance given to education relies on the possibility of promoting behavioral changes in individuals, who can also influence changes in the institutions and the contexts to which they belong. Hence, the academy needs to align with the current times, training the new leaders. They will drive positive change in transitioning the fashion industry to a more sustainable paradigm (D'Itria & Vacca, 2021).

Drawing on work on sustainability education (Bloom & Krathwohl, 2020; Sterling, 2010; Wiek et al., 2011), this article analyzes the knowledge, understanding, skills, and abilities needed to develop a Fashion Design for Sustainability curriculum. As a result of such an analysis, the authors design a new tool to help tutors explore their teaching and knowledge and students' learning about sustainability through reflecting the primary educational taxonomies and critical concepts of the FDfS field.

The selection of these three specific models on education and sustainability was conducted according to several criteria that consider reliability, consistency, and literature and academic accreditation:

(a) Bloom's taxonomy is widely recognized as a fundamental model in educational studies. Due to its long history and popularity, it has been condensed, expanded, and reinterpreted in various ways (Cassar, 2022; Forehand, 2010; Krathwohl, 2002).
(b) Wiek's framework is now widely used in analyzing and designing sustainability courses in higher education as it

relates to comprehensive academic programs in sustainability at the undergraduate and graduate levels (Günther et al., 2022; Lang et al., 2012).

(c) Sterling's model is sketching the conceptual ground for addressing transformative learning and sustainability (Aboytes & Barth, 2020; Ferrer-balas et al., 2008; Michel et al., 2020).

The authors refer to an evolving process in a specific educational context. The goal of providing tutors with such a resource is to improve learning outcomes and enhance the educational experience and its enrichment as an opportunity for developing soft skills.

Methodologically, four main steps were performed to build the knowledge around competencies, capabilities, and skills. (1) Systematizing the taxonomies/models presented and rethinking the skills and abilities of both individuals (tutors and students) and the common ones acquired during the learning experiences, (2) Analyzing the different levels of Bloom's pyramid, the Krathwohl and Anderson's revised version, and the role of the tutors through the FDfS lens. This has led to reformulating the competencies per Bloom's, Anderson & Krathwohl's taxonomies and FDfS key concepts. (3) Putting all three models together and analyzing the "learning experiences" (Sterling), both from the point of view of teachers and students, through Sterling's transformative learning lens. (4) Designing a tool to channel the knowledge produced in the previous stages for supporting tutors and students in codifying their knowledge. They are guided in assessing their teaching and learning activities. The tool offers an intuitive approach to mapping out the planning and execution of their educational curricula.

2.1 On Competences

Bloom's Pyramid is a taxonomy of learning objectives, and it outlines a hierarchy of levels to assess cognitive learning. According to Bloom, learning objectives explain what for the teacher is essential to achieve. A further interpretation of this is proposed in Anderson and Krathwohl's (Anderson & Krathwohl, 2001) revised version of Bloom's pyramid.

Here, they have been instrumental in identifying necessary features about how tutors can guide students to achieve more sophisticated levels of understanding and abstraction in their educational experience. Starting from general settings emerged the specificity of the knowledge necessary for tutors who teach fashion design for sustainability. Its nature consists of various theories and concepts about sustainability (see Benyus, 2002; McDonough & Braungart, 2002; Capra & Luisi, 2014; Fletcher & Grose, 2011; Gwilt, 2011; Salcedo, 2014; Vezzoli & Manzini, 2010), but also skills in performing their task, knowing why they are applying a specific knowledge, and attitudes which is related to their predispositions, values, and beliefs which characterize how this knowledge is applied.

In the FDfS perspective, this translates into fostering conscious knowledge about sustainability through a holistic approach based on the four pillars and helping tutors nurture graduates with a more profound knowledge of fashion design for sustainability.

The identified key competencies are:

- Remembering: "How do I recall this? How would I recognize this?"

Remembering is a cognitive process involving tutors' ability to recall relevant knowledge about fashion design for sustainability through the pillars' lens. This will help tutors in defining their knowledge level concerning the four pillars.

- Understanding: "How much do I know about this? Which is my sustainability understanding?"

Understanding refers to tutors' ability to construct meaning concerning the methods and theory of teaching related to FDfS. It relates to tutors' ability to compare and interpret sustainable contents theories in teaching fashion design for sustainability through the pillars' lens.

- Applying: "How can I apply this knowledge? What types of learning activities can I manage?"

Applying refers to tutors' ability to execute practices and activities in the teaching and learning of FDfS through the pillars' lens.

- Analyzing: "How can I classify and interpret this knowledge? What conclusions can you draw from this?"

Analyzing refers to tutors' ability to examine sustainable knowledge, determine the structure and recognize assumptions to integrate existing teaching or initiate new explorations in the teaching and learning of FDfS.

- Evaluating: "What judgment would you make about this? How could you determine this?"

Evaluating refers to tutors' ability to reflecting on sustainable knowledge expressed through contents, values, and goals to engage tutors in continuous learning.

- Creating: "How could I improve this? What would happen if I rethink this?"

Creating aims to reorganize sustainable contents and elements into a new structure to re-design and enhance teaching practices in the FDfS field through the pillars' lens.

2.2 On Capabilities

A further contribution to exploring the ability to recognize, perform, and apply knowledge is offered by the framework Wiek et al. (2011) developed on identifying critical competencies in sustainability for academic program development.

The proposed framework addresses capabilities and defines them as a functionally linked complex of knowledge, skills, and attitudes that enable successful task performance and problem-solving applied to competencies in sustainability, these are complexes of knowledge, skills, and attitudes that enable successful task performance and problem-solving concerning real-world sustainability problems, challenges, and opportunities (McArthur & Sachs, 2009; Wiek et al., 2011b: 204).

This framework has been functional to define capabilities with a double purpose. On one side, it could help tutors identify and understand the learners' knowledge about sustainability. Putting it in perspective, this could support tutors in setting priorities and goals to be achieved through their learning activities. On the other side, the framework presents students with a reflection on their knowledge and skills around sustainability topics.

In the paper's perspective, this is translated into helping tutors defining and assessing their fashion design for sustainability learning experiences and helping students deepen their knowledge of fashion design for sustainability through a holistic approach based on the four pillars.

The identified key capabilities are:

- Values thinking: "When is it necessary to question the status quo? What factors shape our values?"

Value thinking refers to students' ability to defining in a clear and explicit decision problem that should be addressed and stimulate creative value-focused thinking in modeling the decision and throughout the decision process. This means considering/rethinking values, principles, goals, and targets through sustainable concepts and the pillars' lens.

- Collaboration: "How can I stimulate conversations and joint actions? How can I fill the lack of specific knowledge?"

Collaboration refers to students' ability to combine different expertise, knowledge, and capabilities to build a working team. This attitude helps students use their interpersonal competence to thrive in different types of collaboration, including teamwork and stakeholder engagement, to actively involve other professionals to share the effort in facing FDfS topics.

- Systems thinking: "How could I define what belongs to the system and what does not? Which interactions and relationships define the system structure and drive the system's behavior?"
  Systems thinking refers to students' ability to engage with a holistic approach to address sustainability issues drawing on different domains, processes, and levels of transformation.
- Futures thinking: "How do I identify trends and phenomena that are relevant and meaningful in understanding the future? How do I create alternative views of the future?"

Future thinking refers to students' ability to envision sustainability problems and their evolution and to visualize compulsory new ideas about possible and desirable solutions to achieve sustainable development.

- Strategic thinking: "What is my vision, and what are the associated objectives and corresponding goals to realize it? What will I monitor and measure?"

Strategic thinking refers to students' ability to intentionally and rationally focus on analyzing critical factors and variables that will influence the long-term success of a determinate action. This means students can design and test systemic interventions, transformational actions, and transition strategies toward sustainability.

- Problem-solving: "What steps do I take before deciding on how to solve this? How can I carry out a systemic analysis of this?"

Problem-solving refers to students' ability to determine the problem's source and find an effective solution. This means understanding the sustainable design process, covering the different steps, activities, and tools, and enabling students to design the process by applying different problem-solving frameworks for effect sustainability problems and foster sustainable development.

2.3 Skills Acquired

Sterling's theories are functional in implementing Bloom's and Wieks' theories by addressing the correlation between tutor competencies and student capabilities. Through his theories, Sterling systematizes the levels of transformation that highlight the different types of learning processes managed by tutors and the skills acquired during the educational process by students.

Regarding educational practices, this is represented by learning conditions that explore the potential of transformative learning experiences and their translation into tutors' work. Sterling (2009) argues that learning can ideally be "reflective, experimental, experiential, investigative, participatory, iterative, real-world and action-oriented" (Sterling, 2009, p. 82). This is because learning is understood as a practical act of change toward sustainability and in designing and building sustainable systems instead of a simple act of "learning to know change" or "learning for change" (Sterling, 2009, p. 82), which are instead passive experiences towards a more profound learning response.

From the interpretation of this systemic vision, features concerning the profiles of tutors and students, who work together in the learning path, have been extrapolated from the six inquiring aspects, defined by Sterling (2009), for deepening learning and facilitating transformational experiences, about the learning response to the educational sustainable, transformative system: holistic, critical, appreciative, inclusive, systemic, creative, ethical.

The seven aspects have been considered to evolve gradually according to their active nature and the changes in the learning contexts.

This paper proposed a revised version that focuses on the skills acquired during the transformational educational experiences concerning Sterling's inquiry (Sterling, 2009, p. 82). The authors introduce two main differences: (1) The use of adjectives has remained, but they are linked to the term skill defining the attributes needed to produce a learning response that can enable an educational sustainable, transformative system to thrive; (2) the number of aspects has been reduced from seven to six because the analysis of the aspects related to the ethical sphere showed an overlapping with the themes of the inclusive dimension; it was decided to combine the two aspects into a single skill (Inclusive skill).

In this work perspective, this is translated in considering the transformational dimension of teaching/learning in designing and developing fashion design for sustainability systems, helping tutors in nurturing graduates with a deeper knowledge of fashion design for sustainability, helping tutors in defining and assessing their fashion design for sustainability learning experiences; and helping students in defining and assessing their fashion design for sustainability knowledge through a holistic approach based on the four pillars.

The identified skills are:

- Sustainable holistic skill: "Am I thinking about the big context? How is each pillar interrelated?"

Holistic skill refers to seeing and thinking of the world in two broad dimensions: interconnected and evolving systems, dealing with or treating sustainability as a whole and not just a part, and thinking about the larger context.

- Sustainable critical skill: "Why are things this way? Whose interests are affected by sustainability?"

Critical skills refer to thinking clearly and rationally, understanding the logical connection between ideas, and engaging in reflective and independent thinking, all reasoning about why things are this way and in whose interests, compared to sustainability issues.

- Sustainable appreciative skill: "What already works well in the system that should be developed and built upon? How could this be implemented?"

Appreciative skill refers to determining the system's significance, worth, or quality, understanding and valuing what is good and positive, and what already works well in a sustainable system.

- Sustainable inclusive skill: "How important is diversity to me? What am I doing to make sure everyone feels included?"

Inclusive skills emphasize the contribution of understanding diversity to informing decisions, covering variation in capabilities, needs, and aspirations. This refers to considering who/what is being heard, listened to, and engaged in sustainability matters.

- Sustainable systemic skill: "What can trigger a reaction in the system? What implications can I see about my sustainable practice?"

Systemic skill refers to the ability to organize something according to a rationale that considers the context and possible effects of specific actions on the system itself: reasoning about what are or might be the consequences and effects of the sustainable actions.

- Sustainable creative skill: "How do I define innovation? How might I implement this?"

Creative skill refers to producing original thoughts and implementing them to create something entirely different from their original form, thinking about what innovation might be required in a sustainability scenario.

3 The CCS Tool: Competences, Capabilities and Skills

The CCS tool (Competences, Capabilities and Skills tool) aims at improving educational processes in a FDfS context through the exploration of teaching and learning features that emerge from reflecting on the main taxonomies about processes and objectives related to the field of education (Bloom & Krathwohl, 2020; Anderson & Krathwohl, 2001), education for Sustainability (Wiek et al., 2011), the theories about the Education for Sustainable development pedagogical levels of transformation (Sterling, 2010) and the fundamental concepts of the FashionSEEDS project.

A holistic approach has been applied to the knowledge about sustainability. Systematizing sustainable approaches via four founding pillars—Environment, Economy, Society, and Culture (UNESCO, 2010; FashionSEEDS, 2019) representing a combination and assembly of the four founding pillars to offer elementary knowledge through a series of overarching topics.

This considers the learning experience as a gradual path, from novice to expert, translated in transformative levels called scales of transformation (Sterling, 2010). This progression is related to Sterling's (2010) discussion around a change in behavior and practices, understanding and thinking, and underlying beliefs.

- Level 1: Cognition. This introduces new inclusive teaching and learning styles, and teaches the connection between sustainable futures and how we think, live, and work. It also connects learners to their local and global spheres of influence.

- Level 2: Meta-cognition. This critically assesses change processes in society and envisions a sustainable future; facilitates the emergence of new worldviews that address sustainable development; encourages negotiation of alternative futures; stimulates willingness to take considered action even in situations of uncertainty.
- Level 3: Epistemic learning. This assesses learning outcomes regarding changes and achievements in sustainable development; facilitates participatory and learner-centered education that develops critical thinking and active citizenship; challenges unsustainable practices across educational systems, including at the institutional level; and stimulates to be willing to challenge assumptions underlying unsustainable practice.

According to that, the CCS tool is based on radar charts to immediately focus on the three topics that most need attention: competencies, capabilities, and acquired skills. Each topic is identified through a set of variables (axes) and the level of preparation regarding the proposed topics (scales of transformation). The data from each sustainability pillar observation are plotted along each axis and connected to form a polygon (see Fig. 10.1).

3.1 Competences Radar

The Competences Radar (Fig. 10.1) has been developed from Bloom's taxonomy of learning (Bloom & Krathwohl, 2020) and its revised version by Anderson and Krathwohl (Anderson & Krathwohl, 2001) that outlines a hierarchy of levels to assess cognitive learning. Here they have been instrumental in informing the identification of necessary features about achieving more sophisticated levels of understanding and abstraction in the educational experience. Both taxonomies have been functional in defining the phases of learning, building the educational process, and then defining the procedure of acquiring competencies out of the learning experience.

This radar aims at supporting tutors to self-assess their knowledge through a holistic approach based on the four pillars of sustainability. This tool can guide tutors in the following:

1. Reflecting on the sustainable level offered in the course/module/unit in which they perform;
2. Adopting a holistic approach or focus vertically on a specific pillar;
3. Setting priority and future goals and defining what tutors want to achieve regarding the learning experience offered to students.

Specifically, educators are guided in the following:

- Developing their perspective on conscious knowledge about sustainability through a holistic approach based on the four pillars
- Nurturing graduates with more profound knowledge of fashion design for sustainability.

3.2 Capabilities Radar

The Capabilities Radar (Fig. 10.2) has been developed from the model proposed by Wiek et al. (2011). This work implements Bloom's taxonomy by addressing capabilities and defining them as knowledge and skills related to each other and allowing successful performing tasks and solving problems. Here, the model is considered in the Hasling and Ræbild interpretation (Hasling & Ræbild, 2017), which structures the competencies in a hierarchical order considering their cognitive and collaborative complexity.

The radar has been designed to be used as an initial activity offered by tutors to students. Students are asked to navigate the Capabilities Radar to self-assess their knowledge of FDfS according to the four pillars. This tool has a double level of analysis. On the one hand, students are stimulated to reflecting on their capabilities according to sustainable topics. On the other hand, it can help tutors understand the sustainable level of learners to set priorities and goals to be achieved.

From tutors' perspective, they are supported in defining, assessing, and helping:

1. Defining and assessing their fashion design for sustainability learning experiences.
2. Helping students deepen their knowledge of fashion design for sustainability through a holistic approach based on the four pillars.

3.3 Skills Acquired

The Skills Acquired Radar (Fig. 10.3) has been developed from Sterling's theories (Sterling, 2010) which are functional to implement Bloom's and Wiek's ones (Bloom & Krathwohl, 2020; Wiek et al., 2011) by addressing the correlation between tutor competences and student capabilities. Through his theories, Sterling systematizes the levels of

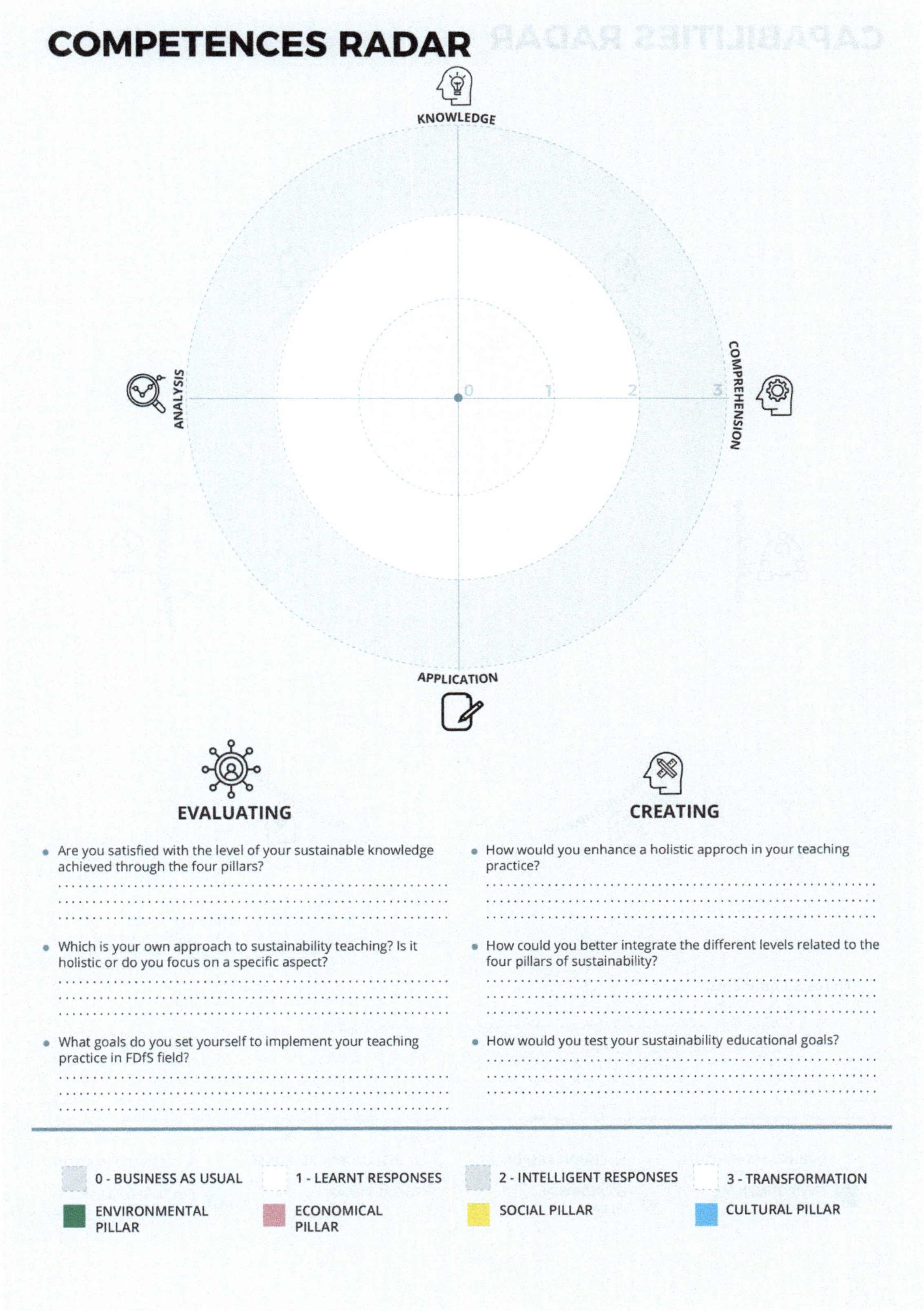

Fig. 10.1 The competences radar

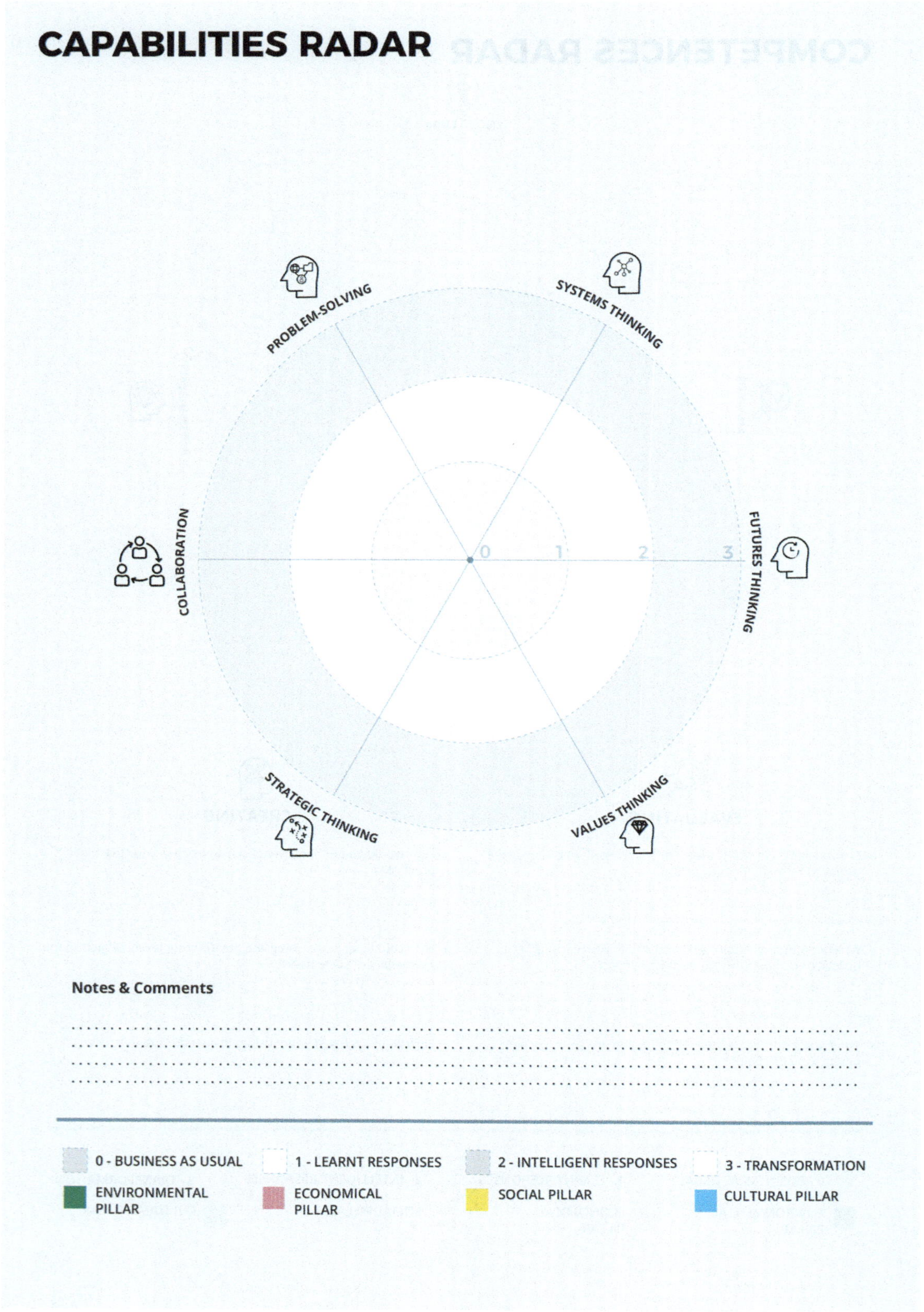

Fig. 10.2 The capabilities radar

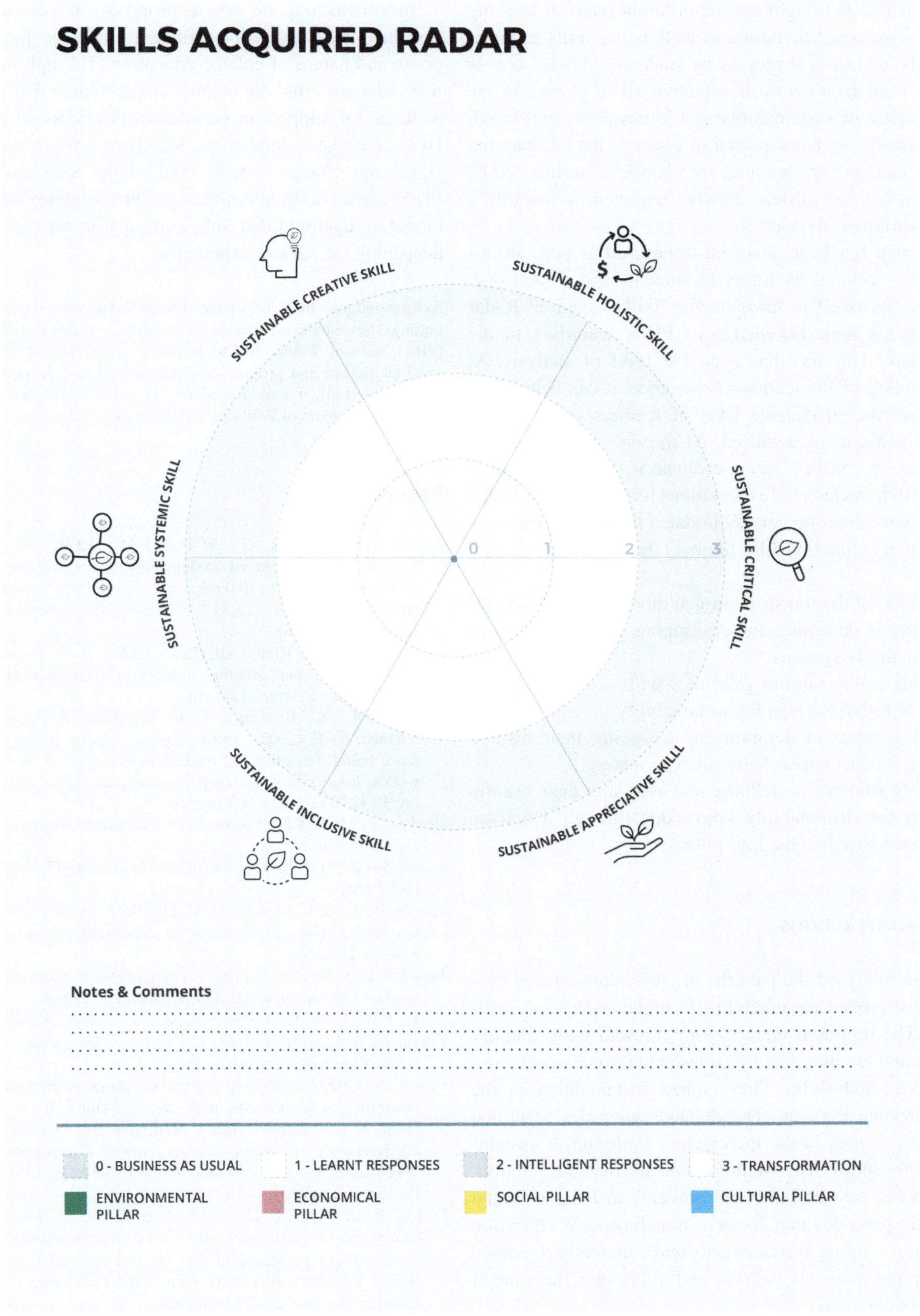

Fig. 10.3 The skills acquired radar

transformation that highlight the different types of learning processes managed by tutors, as well as the skills acquired during the educational process by students. This is because learning is understood as an effective act of change in the active pursuit of sustainability and in designing and building sustainable systems instead of a simple act of "learning to know change" or "learning for change" (Sterling, 2009, p. 82), which are instead passive experiences towards a deeper learning response.

The radar has been designed to be used as both an initial activity offered by tutors to students and a final one. Students are asked to navigate the Skills Acquired Radar to self-assess their knowledge of FDfS according to the four pillars. This tool has a double level of analysis. At the beginning of the learning experience, it can help tutors understand the sustainable level of learners to set priorities and goals to be achieved. At the end of the learning experience, it can help tutors evaluate if there has been an increase in knowledge if expectations have been met, and if students have appropriately assimilated the knowledge.

This tool is functional for tutors in the following ways:

1. Considering the transformational dimension of teaching/learning in designing and developing fashion design for sustainability systems.
2. Helping tutors nurture graduates with a deeper knowledge of fashion design for sustainability.
3. Helping tutors in defining and assessing their fashion design for sustainable learning experiences.
4. Helping students in defining and assessing their fashion design for sustainability knowledge through a holistic approach based on the four pillars.

4 Conclusions

The paper analyzed the patterns of curriculum change processes that embed sustainability in higher education institutions. The proposed methodology allowed us to identify teaching and learning features translated into competencies, capabilities, and skills. They embed sustainability in the HEIs curricula systems. The defined categories identified a progressive path in the incremental evolution of sustainability knowledge. The findings indicate that the relationship between sustainability and academy in HEIs is linked to holistic practices that foster a transformative education model—responding systematically and coherently to embed sustainability concepts, values, and skills into the student learning experience.

Further research will now focus on testing product tools to verify their validity on the field. Subsequently, it is imagined that it will also be tested in different contexts, such as disciplines, sizes of HEIs, or geographical areas.

In conclusion, the article proposes and discusses that sustainability implies a significant challenge for the purposes and nature of college education. Through the indicative radar patterns, the authors suggest tools that can point to ideas for support in curriculum development that each HEI can begin to implement, ideally as a precursor to more significant change, which could help academics design FDfS curricula. In addition, a gradual learning and change model is suggested that links curricular transformation with deepening the student experience.

Acknowledgements This paper results from common research and findings; nevertheless, Federica Vacca edited sections 1, 4 and Erminia D'Itria sections 2 and 3. The methodology presented results from research studies and projects undertaken in recent years by Fashion in Process (Fip), a multidisciplinary research laboratory within the Design Department of Politecnico di Milano.

References

Aboytes Rodríguez, J. G., & Barth, M. (2020). Transformative learning in the field of sustainability: A systematic literature review (1999–2019). *International Journal of Sustainability in Higher Education, 21*(5), 993–1013. https://doi.org/10.1108/ijshe-05-2019-0168

Anderson, L. W., & Krathwohl. (Eds.). (2001). *A taxonomy for learning, teaching, andassessing: A revision of bloom's taxonomy of educational objectives*. Longman.

Baartman, L. K., Bastiaens, T. J., Kirschner, P. A., & van der Vleuten, C. P. (2007). Evaluating assessment quality in competence-based education: A qualitative comparison of two frameworks. *Educational Research Review, 2*(2), 114–129. https://doi.org/10.1016/j.edurev.2007.06.001

Benyus, J. M. (2002). *Biomimicry: Innovation inspired by nature*. Harper Perennial.

Black, S. (2008). *Eco-chic: The fashion paradox* (Illustrated). Black Dog Press.

Bloom, B. S., & Krathwohl, D. R. (2020). *Taxonomy of educational objectives: The classification of educational goals. Cognitive domain*. Longman.

Bowden, J., & Marton, F. (1998). *The university of learning: Beyond quality and competence in higher education*. Kogan.

Burke, J. W. (1989). *Competence-based education and training*. Falmer.

Capra, F., & Luisi, P. (2014). *The systems view of life: A unifying vision*. Cambridge University Press.

Cassar, C. (2022). Designing curriculum about governance and sustainability in higher education: A case study. In S. Grima, E. Özen, & I. Romānova (Eds.), *Managing risk and decision making in times of economic distress, part A. Contemporary studies in economic and financial analysis* (vol. 108A, pp 1–35). Emerald Publishing Limited.

D'Itria, E., & Vacca, F. (2021). *Fashion design for sustainability. A transformative challenge across the European fashion education system*. Paper presented at the 7th International Conference on Higher Education Advances, Universitat Politècnica de València, online conference, 22–23 June 2021.

Ferrer-Balas, D., Adachi, J., Banas, S., Davidson, C. I., Hoshikoshi, A., Mishra, A., & Ostwald, M. (2008). An international comparative analysis of sustainability transformation across seven universities. *International Journal of Sustainability in Higher Education, 9*(3), 295–316.

Fletcher, K. (2008). *Sustainable fashion and textiles—Design journeys*. Earthscan.

Fletcher, K., & Grose, L. (2011). *Fashion and sustainability: Design for change*. Laurence King Publishing.

Forehand, M. (2010). *Bloom's taxonomy. Emerging perspectives on learning, teaching, and technology* (vol. 41, no. 4, pp. 47–56).

Fornasier, C. B. R. (2011). *Sistema de integração do conhecimento organizacional pelo design thinker*. Dissertation, Universidade Federal de Santa Catarina.

Günther, J., Overbeck, A. K., Muster, S., Tempel, B. J., Schaal, S., Schaal, S., & Otto, S. (2022). Outcome indicator development: Defining education for sustainable development outcomes for the individual level and connecting them to the SDGs. *Global Environmental Change, 74*, 102526.

Gwilt, A. (2011). Producing sustainable fashion: The points for positive intervention by the fashion designer. In A. Gwilt & T. Rissanen (Eds.), *Shaping sustainable fashion: Changing the way we make and use clothes* (pp. 59–74). Earthscan.

Hasling, K. M., & Ræbild, U. (2017). Sustainability cards: Design for longevity. In *Proceedings of PLATE 2017—Product Lifetimes and the Environment*.Delft, the Netherlands.

Krathwohl, D. R. (2002). A revision of Bloom's taxonomy: An overview. *Theory into Practice, 41*(4), 212–218.

Lang, D. J., Wiek, A., Bergmann, M., Stauffacher, M., Martens, P., Moll, P., Swilling, M., & Thomas, C. J. (2012). Transdisciplinary research in sustainability science: Practice, principles, and challenges. *Sustainability Science, 7*(S1), 25–43. https://doi.org/10.1007/s11625-011-0149-x

Loorbach, D., & Rotmans, J. (2006). Managing transitions for sustainable development. In X. Olsthoorn & A. J. Wieczorek (Eds.), *Understanding industrial transformation* (Vol. 44, pp. 187–206). Springer.

McArthur, J. W., & Sachs, J. (2009). Needed: A new generation of problem solvers. *The Chronicle of Higher Education, 55*(40), A64–A66.

McDonough, W., & Braungart, M. (2002). *Cradle to cradle*. North Point Press.

Michel, J. O., Holland, L. M., Brunnquell, C., & Sterling, S. (2020). The ideal outcome of education for sustainability: Transformative sustainability learning. *New Directions for Teaching and Learning, 161*, 177–188.

Niinimäki, K. (2010). Eco-clothing, consumer identity and ideology. *Sustainable Development, 18*(3), 150–162.

Orr, D. (2002). Four challenges of sustainability. *Conservation Biololgy, 16*, 1457–1460.

Salcedo, E. (2014). *Moda ética para um futuro sustentável*. GGBrasil.

Setó-Pamies, D., & Papaoikonomou, E. (2016). A multi-level perspective for the integration of ethics, corporate social responsibility and sustainability (ECSRS) in management education. *Journal of Business Ethics, 136*(3), 523–538.

Spady, W. G. (1994). *Outcome-based education: Critical issues and answers*. American Association of School Administrators.

Sterling, S. (2009). Towards ecological intelligence. *International Journal of Sustainability in Higher Education, 11*(1), 77–83.

Sterling, S. (2010). Transformative learning and sustainability: Sketching the conceptual ground. *JLTHE, 2011*(5), 17–33.

UNESCO. United Nations Educational, Scientific and Cultural Organisation (2010). The UNESCO climate change initiative: Climate change education for sustainable development. Retrieved from https://unesdoc.unesco.org/ark:/48223/pf0000190101_eng. [14 September 2023].

Vezzoli, C., & Manzini, E. (2010). *Design for environmental sustainability*. Springer Verla.

Von Busch, O. (2009). *Post-script to Fashion-able, or a methodological appendix to activist design research*. Lulu.com.

Voorhees, R. A. (2001). Competence-based learning models: A necessary future. *New Directions for Institutional Research, 110*, 5–13.

Wiek, A., Withycombe, L., & Redman, C. L. (2011). Key competencies in sustainability: A reference framework for academic program development. *Sustainability Science, 6*(2), 203–218.

Williams, D. (2016). Transition to transformation in fashion education for sustainability. In W. Leal Filho, & L. Brandli (Eds.), *Engaging stakeholders in education for sustainable development at university level* (pp. 217–232). Springer Nature.

Williams, D., Baldwin, N., & Fletcher, K. (2009). *Volume 3.0: Centre for sustainable fashion: Tactics for change*. Centre for Sustainable Fashion.

Project-Based Learning (PBL): Student Creativities in the Upcycling Projects

Ninik Juniati, Viofelita Gunawan, Aluisius Hery Pratono, Veny Megawati, Hari Hananto, and Njoto Benarkah

Abstract

Fashion waste has become a global issue because it is the second-largest contributor to waste after oil and gas. Thus, every sector in the fashion industry, especially the education sector, has begun to think about a zero-waste learning curriculum that can form the competence of fashion designers who can be creative wisely without producing any waste. However, Fashion waste in fabric trashes and used clothes today requires serious handling. Designers' expertise is needed to turn this waste into a unique, exclusive, useful product. Fabric waste upcycling needs a wide variety of competencies, such as sorting fabric trash and choosing the proper technique according to the type of fiber. The most important thing is finding the right market for the product. A community service program (PKM) has seen an opportunity to develop various upcycled products according to market needs, namely a guest house that needs multiple upcycled products ranging from bed linings to home décor. Students are also involved in the Project-Based Learning (PBL) through this community service program (PKM). This project-based learning (PBL) study aims to increase student's creativity in project upcycling for fabric waste. This research used qualitative methods with descriptive analysis, the data collection methods used observation techniques, and the interviews of students in textile manipulation courses—data analysis used the theory of Project-based Learning (PBL). The study's results showed that PBL had produced variations in student creativity in the products made, originality in describing problems, and creativity in finding solutions. The upcycling project carried out by students in PBL has become a solution for two institutions related to the waste problem. The first institution is the classroom, which will always produce fabric waste from course projects every semester. The second institution is a partner in the study. Student creativity in the form of upcycling products such as bed linings and home décor contributes creative ideas for Ketapanrame tourism villages so that they can improve the creative economy in the town.

Keywords

Project-based learning (PBL) · Eco-art education · Upcycle · Fabric waste

1 Introduction and Literature Review

A textile manipulation course is a course that aims to provide an understanding of fabric manipulation techniques for new forms or types of textiles. It covers several techniques of processing fabric surfaces, such as natural dyeing and making new types of materials such as felting, warm water-soluble (WWS), and tapestry. Along with individual projects of processing fabric waste and making various wearable art forms during the effectiveness of

N. Juniati (✉) · V. Gunawan
Product Design Study Program, Faculty of Creative Industries, University of Surabaya, Surabaya, Indonesia
e-mail: ninik.juniati@staff.ubaya.ac.id

A. H. Pratono
Economics Development Study Program, Faculty of Business and Economics, University of Surabaya, Surabaya, Indonesia
e-mail: hery_pra@staff.ubaya.ac.id

V. Megawati
Management Study Program, Faculty of Business and Economics, University of Surabaya, Surabaya, Indonesia

H. Hananto
Accounting Study Program, Faculty of Business and Economics, University of Surabaya, Surabaya, Indonesia
e-mail: hananto@staff.ubaya.ac.id

N. Benarkah
Informatics Engineering, Faculty of Engineering, University of Surabaya, Surabaya, Indonesia
e-mail: benarkah@staff.ubaya.ac.id

M. Nawaz Tunio et al. (eds.), *Sustainability in Creative Industries*,
Advances in Science, Technology & Innovation, https://doi.org/10.1007/978-3-031-48453-7_11

this textile manipulation course, I initially used a learning method intended for active student activities (student-centered). This kind of learning makes students more creative, the learning process is more effective, and the classroom atmosphere becomes more pleasant (Azizah, 2011). In this learning method, everything is student-centered, starting from the selection of the form of wearable art to be made, the concept, and the choice of supporting materials needed in processing fabric waste, to the completion technique. However, the observations made by researchers in previous semester classes show that the results of wearable artwork produced in processing fabric waste are still less than optimal. However, students have the freedom to compile concepts and choose the techniques used until the application becomes wearable art. One of the reasons that caused suboptimal work results was the absence of continuous use of wearable artwork. The continuation in question is this wearable art's end user and customer.

Based on the results of these observations, it requires the involvement of the user community who wants wearable art derived from fabric waste, so a Project-based Learning (PBL) method involving partners is needed. According to Milter and Stinson (Sulisworo, 2019), the Project-based Learning (PBL) method is the same as problem-based learning, except that in PBL, the problems consist of multiple subproblems and longer activity times. Generally, problems used as materials in problem-based learning methods are solved in one or two sessions. Meanwhile, in Project-based Learning (PBL), problems must be solved within one semester, so in specific courses, one type of project problem is given for one semester. The principles of Project-based Learning (PBL) are (1) Fostering interest in learning; (2) Building a solid group through various activities; (3) Training students to plan and manage learning projects; (4) Carrying out learning activities following predesigned projects; (5) Ensuring the progress of learning. To be able to see the progress of this learning, students are asked to discuss, report and present their progress, so that improvements can be made to the project in the future.

Based on the definition and principles of the Project-based Learning (PBL) method, a project partner who can be part of the learning process and function as the originator of concept ideas and evaluators of the work is needed. The learning partner is Bumdesa Ketapanrame, a tourism village, in Trawas District, Mojokerto Regency, East Java. Bumdesa Ketapanrame and The University of Surabaya develop educational tourism through the Matching fund program, which consists of three main projects: (1) The use of digital technology to help develop educational tourism (facilitate and expand the delivery of Education) while helping manage tourism management. (2) The development of Edu-tourism programs. (3) The development of waste processing and utilization. As a solution to the consequences of the development of tourist areas that are increasingly developing, this third program is an improvement in terms of the amount of waste. The Matching Fund program is intended as one of the efforts to solve the waste problem, be directed as an alternative educational material, and improve community welfare (Hananto et al., 2022).

The Ketapanrame village development program above was also an effort to provide opportunities for lecturers and students in the Textile Manipulation course to conduct joint research in developing waste treatment techniques into new products. The products, such as bed lining and home décor, complement the village's existing educational facilities in the form of four guest house units. In developing this fabric waste materialized product concept, students individually designed product manufacturing, from the design concept of fabric waste and embodiment to product completion techniques. The products included bed covers, chair pillowcases, tissue holders, window curtains, lamp caps, and multipurpose baskets. The plan of upcycling products to be made refers to the research carried out by Singh and Rani (2021), this study explored many ways to recycle and upcycle textiles traditionally practiced in various states in India, such as Kantha embroidery from West Bengal, Panja Dhurrie technique from Haryana state, Kasuti embroidery originating from Karnataka, Sujani Bihar state, Chindi and Chaklo, Gabba, and Namda carpets, as well as Patchwork and Kathputlis techniques from various Indian states. It was found that these traditional techniques from various Indian states contribute passively to the sustainable future of the Indian textile industries. Recycling old fabrics not only improves the life cycle of products but also provides jobs for millions of people in India. The artisans are gradually adapting to selling recycled textile items through e-commerce websites for better reach to targeted consumers in different Indian states. From the research, the product plan to be made uses the most patchwork techniques on the grounds that the technique is easier to follow, and the availability of large patchwork waste is quite a lot.

There was another research related to the treatment of used clothing waste which is also a reference in this class. This research related to recycling and upcycling on used clothes was also carried out by Teli et al. (2015). This study discussed the problem of converting used clothes by restoring and adding additional details, such as screen-printing motifs. This research was a type of experimental research applied to materials purchased from used clothing outlets in Mumbai, India. The results of this study stated that 72 samples were recycled, 29 of which provided a profitability of 99.34%, but the rest (43 samples) gave a profitability of 821.93%. However, all prices became lower when respondents knew that the product came from recycled used clothing. This shows the influence of psychological barriers on their purchase price. In relation to the quality of recycled products and the selling price, in the absence of knowledge that the product comes from used clothing, the selling price can be much higher, and in this case, the initial profitability ranges from 100 to 893%, which increased many times. In other words, upcycling is not only environmentally friendly but also sustainable, and it can open other avenues for entrepreneurs to get a high level of profitability, and someone can work on it as a solid business model. These two studies can be a reference to develop student creativity related to the solutions needed in the treatment of fabric waste produced in the pattern and sewing studio every semester. Not only student's creativity is developing, but also other possibilities can be developed through this upcycling project, such as the concept of sustainable design curriculum and eco-art entrepreneurship.

In line with the program developed by both parties, the Project-based Learning (PBL) method used in assisting students in developing creativity in processing fabric waste is a form of eco-art Education. Eco-art Education is a method that integrates art with the environment to build awareness and interaction with environmental concepts and issues, such as conservation, preservation, restoration, and sustainability of the natural environment (Inwood, 2008).

A study on eco-art Education stated that eco-art Education is increasingly important for schools as art education becomes more significant in the environment. The results showed that students prefer to give up their interest by formulating a better ecological perspective. On the plus side, only a few students experience frustration during programs and activities. However, students have given positive feedback to the program in favorable terms, such as "fun," "interesting," and "cool," to express the experience gained through class activities (Sunassee & Bokhoree, 2021). Based on the definition of eco-art Education, it can be said that the results of products produced by students and their creativity must refer to environmental concepts and issues and must consider the balance of nature.

Furthermore, product results must be effective as a solution for the community regarding environmental issues. Materials derived from fabric waste become various products with serviceability and certainly have high economic value. The most important thing is that fabric waste must be conducive, and the learning process should be fun in realizing processed products.

Based on the database, PBL research is ideally quantitative approach research, such as in research conducted by Hutasoit, who aimed to analyze the influence of conventional (teacher-centered learning) and Project-based Learning (PBL) models. The study developed scientific performance values and reviewed student character on daily product materials made from chemistry (Hutasoit, 2021). Sukmawati's research tested the effectiveness of project-based collaborative learning (PBCL) applied to biological materials that used social abilities (Sukmawati et al., 2019). Other research measured the improvement of learning outcomes using a Project-based Learning (PBL) model to improve students' comprehension ability to understand phrase material (Latifah, 2017). The textile manipulation course cannot carry out experimental methods because the projects for each student are done differently. The design concept they develop certainly has a level of difficulty in the process of embodiment, also it also varies according to current environmental issues. This research focused on analyzing the creativity of upcycling projects students in learning Project-based Learning (PBL), which refers to eco-art Education ranging from student creativity in product development to problem-solving.

2 Method

This research was descriptive qualitative research. Data collection using interview techniques, observations to explore information related to student creativity while working on projects, and documentation techniques were used to complete the presentation of research data. The source of research data were all students who were registered in the textile manipulation courses. The data analysis used Project-based Learning (PBL) theory, which refers to eco-art Education. Based on the principles of Project-based Learning (PBL), the analysis included (1) the interest in learning; (2) the ability to build a solid group; (3) the ability to plan and manage learning projects independently; (4) the ability to carry out learning activities following pre-designed projects; (5) the learning progress. The research design consisted of two cycles: (1) The activities in the first cycle were carried out together, namely sorting fabric waste. This sorting of fabric waste grouped the origin of the fibers, colors, motifs, and sizes of fabric waste. (2) The activities in the second cycle were carried out individually, consisting of three stages: design, production process, and product completion.

The success of this project was based on the assessment of each product carried out by a team of lecturers. The assessment carried out included the suitability of the material techniques used, the quality of control over product quality, such as neatness and cleanliness, and the total look produced. This assessment was carried out every week to avoid unsuccessful product manufacturing considering the limited processing time.

3 Result of Project-Based Learning (PBL) and the Output

The number of students involved in this upcycling project was 11 students. The products from the treatment of upcycling project were bed linings and home décor using geometrical patchwork techniques, random patchwork, rope bowls, Yoyo, and warm water-soluble (WWS), which can be seen in Table 11.1:

Based on the tables above, further discussion related to research data based on grouping techniques was used because they had similar difficulty levels. Based on the principles of Project-based Learning (PBL), there should be (1) The interest in learning; (2) the ability to build a solid group; (3) the ability to plan and manage learning projects independently; (4) the ability to carry out learning activities following pre-designed projects; (5) There is learning progress. The PBL principles were described in the following research design, consisting of two phases: the sorting and embodiment phases.

3.1 Sorting Phase

The activities in this phase were carried out together, namely sorting fabric waste. This group's activities were sorting the fibers, colors, motifs, and sizes of fabric scraps. This phase explored information related to the ability to build a solid group because, in the sorting step, fabric waste must be carried out together.

Questionnaire data stated that 63% of the students struggled in the sorting phase. One of the things that caused difficulties was that students were too picky in mixing fabric scraps that fit their product design concept. Meanwhile, the fabric waste they want was limited, adhering to the eco-art idea that all available fabric waste, as much as possible, must be used and not wasted. Large fabric scraps must first be divided evenly according to the needs of bed covers and curtain window products using geometrical patchwork techniques. However, 37% of students answered that they had no difficulty in this fabric waste sorting phase because they processed the remains of small scraps using Yoyo, random patchwork, warm water-soluble (WWS), and rope bowls. This upcycling project complemented the four guest house rooms in one upcycle theme so that the unity of this theme must be seen even though the products developed were done with various techniques. The steps taken to achieve harmony in the composition were that all products that students worked on must have the same tone, color, and scraps. Therefore, the division of fabric scraps had to be evenly distributed even though there were often struggles between them. Specific fabric waste available was quite limited, but the need for patchwork techniques was quite a lot in bed cover products, so the lecturer team often helped this class by dyeing plain fabric waste to adjust the tone so that the unity of the theme could be achieved.

Table 11.1 List of upcycling projects dan techniques used

Upcycle products	Amount (pieces)	Fabric manipulation techniques used	Students who create projects (person)
Bed covers	3	Geometrical patchwork	3
Bed cover	1	Random patchwork	1
Tissue boxes	4	Geometrical patchwork	1
Snacks bowls	4	Rope bowl	1
Lamp caps	4	Warm water soluble (WWS)	1
Curtain windows	8	Geometrical patchwork	2
Chair pillows	4	Yoyo	1
Pillows	4	Geometrical patchwork	1

3.2 Embodiment Phase

This embodiment consisted of the design, production, and completion of the product phase, which unearthed related information:

- Interest in learning

Questionnaire data stated that 100% students were familiar with sustainable design and global issues before this upcycling project was carried out and knew several fabric manipulation techniques, so this did not make it difficult for them to conduct research related to upcycling projects. Through this PBL, students could understand that waste treatment management was to use fabric waste according to their needs, not to throw small fabric scraps or give the remaining fabric scraps that are no longer needed to friends in need. They returned the fabric waste that was no longer required to the fabric waste storage in the classroom.

- Ability to plan and manage learning projects independently

Students could prepare and manage learning projects independently. Based on the results of observations made by the lecturer team, students could calculate the need for patchwork and supporting materials for each product design they would make. They could determine manufacturing and completion techniques based on the research they had done before.

- Ability to carry out learning activities following predesigned projects

All students could carry out the process of doing learning projects independently, but almost all required the assistance of a team of lecturers for the consultation of progress which was mainly related to using supporting materials other than fabric waste and when fabric waste was limited. Students did not experience difficulties procuring materials because they only processed materials, fabric waste, and support following their product design concepts. The procurement of materials was carried out by a team of lecturers and laboratory workers to maximize student performance in their work. Figure 11.1 shows an example of a patchwork lay-outing design and the division of patchwork pieces on curtains. Figure 11.2 shows pieces of fabric waste in a small size.

- Progress in the learning process

This upcycling project was carried out for 11 weeks, from selecting the idea of home décor and bed lining products, designing concepts, sorting fabric waste, making products, to finishing. Based on the results of observations made by the lecturer team, the developments carried out by students using geometrical patchwork techniques, random patchwork, and Yoyo, progress every week was not problematic. Upcycling always showed progress even though students experienced problems, mainly working on bed covers. The obstacle experienced by students was that the field being done was very large, namely a bed cover 200 cm × 200 cm size, causing students' space to move was limited. Another obstacle was adjusting the sewing machine settings because of the thickness of the bed cover being worked on. The following is documentation of the PBL's output. The upcycling products utilized small patchwork; Fig. 11.3 is a bed cover that used a random patchwork technique in the middle of the bed cover, while Fig. 11.4 is a pillow that used the yoyo technique.

Two of the 11 students involved in this upcycling project experienced slow progress. This slow progress was due to the difficulty of waste treatment techniques that differ from other methods. For example, in rope bowls and warm water-soluble (WWS) techniques, students often had to repeat experiments to get the shape, surface, and neatness according to the design concept. The difficulties felt by students were also caused by inability to share experiences related to these two kinds of techniques, so they had to research and find their solutions. The following is product documentation, and Fig. 11.5 is a photograph of a lamp cap product that used the warm water-soluble (WWS) technique. And Fig. 11.6 shows snacks bowl using the rope bowl technique. It also shows other products such as tissue boxes and pillows that use geometrical patchwork techniques.

Fig. 11.1 .

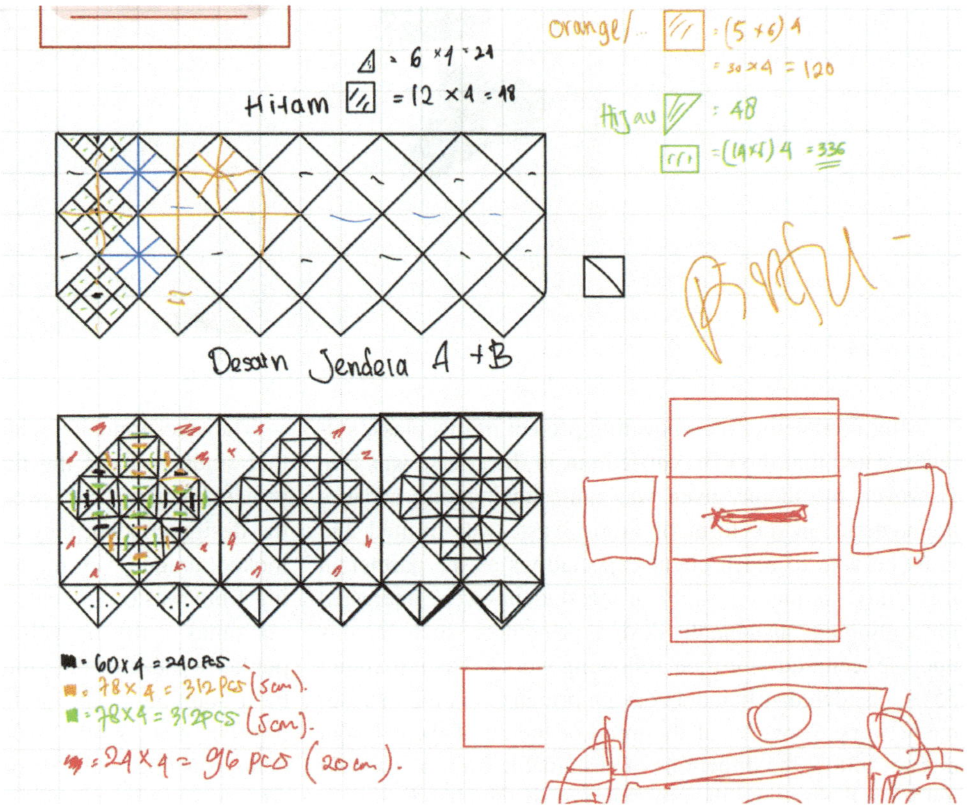

Fig. 11.2 .

After completing the upcycling project in this class, students stated that the PBL carried out in the course was fun. However, the theme given was quite challenging because the material in the form of fabric waste was limited and must be able to reach client expectations as the partner in this study. As many as 45% of the students were interested in continuing sustainable design principles, such as this upcycle technique for their following works. The remaining 55% considered sustainable design principles in their subsequent work. However, if the trend or theme of their design did not follow the directions of sustainable design, it would not force them to use the principles of sustainable design.

The questionnaire related to the finished results stated that students liked the finished products of the upcycling project because they were following the expectations of the design concept they designed. The products they made looked unique yet modern without being seen as coming from waste. Students also had high confidence that the outcomes of this upcycling product were in demand by the public because of the uniqueness produced, and they could be a solution to the problem of fashion waste which has now become a global issue. The high confidence of the students was due to the response of the partners in this study, who were very satisfied with the results of the student's

Fig. 11.3 .

Fig. 11.4 .

work in this upcycling project. The partner was impressed with the students' accuracy in mixing various motifs and colors from batik fabric waste. The good appreciation from partners for the work of these students could also be seen directly by students when they visited to enjoy the tourist services offered by this village. Figure 11.7 shows the students when visiting to hand over their product while traveling.

Furthermore, the partner gave suggestions to develop other products that are needed by the community in general, such as hotel sandals, prayer mats, and *mukena* (praying cloth for Indonesian Muslims) with detailed patchwork

Fig. 11.5 .

or yoyo decorations. Figure 11.8 is the documentation of the guest house room arranged using the upcycling products' work. Figure 11.9 is the documentation of the handover of the upcycling product from the team of lecturers to the village head as a representative of this research partner Bumdesa Ketapanrame, Trawas District, Mojokerto, East Java.

The results of the research that had been described related to the learning process followed the principles of PBL. Students were interested in PBL, shown by their interest in using sustainable design principles to continue their work. Students could plan and manage the process of making products according to their design concepts. Students could carry out learning activities and show the development of the learning process every week, making it easier for the team of lecturers to guide and direct students if they experienced problems in the process. Student individual or group learning activities had been running conductively. In group activities, it could be seen that although most students experienced difficulties, they could cooperate in sorting fabric waste. Students who had experienced slow progress had also completed the upcycling project product well. Despite the problems, students could find the solutions by doing more experiments than the others. From each process of making this upcycling product, students also understood that the processed waste patchwork must be utilized as much as possible. The remaining fabric waste was still returned and collected in the storage class for future projects.

The most important thing was that treating waste still required the same capabilities and attention as processing non-waste materials. The finished upcycling products must be of high quality so that the public remains interested even though the products were made from waste. Creativity and high quality could be achieved because quality control was carried out every week for each manufacturing

Fig. 11.6 .

Fig. 11.7 .

Fig. 11.8 .

process. The role of the lecturer team was very important during quality control because lecturers other than teachers were also representatives of research partners of this class related to product quality. In addition, the assessment of the final results of this learning was also based on the results of quality control. The research conducted by Teli et al. (2015) showed the price lower when the respondents knew that the product came from used materials/waste. This study shows the influence of psychology on buying prices. However, in this PBL, Bumdesa Ketapanrame's interest in upcycling products by students was quite high, and it could even inspire villagers, especially housewives, to collect used fabrics and used clothes to be used as useful upcycling products. This stimulated creativity that could lead to entrepreneurship for the residents of Ketapanrame village.

Project-based Learning (PBL) carried out in textile manipulation classes had been successful if seen from the positive partner's response to the results of upcycling products made by students in this learning. The partner's need for unique bed linings and home décor products had been achieved. The output of this PBL was also performed in a conducive learning process following the principles of PBL and student learning outcomes in the form of increasing student knowledge related to upcycling and unique product

Fig. 11.9 .

results made optimally by students during this PBL. The success of PBL was also in line with Eco-art Education's principle that creating a work must consider the conditions of the community and the surrounding nature and its problems. One of the programs being developed by Bumdesa Ketapanrame as a tourism village manager was to create a waste processing program due to the waste problem as an effect of tourism. Students' product from the upcycling project contributed to creative ideas for the village using fabric waste. Therefore, it is hoped that through these innovative products, they will be able to bring out new works from the use of waste from the Ketapanrame village community, whose continuation can make the creative economy in the tourist village be improved.

4 Conclusion

The upcycling product made by students in PBL was a unique but modern product without being seen coming from waste. Students also had high confidence that the outcomes of this upcycling product were in demand by the public because of the uniqueness produced, and it can be a solution to the problem of fashion waste which has now become a global issue. The partner was very impressed with the students' accuracy in mixing various motifs and colors from batik fabric waste and suggested developing other general products that the community needs. The upcycling project carried out by students in PBL has become a solution for two institutions related to the waste problem. The first institution is the classroom, which will always produce fabric waste from other course projects every semester. The second institution is the partner in the study. Student creativity in the form of upcycling products such as bed linings and home décor contributed creative ideas for Ketapanrame tourism villages, which can further develop the results of village products made by the Ketapanrame tourism village community so that they can improve the creative economy in the village. The results of PBL in this qualitative research are the first step to finding a roadmap related to eco-art Education that can be developed in every course in the fashion department as a solution in the field of Education to the issue of global fashion as the second largest contributor to waste in the world.

References

Azizah, L. F. (2011). Implementasi Pembelajaran Kimia Pokok Bahasan Minyak Bumi Berorientasi Pembelajaran Terpadu Tipe Immersed. *LENSA (Lemtera Sains): Jurnal Pendidikan IPA, 1*(2), 1–7. https://doi.org/10.24929/lensa.v1i2.138

Hananto, H., Dewi, H. P., Pratono, A. H., Tjahjono, H., & Pandjaitan, L. N. (2022). Proposal Usulan matching fund. In *Matching fund Kedaireka*.

Hutasoit, S. A. (2021). Pembelajaran teacher centered learning (Tcl) dan project-based learning (PBL) dalam Pengembangan Kinerja Ilmiah dan Peninjauan Karakter Siswa. *Jurnal Pendidikan Indonesia (Japendi), 2*(10), 1775–1799.

Inwood, H. (2008). Mapping eco-art education. *Canadian Review of Art Education, 35*, 57–73.

Latifah. (2017). Penggunaan model Pembelajaran problem-based learning pada Mata Kuliah Sintaksis sevagai Upaya Untuk Meningkatkan Kemampuan Pemahaman Mahasiswa STKIP Siliwangi Dalam Memahami Materi Frase. In *International conference Riksa Bahasa XI "Penguatan Pendidikan Bahasa Indonesia Pada Abad Ke-21"* (pp. 594–599).

Singh, S., & Rani, J. (2021). Traditional Indian textile techniques used to upcycle and recycle textile waste traditional Indian textile techniques used to upcycle and recycle textile waste. *Textile & Leather Review*, 2623–6281. https://doi.org/10.31881/TLR

Sukmawati, F., Setyosari, P., Sulton, & Purnomo. (2019). The effect of project-based collaborative learning and social skills on learning outcomes in biology learning. *Journal for the Education of Gifted Young Scientists*, *7*(4), 1325–1344. https://doi.org/10.17478/jegys.630693

Sulisworo, D. (2019). *Konsep Pembelajaran project based learning* (N. Setyaningsih, Ed.). ALPRIN.

Sunassee, A., & Bokhoree, C. (2021). A designed eco-art and place-based curriculum encouraging students' empathy for the environment. *Ecologies, 2*(3), 248–267. https://doi.org/10.3390/ecologies2030014

Teli, M. D., Valia, S. P., Maurya, S., & Shitole, P. (2015). Sustainability based upcycling and value addition of textile apparels. *International Journal of Applied and Physical Sciences*, *1*(3), 55–60. https://doi.org/10.20469/ijaps.50002-3

Upcycling the Abandoned Students Artwork with Bateson's Type of Learning in Entrepreneurship Course

Monica Hartanti, Elizabeth Wianto, and Miki Tjandra

Abstract

Many final artworks were ignored among many design courses offered in higher education. Each final state of the assignment can be spelled out through the design process, thereby placing valuable thoughts from the insights of students and teachers. Unfortunately, the end state stops developing once the final point was set, and students started from scratch to come up with new ideas all the time. This study recruits junior students from the Entrepreneurship course, majoring in visual communication design at a private university in West Java, Indonesia. They reuse and recycle previous assets from their colleagues or themselves and explore the possibility of making them suitable for certain targets using Gutman's Means-End theory. The educational method used is the Bateson Learning type. This research will determine whether the educational process can be understood by students or they are just learning to pass. It's a shame that even though the design students are working on it, upcycling the abandoned design art to make it more valuable and have a selling value has not been maximized, it is still affected by the effectiveness of the assignment, so the products being sold have not yet reached the 'desirable of existence'. Working in groups in the Entrepreneurship course has weaknesses, namely the understanding of the material is uneven and not comprehensive for each group member, especially for students who do not try to learn this material; it can be said that the stereotype of students belonging to Bateson's Learning type 0. Entrepreneurship courses at the undergraduate level for visual communication design students have not made them interested in becoming new entrepreneurs. The real project-based course in this learning is only limited to providing new experiences in learning things related to entrepreneurship, this applies to stereotypes of students belonging to Bateson's Learning type 0–2.

Keywords

Entrepreneurship course · Real project-based · Student's artwork · Upcycling

1 Introduction

As a vocational education, design in higher education is more practical than theoretical, resulting in various artworks and design. The work has gone through a process with assistance between students and lecturers (Ellmers, 2015). Those artworks and designs are overwhelming, and eventually become piled up and worthless. This is very unfortunate, especially in the field of art and design, which is one of the subsector for nation's creative economy. Recognizing the significant contribution of the creative economy sector to Indonesia's Gross Domestic Product (GDP) as highlighted in the 2019 Opus Creative Economy Outlook, thus the government actively supports its development, placing Indonesia in the third position globally. With such a substantial impact, the creative economy is projected to play a crucial role in driving Indonesia's economic growth by 2025 (Kemenparekraf, 2021). In correlation with above statement, Zamzami and Hastuti (2018) mentioned

M. Hartanti
Bachelor Program of Visual Communication, Maranatha Christian University, Bandung, Indonesia
e-mail: Monica.hartanti@art.maranatha.edu

E. Wianto (✉)
Doctoral Student of Industrial Design Department, Faculty of Design and Planning, National Cheng Kung University, Tainan, Taiwan
e-mail: elizabeth.wianto@art.maranatha.edu

E. Wianto · M. Tjandra
Faculty Member of Faculty of Arts and Design, Maranatha Christian University, Bandung, Indonesia

M. Nawaz Tunio et al. (eds.), *Sustainability in Creative Industries*,
Advances in Science, Technology & Innovation, https://doi.org/10.1007/978-3-031-48453-7_12

six advantages of entrusting the fields of art and design as the cornerstone of the progress of the creative economy in Indonesia are as follows: (1) Economic contribution capacity; (2) Creating a positive business environment; (3) Building national image and identity; (4) Developing an economy based on renewable resources; (5) Fostering competitive innovation and creativity; and (6) Providing social impact that stimulates the birth of new ideas.

One of the efforts made by the university to ensure creative economy development was by making Entrepreneurship courses as a mandatory courses. This course has become a major area of interest in public policy, industry, society, and education. Entrepreneurship is also believed to be the key to success in developing job creation and economic growth (Mawardi & Sahputri, 2022), and introduced by university to broaden student knowledge as well as to expanding their perspective of work opportunities (Anjum et al., 2018). In addition, increasing the ratio of entrepreneurs to Indonesia's total population is essential to enhance competitiveness and compete with other countries. For instance, Singapore has 7.2% entrepreneurs, Malaysia 2.1%, Thailand 4.1%, South Korea 4.0%, and the United States 11.5% entrepreneurs in relation to their respective populations. Encouraging entrepreneurship offers an alternative solution to societal challenges, including poverty, social inequality, unemployment among the productive age group, and diminishing energy resources, necessitating creative and innovative approaches (Marti'ah, 2017).

The Entrepreneurship course introduces one application of visual communication design results in the form of entrepreneurial project planning through a simulation of the establishment of small and medium businesses in the creative field which is still being pioneered and can be developed further. It can be said as a start-up company that is still in the development or research phase to continue to find markets and develop products (Nugraha & Wahyuhastuti, 2017). Based on real projects, the completion of assignments in this course is carried out in groups of three people and the final result is a business project book with five generic components equipped with STP (segmentation, targeting, positioning) analysis, artworks product development and design using a means-end theory approach, business canvas model and sales activity recap carried out.

Means-end chain model theory, is one of the classical theories proposed by Gutman (1982) with the aim of assigning value to objects or activities, thus reaching the stage of "desirable of existence". With a special theme to upcycle previous products that have been produced in the courses that have been taken previously, each group of participants is trained to assess the potential for developing works that are tailored to their chosen target.

Technological developments provide convenience in the creation of entrepreneurs today (Susan & Novianti, 2019). Those without formal business knowledge can also self-taught trying to sell online on platforms that have provided the infrastructure (Finkle & Olsen, 2019). This is because the Internet, social media, mobile applications, and other digital communications have become part of the daily lives of billions of people around the world (Meyen et al., 2010). According to Statista (2020), 4.54 billion people are active internet users, accounting for 59% of the global. The self-taught process of selling online can be found out easily, and put into practice immediately. Moreover, the capital required is not large. The utilization of the Internet and social media has revolutionized consumer behavior and transformed business operations (Yogesh et al., 2021), driven by technological innovations and widespread adoption of handheld devices. This has significantly impacted the way consumers engage in social commerce, influencing their decision-making process and online shopping habits. Moreover, the growing prevalence of digital marketing and social media has had a positive influence on consumer attitudes towards online shopping, resulting in a rising market share for e-Commerce-focused organizations (Abou-Elgheit, 2018; Alam et al., 2019; Komodromos et al., 2018). The online phenomenon has merged with the daily lives of most people in this world, buying and selling interactions are also included in it. By taking advantage of this phenomenon, art and design products can be promoted and sold online as well. In the end of this course, students were required to try selling their products online.

Referring to the Oxford Languages dictionary: "upcycling means reusing discarded objects or material in such a way as to create a product of higher quality or value than the original". The original value of the original state which of "product" has been gone; therefore, it consider as "trash". However, the state of "trash" in this beginning stage of upcycling material which has been abandoned and worthless eventually will be "shifted". Together with online phenomena that are closely related to technological developments and Entrepreneurship learning materials are the three main things that are discussed in this scientific article.

During the first semester, students are given learning materials to project a business from art and design products, seek to reuse artwork and design and recycle previous assets from the work of their colleagues or themselves and explore the possibility of making it suitable for certain targets using Gutman's Means-End theory. The final goal of this study is to find out the stereotype of zero-hyper-complexity learning students based on the educational process that has been carried out for one semester and what factors influence the formation of these stereotypes. Do students who attend and complete this Entrepreneurship course just want to graduate or simply get rewarded by getting good grades or do they want to understand how to be an entrepreneur and are even interested in becoming a new entrepreneur? The level of complexity is also added by the need for coordination between group members so that it is not only in the form of completing the final task, but the soft skills of negotiation between students can be further discussed related to attitudes that may arise from the initial selection of works to the willingness to complete this real project-based course.

2 Research Method

Real projects were given in the Entrepreneurship course in visual communication design at one private university in Bandung with the purpose to introduce how to curate the product value to initiate the student to familiarize the online shopping platform. For this reason, the entire population of 48 students taking Entrepreneurship courses (24 males and 24 females) was instructed to form groups independently, so 16 groups were formed.

To determine the effectiveness of giving courses that end with an offer to open a new business, this study evaluates the final score given by the lecturer and sets the final grades obtained into three groups as follows: (1) Group A is for those who get C and C+ grades; (2) Group B is for those who get B, B+ and A grades; while (3) Group C is for those who also get B, B+ and A grades, but express their concern about the unfavorable condition of the group and the uneven contribution made by their group members during the assignment period. The grouping carried out will be validated in this study, with the assumption that in the stages or levels of learning based on Bateson, even though they receive the same method and treatment, in the end, they learn things and matter for different purposes. Group A was assumed to be stereotyping those who followed the process but had no changes in abilities and attitudes afterward (Learning Type 0). Group B was assumed to be stereotyping those who follow the process and are serious about learning to gain an understanding of the new ability or knowledge itself (Learning Type 1). Group C was assumed to be stereotyping those who follow the process, seeking to gain new understanding or abilities, but more concerned with the value obtained as a reward in the higher education learning system (Learning Type 2). To examine in depth whether the type of learning is relevant to the results obtained and the process of undergoing it, semi-structured interviews were conducted, with the grouping of questions arranged per cluster, as written in Table 12.1.

Semi-structured interviews were proposed to students who stated their willingness to be interviewed, using purposive sampling method, each representing the evaluation group, so that the factors explored could then be compared between each student included in the three types of learning. The category of structured questions is divided into the four questions above (i–iv).

Students who have expressed their willingness, based on the evaluation group they represent, will each be faced with the following structured question:

- Sample from Group A (assumption learning type 0): question cluster i, if participants could answer question i, proceed to questions ii, iii, however, if participants could not answers, then stop and be classified into the learning type 0 category.
- Sample from Group B (assumption learning type 1): question cluster i, ii, iii, however, if participants could not answers question in cluster i, then validate as type 0 learning type, and if able to answer satisfactory, fall to type 1 learning type.
- Sample from Group C (assumption learning type 1): question cluster i, continue to question ii and iii, however, if participants could not answers question in cluster i, then validate as type 0 learning type, or if able to answers satisfactory, fall to type 2 learning type because they commented about unequal workload.

Furthermore, the results of collected answers that have been grouped will be coded in stages, and after reaching saturation, it is declared that the expected sampling has been achieved. Finally, related to the Guttman Means-End Theory, it will also be sought whether students' motivation to understand the material is affected by the initial selection of products to be sold.

3 Results

This section contains a summary of in-depth interview data on group representatives belonging to the category levels 0, 1 and 2 according to Bateson learning. In the end, there is a discussion and analysis of the data collected.

1. **The first part, questions related to Learning Motivation and Product Development**

Students who are assumed to be in the level 0 category because their grades are C or C+, and their product development analysis is not carried out correctly. They do not master the learning material provided. Students can briefly describe what they learned in Entrepreneurship class, but most do not remember it in detail. All good students who finally validated the learning type category 0–2 can remember where the selection of artwork came from as the product chosen. However, most category 0 types admit that

Table 12.1 Cluster grouping

Cluster	Group A Indicated as: Learning tipe 0 Grades C, C+	Group B Indicated as: Learning tipe 1 Grades B, B+, A	Group C Indicated as: Learning tipe 2 Grades B, B+, A (Students who complain that their members are not doing well)
(i) Related to learning motivation dan product development	Who and where did the artwork selection come from for the products used in this course? Remember what you have learned in this entrepreneurship class?		
(ii) Related to course material	Which is the most recent part that you can fill in the BP (Business Plan) book that has been done for 1 semester, besides the recap? Why the reason? Common questions for which there are not in the learning materials: Why are business projections necessary for investors? Why should you try to promote and sell online? Why is the organizational structure within BP that you have suggested to be as lean as possible? When deciding which products to sell, what do you need to think about? Questions in the learning materials: Why should depreciation expense be taken into account when making BP? Mention the 3 most important roles in this business created What is Break Even Point?		
(iii) Related to optimizing the opportunity to continue the project	Is it possible to continue this effort? What's the reason? If you were permitted by your parents to do business online while studying, would you be interested? Will you continue this planned effort? Is the product attractive enough to sell? Or is there something else?		
(iv) Related to simulation in the complexity of group work	–	Imagine that you are grouped into one course with self-selected members. Everyone will get tasks that support each other, but at the time of implementation not all members work with the same load. Do you think the grades obtained should be averaged? What is the reason? What if the workload of the group is not evenly distributed because one of the members has a misfortune? Should the final grade be averaged? In your opinion, what are the causes of the unbalanced workload in a group?	

it is not their artwork that is chosen. It is often seen that if the choice of artwork comes from them, then students are automatically categorized in learning type 1 or 2. It is alleged that they become more responsible for completing this assignment. There are several reasons students determine product selection. First, some groups have already sold these products before joining Entrepreneurship class; Second, due to the mutual agreement of the entire group; Third, there is also the fact that the product is ready in digital form so it doesn't need to be processed anymore.

2. **Part Two, Question Related to Course Material**

In this section, some questions are general knowledge and some are related to the learning materials that have been given. All students quite understand the function of business projections for investors, but they cannot explain it in detail, only so that investors have an idea of what kind of business this business will develop. The majority of the participants were unable to correctly pronounce terms that did not fall into the category of visual communication design disciplines (i.e., depreciation cost, break-even point), but were still able to describe more or less what the term meant, along with the need for preparation or logically necessary calculations. Of course, students who are categorized as learning type 0, are not too able to state in detail the components compiled in the business plan. In general, groups that divide work separately and do not coordinate, are more likely to not fully understand the components being taught. They were only able to re-describe what they had done during group work.

Students already understand the benefits of promoting and selling online, they understand that online media is relevant today, widely used, and has a very wide reach. The prospect of an online business is better, small capital can be done from home, it is estimated that it will sell better and the way to shop is practical. They also think that the task of selling directly online will give them a hands-on experience. Only some students understand that the business projection they make is a lean start-up business from an organizational perspective. Students already understand the things that need to be considered when deciding which products to sell but from their point of view. They will look for the right STP, product functionality for the community, attractive designs, whether this business is sunset or sunrise, product innovation, selling prices, and promotions. The second part of the question is in the learning material. Students do not understand what depreciation costs are and why they need to be taken into account when making business projections. Most of them also do not understand BEP (Break Even Point), moreover how to calculate it on the grounds of working in groups so that they divide tasks. Students also forget the three most important roles in the structure of a start-up business that is designed.

3. **Part three, Question Related to Optimizing the Opportunity to Continue the Project**

Students have been given learning materials to design business projections for one semester by selling products directly at the end of the lecture. However, they are not yet willing to continue the business they made for several reasons, namely, because the product still needs to be developed, there are still many ongoing lectures, no capital, and team shortage. Some students are interested in trying but according to their expertise as graphic design freelancers. About the attractiveness of the artwork and design products being sold, some think that they are too specific, but some of them think that art and design products have their niche market, especially for art and design lovers.

4. **Part four, Question Related to Simulation in the Complexity of Group Work**

All students agreed that even though the group was chosen by themselves, when the workload was not the same, the results or grades obtained should not be generalized. Except when some obstacles or misfortunes befall the members of the group who are not working, they are willing to help but of course, they need evidence. They realize that the things that cause uneven workloads can be due to lack of coordination, unequal initiative in each group member, and differences in standards of excellence in finished work so that sometimes work results become unused.

4 Discussion

The initial assumption of the research team, related to categorization according to Bateson learning about students' understanding of the learning materials for entrepreneurial business projects that they have completed, is that there are three categories, namely Category level 0 is for groups that get C, C+, level 1 and 2 categories for students who got the grades B, B+, A.

After conducting in-depth interviews with representatives of these categories, it was found that for the students who scored C or C+, it was true that the learning level was 0, they did not understand the Entrepreneurship learning material. At the time of doing the given task, they depend on the handout learning material uploaded on the learning web system provided by the university. All stages in making a business projection book are 70% done, but the content is very shallow and instant. This type of student when interviewed about understanding the material also cannot give the correct answer. However, regarding group collaboration, these students did not experience any problems at all, their tolerance for working in groups was very high, and all group members even though they got C and C+ scores were willing, it could be said that they just graduated from this course. From the point of view of the supporting lecturer, because this course is given full online, there is tolerance in giving grades, so it only measures the completeness of the contents of the business plan book.

Students who get grades B, B+ or A in terms of understanding the material, understand them in general, why it is necessary to carry out the stages in making a business projection book, and what are their uses. However, when asked about the theories, only part of them are still remembered. This category of students is dominant in their group. In terms of group collaboration, they also do more work than their group mates. If there are group members who do not work, they expect the lecturer to give a different assessment as well, but they still strive to get good grades.

This is due to several external factors. When it comes to online learning, entrepreneurship learning materials have quite a lot of processes and stages of work. Every week there is a different new material. This requires students to reread the material that has been given and maybe look for other sources to complement the material so that they can do the assignments given. This cannot be generalized to each student. Those who have good learning facilities, adequate time, and a good internet network will make it easier to do this.

Assignments in this Entrepreneurship Course are intentionally made to have group complexity because in the real world of work the ability to work together in groups is required (Chinoy et al., 2022). In addition, the goals are made per group because there is a lot of work and it trains how to behave from each group member when completing the given task. In designing business projections, it can only be made with a comprehensive understanding of the material. For students who do not try to learn the whole material, then after the learning is completed their understanding is not thorough. So, even though the assignments are given in groups, individually they need to understand the whole material (Colbeck et al., 2000).

Entrepreneurship course is a new material in the context of design. Design students are invited to learn something new outside the field of design. In line with Vogler's research, this interdisciplinary project provides opportunities for students to practice skills that contribute to success outside the field of design, and of course add a new dimension to learning (Vogler et al., 2018).

Design is an applied field of science, so the task of designers is not only to create products. Designers also need to think about their design work being upcycled which reaches the stage of a desirable existence, so that entrepreneurship learning materials from economics also need to be understood by designers. This is because design thinking and entrepreneurship are complementary processes, and by mastering entrepreneurial skills students will be able to seize opportunities to develop commercial businesses after graduation (White & Kennedy, 2022).

5 Conclusions

Stereotypes of student understanding of learning materials can be formed from internal and external factors (AL-Muslimawi & Hamid, 2019). Internal factors are courses, lecture materials, lecture time, and course lecturers. External factors are students who take courses, how to do assignments, environmental conditions when doing assignments, ambitions or student targets to get rewards in the form of grades, and student interest in courses. In this course, there are only two categories of Bateson learning, namely types 0 and 1. The task of designing business projections and selling real upcycling art and design products in the Entrepreneurship course is carried out which is completed for one semester fully online is not easy for students as a whole to directly understand in detail the lecture material that is not designed material. Coupled with the existence of group assignments that have their complexity. Group assignments are perceived negatively by students who not interest to understand the whole lesson. This leads to uneven and incomplete understanding of the subject matter.

Stereotype level type 0, have a low understanding of the material, they choose group members according to their character, do not pursue grades, assignments are done poorly, friendships remain good and the important thing is to pass the course. Type 1 level stereotypes have a fairly good understanding and some are very good, they are looking for group members who have a cooperative rhythm or who they think will be the same. They pursue the value, trying to get the best possible outcome, they are even willing to work individually if the other members do not work with the same portion. However, they were not willing and reported to the lecturer so that the member who did not work would get a lower score. On the positive side, the stereotype of ambitious individuals with a proactive attitude towards success motivates them to make sincere efforts in comprehending the entirety of the learning material, even within a group assignment context. However, on the negative side, the dominance exhibited by the most prominent group members can inhibit and restrict the creative contributions of other members who may be perceived as less equal. However, both groups still passed the course. This needs to be a concern for study programs, especially regarding the lack of knowledge from students about learning materials but still being able to graduate with a C grade. Thus a C grade in higher education cannot be interpreted as sufficient but lacking.

Student have not been optimal or optimized product development. The product development steps with the mean-ends theory that have been described in the learning materials are carried out but are not evenly distributed and have not been maximally developed in all groups. Upcycling's efforts to provide value to artworks and designs to make them more valuable and have a selling value are still affected by the effectiveness of the assignment, not yet on the idea of how to turn these abandoned artworks and designs into selling value products, so that the products they sell have not yet reached the stage "desirable of existence". In conclusion, the Entrepreneurship course for undergraduate visual communication design students fails to generate sufficient interest in entrepreneurship as a career option. While the course incorporates project-based learning, it primarily serves as a means of introducing students to entrepreneurship rather than fostering a genuine entrepreneurial inclination.

6 Limitation of the Study

This research discusses findings based on a limited number of participants drawn from regular classroom activities from a private university located on Java Island, Indonesia. Therefore, the opinions generated have limitations and cannot be generalized to the entire population of college students. However, the exploratory nature of this study allows for potential development and adaptation based on the unique cultural context of the local region.

Acknowledgements Acknowledgments to Maranatha Christian University for funding the publication of this research.

References

Abou-Elgheit, E. (2018). Understanding Egypt's emerging social shoppers. *Middle East Journal of Management, 5*(3), 207–270.

Alam, M. S. A., Wang, D., & Waheed, A. (2019). Impact of digital marketing on consumers'_impulsive online buying tendencies with intervening effect of gender and education: B2C emerging promotional tools. *International Journal of Enterprise Information Systems, 15*(3), 44–59.

AL-Muslimawi, I. A. J., & Hamid, A. A. (2019). External and internal factors affecting student's academic performance. *The Social Sciences, 14*(4), 155–168. https://doi.org/10.36478/sscience.2019.155.168

Anjum, T., Ramzani, S. R., Nazar, N., Shahzad, I. A., & Salman, S. (2018). Entrepreneurial intention: Does entrepreneurial education matter in Pakistan. *International Journal of Human Resource Studies, 8*(3), 147–161. https://www.macrothink.org/journal/index.php/ijhrs/article/view/13213

Chinoy, N., Stoub, H., Ogrodzinski, Y., Smith, K., Bahal, D., & Zubek, J. (2022). Assessing student desire for professional skills development within the undergraduate science curriculum: A focus on teamwork. *Advances in Physiology Education, 46*(1), 179–189. https://doi.org/10.1152/advan.00051.2021

Colbeck, C. L., Campbell, S. E., & Bjorklund, S. A. (2000). Grouping in the dark. *The Journal of Higher Education, 71*(1), 60–83. https://doi.org/10.1080/00221546.2000.11780816

Dwivedi, Y. K., Ismagilova, E., Laurie Hughes, D., Carlson, J., Filieri, R., Jacobson, J., Jain, V., Karjaluoto, H., Kefi, H., Krishen, A. S., Kumar, V., Rahman, M. M., Raman, R., Rauschnabel, P. A., Rowley, J., Salo, J., Tran, G. A., & Wang, Y. (2021). Setting the future of digital and social media marketing research: Perspectives and research propositions. *International Journal of Information Management, 59*.

Ellmers, G. (2015). The graphic design project: Employing structured and critical reflection to guide student learning. *Communication Design, 3*(1), 62–79. https://doi.org/10.1080/20557132.2015.1057376

Finkle, T. A., & Olsen, T. (2019). Entrepreneurship in the digital era: Creating your own online business. *Entrepreneurship Education and Pedagogy, 2*(2), 133–150. https://doi.org/10.1177/2515127418820680

Gutman, J. (1982). A means-end chain model based on consumer categorization processes. *Journal of Marketing, 46*(2), 60–72. https://www.jstor.org/stable/3203341

Kemenparekraf. (2021). Ekonomi kreatif jadi garda terdepan pemulihan ekonomi nasional. Kementrian Pariwisata dan Ekonomi Kreatif, https://kemenparekraf.go.id/ragam-ekonomi-kreatif/Ekonomi-Kreatif-Jadi-Garda-Terdepan-Pemulihan-Ekonomi-Nasional

Komodromos, M., Papaioannou, T., & Adamu, M. A. (2018). Influence of online retailers' social media marketing strategies on students' perceptions towards e-shopping: A qualitative study. *International Journal of Technology Enhanced Learning, 10*(3), 218–234.

Marti'ah, S. (2017). Kewirausahaan Berbasis Teknologi (Technopreneurship) dalam Perspektif Ilmu Pendidikan. *Jurnal Ilmiah Edutic, 3*(2), 2017.

Mawardi, M. K., & Sahputri, R. A. M. (2022). The effect of entrepreneurship education and family entrepreneurship on student entrepreneurship intention. *Jurnal Aplikasi Manajemen, 20*(2). https://doi.org/10.21776/ub.jam.2022.020.02.20

Meyen, M., Pfaff-Rüdiger, S., Dudenhöffer, K., & Huss, J. (2010). The internet in everyday life: A typology of internet users. *Media, Culture & Society, 32*(5), 873–882. https://doi.org/10.1177/0163443710374792

Nugraha, A. E. P., & Wahyuhastuti, N. (2017). Start up digital business: Sebagai Solusi Penggerak Wirausaha Muda. *Jurnal Nusantara Aplikasi Manajemen Bisnis, 2*(1), 1. https://doi.org/10.29407/NUSAMBA.V2I1.701

Statista. (2020a). Global digital population as of January 2020. Available at https://www.statista.com/statistics/617136/digital-population-worldwide/. Accessed on 9 April 2020.

Susan, A., & Novianti, W. (2019). Benefits of technology for business. *IOP Conference Series: Materials Science and Engineering, 662*(3), 032036. https://doi.org/10.1088/1757-899X/662/3/032036

Vogler, J. S., Thompson, P., Davis, D. W., Mayfield, B. E., Finley, P. M., & Yasseri, D. (2018). The hard work of soft skills: Augmenting the project-based learning experience with interdisciplinary teamwork. *Instructional Science, 46*(3), 457–488. https://doi.org/10.1007/s11251-017-9438-9

White, P. J., & Kennedy, C. (2022). Designing a module in entrepreneurship for product design students. *Industry and Higher Education, 36*(2), 217–226. https://doi.org/10.1177/09504222211013742

Zamzami, Z., & Hastuti, D. (2018). Determinan penerimaan daerah dan pertumbuhan ekonomi terhadap pengembangan ekonomi kreatif di Provinsi Jambi. *Jurnal Paradigma Ekonomika, 13*(1).

Student Perception and Behavioral Changes in Blended Learning Implementation

Rezart Prifti and Ana Shkreta

Abstract

In this paper, we investigate students' behavior and attitude toward blended learning by comparing first-time and second-time users. A sample of 202 first-time users and 60 experienced users has been analyzed. As derived from the data analysis, there are significant differences on students' perceptions and attitudes toward blended learning usage when comparing novice and experienced users. Moreover, we identify that the three external variables multimedia instruction, content quality, and learning management system (LMS) self-efficacy have a positive effect on perceived usefulness and ease of use for the first-time users, meanwhile for experienced users, perceived usefulness is affected only by content quality and LMS self-efficacy, and content quality and multimedia instructions affect ease of use.

Keywords

Blended learning · Perceived usefulness · Perceived ease to use · Behavior intention · TAM model

1 Introduction

Blended learning has become increasingly important, as a methodology that breaks down traditional walls of teaching and provides high-quality learning experiences (Boelens et al., 2018; Broadbent, 2017; Hameed et al., 2008; Moskal et al., 2013; Oh & Park, 2009). Many authors have considered it as "the new normal" of the teaching process (Dziuban et al., 2018; Norberg et al., 2011; Ross & Gage, 2006). Technological improvements, widespread integration of IT infrastructure, new approaches toward digital skills, and internet usage have changed learning and teaching styles rapidly (Cahya et al., 2018; Sethy, 2008). Azizan (2010) argues that technology usage in classroom enhances students' competences and confidence and improves learning quality, given that it can offer extra learning resources. The term "blended learning" represents a wide spectrum of delivery options, tools, and pedagogies but conceptually refers to distance education that emphasizes flexibility of time, place, and pace of student learning. Staker and Horn (2012) define blended learning as "*any time a student learns at least in part at a supervised brick-and-mortar location away from home and at least in part through online delivery with some element of student control over time, place, path, and/or pace.*" Despite its importance and substantial growth, this approach has gained in recent years, and education institutions still face a number of challenges in relation to blended learning methodology. The literature shows that the challenge education institutions face mostly is users' acceptance and delivery of learning materials (Lee & Hung, 2015; Park, 2009). Interactivity in online platforms, quality of learning material, and interaction among students and teachers are also claimed as challenges (Gómez-Rey et al., 2016; Kearsley, 2000; Kent et al., 2016; Lear et al., 2010; Zain et al., 2016). Hence, while designing an effective online learning environment, there are two critical dimensions to be considered, instructor and student characteristics (Campbell, 2006; Liaw, 2004; Mbengo, 2014). Understanding their perceptions and attitude toward such a methodology helps institutions understand how to use the approach appropriately. Precisely to study these interactions in this research paper, we have used the technology acceptance model (TAM) developed by Davis et al. (1989) and later adopted by many authors to investigate factors that influence the usage and adoption of blended learning (Padilla-meléndez et al., 2013; Pituch & Lee, 2006; Tselios et al., 2011). This model is

R. Prifti (✉) · A. Shkreta
Department of Economy, Entrepreneurship and Finance, Faculty of Economy, Governance and Law, Barleti University, Tirana, Albania
e-mail: r.prifti@umb.edu.al

A. Shkreta
e-mail: a.shkreta@umb.edu.al

M. Nawaz Tunio et al. (eds.), *Sustainability in Creative Industries*,
Advances in Science, Technology & Innovation, https://doi.org/10.1007/978-3-031-48453-7_13

used to understand factors that influence usage and adoption of blended learning in university premises. The model has been supported in the literature by numerous theoretical and empirical studies (Hsu & Chang, 2013; Martín-García et al., 2019; Wong et al., 2017). This model analyzes and describes how users accept and use a particular technology taking into account such factors as: system characteristics, perceived usefulness, perceived ease of use, and attitudes toward using technology. Through this research study, we analyze how external factors like multimedia instructions, content quality, and learning management system self-efficacy affect perceived usefulness and ease of use. In addition, we investigate whether their perception toward usefulness and easiness of the online learning platform affect their intention and actual use. Moreover, we find relevant the differences in vantage points related with perception and attitudes between new and experienced time users. The contribution we aim to make in this paper is to investigate the underlying factors that influence adoption and usage of blended learning approach, and, in specific, we investigate how student's perceived usefulness and ease of use affect their behavior intention toward blended learning and the overall use of this methodology. Through this analysis, we aim also to compare how these factors change between new users and experienced users.

Behavioral intention influences a number of important factors that determine the effectiveness of blended learning methodology; it is one of the key determinants of technology incorporation in the classroom (Fishbein & Ajzen, 1980), which is influenced by such factors as, self-efficiency and motivation (Liaw & Huang, 2011; Thai et al., 2017), compatibility, complexity, observability, relative advantages, Lee et al. (2011), ease of use and usefulness (Yi & Hwang, 2003). Students' behavioral intention to use e-learning is also influenced by perceived satisfaction and perceived usefulness from the LMS that directly affect the last one (Damnjanovic et al., 2015). For the purpose of this paper, perceived usefulness and ease of use are defined according to Davis et al. (1989), who describes perceived usefulness as the level in which students believe that using blended learning approach will enhance their final results and perceived ease of use as the level in which students believe that using blended learning approach will facilitate their learning process. The structure of this paper is organized as follows: The next section describes the methodology used to analyze gathered data, followed by their analysis and lastly research questions. After data analysis, we conclude with main results.

2 Methods

This study was carried out in university premises, with business administration students at the first and third year of their studies. The sample consists of 262 business administration, bachelor-level students, from which 202 are new users and 60 are experienced users. The online course is offered as supplementary of face-to-face learning. Students were asked to attend the online learning once a week to find additional learning information, video explainers, illustrations, assignments, and quizzes. The online curricula were adapted as complementary of in-class learning following instructional design principles (Ghirardini, 2011) in that way to create an appropriate blended learning environment. Students have been part of this methodology at the first year of their studies and at the third year of their studies. The learning methodology has not changed, only subjects.

The aim is to understand whether there are differences in their behavior and how these change with time. How do student priorities change while using blended learning through an online platform? Are their behavioral intentions affected the same way and by the same factors as first-time and experienced users? Are usefulness and ease of use affecting the same way first-time and experienced users? While students gain experience, do they favor multimedia instructions over content quality over LMS self-efficacy, or the other way around?

Students attended a full-fledged blended learning course for one semester, and afterward, a five-scale Likert questionnaire was distributed online. The questionnaire was designed based on the work of Liaw (2008), Liaw and Huang (2013), and Cigdem and Ozturk (2016a).

As external factors, Lau and Woods (2009b) used multimedia instructions, content quality, and learning management system self-efficacy to understand how students' perceived usefulness and ease to use of the platform is affected. System quality and content quality are considered as important factors that influence learners' usage and satisfaction (Dorobat, 2014). Pedagogical and technical aspects constitute the infrastructure and quality of the learning process (Ouadoud et al., 2018), and the content provided in the online platform make it eligible or not to learners (Vonderwell & Zachariah, 2005). Studies have shown that the easier the usage of the platform and the general impression it creates, the greater the acceptability (Aristovnik et al., 2016). Referring to the TAM Fig. 1 model and supported the literature, the following research questions were raised:

1. Can multimedia instruction, content quality, and LMS self-efficacy impact perceived usefulness?
2. Can multimedia instruction, content quality, and LMS self-efficacy impact perceived ease of use?
3. Can perceived usefulness and perceived ease of use impact behavior intention to use blended learning approach?
4. Can behavior intention to use blended learning impact actual use of the online platform?

Fig. 1 Theoretical framework derived by technology acceptance model developed by Davis et al. (1989)

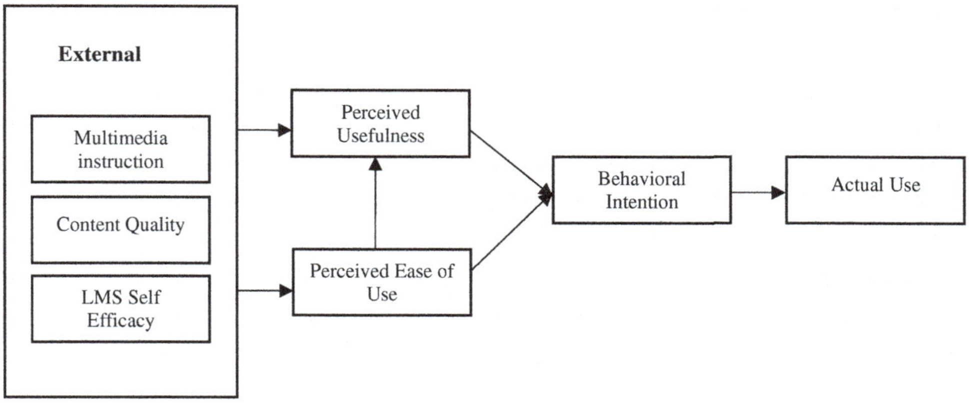

3 Data Collection and Analysis

The data for the study were collected during the first semester of the first year of studies and the second semester of the third year of studies of undergraduate course. During the semesters, the participants attended the blended learning for 15 weeks, 2 h per week. At the end of the semester, an online questionnaire has been distributed to them, asking for their perception and attitude toward their experience with the blended learning methodology.

The data were collected by the questionnaire composed by seven constructs representing multimedia instructions, quality of learning materials, LMS self-efficacy, perceived usefulness and ease to use, behavior intention to use blended learning approach, and actual use of the learning management system (Cigdem & Ozturk, 2016a; Liaw, 2008; Liaw & Huang, 2013). All items are presented in Table 1.

We have computed Cronbach's alpha to determine the internal consistency of each construct and investigate whether the items used formed a reliable scale. Cronbach's alphas for each construct regarding the "new user" model and the "experienced user" model are very reliable and within statistical measurement standards, revealing reasonable levels of reliability Table 2.

To be able to generalize the sample, we have analyzed data through inferential statistics, given that we use a random sample of data taken from a population. Furthermore, Pearson correlation coefficients and linear regression analyses were carried out to investigate relationships and correlations between constructs within the research design. The same analyses has been performed for the two models in the way to examine construct validity. Table 3 presents Pearson correlations among factors with a p value of less than 0.01 in order to control the Type 1 error across the correlations.

As shown in Table 3, all correlations were all less than 0.80 and there was no evidence of correlation ($rxy>0.90$) among predictors, and for this reason, it is assumed there is no multicollinearity problem among predictor variables of the regression analyses. It seems that all the seven factors were significantly correlated with each other. The strongest correlation is between LMS self-efficiency and perceived usefulness (0.601), LMS self-efficiency and perceived ease of use (0.552), content quality and perceived usefulness (0.546), and the lowest correlation is between LMS self-efficiency and actual use.

In "experienced user" model, some correlations have changed as referred to Table 4. There is a negative correlation between multimedia instruction and actual use, and this correlation has been weak in "new user" model; the relation between these two factors has been analyzed by many authors that argue the positive impact of multimedia instructions on the intention to use the LMS (Al-Busaidi, 2012; Cigdem & Ozturk, 2016b). There are also authors that suggest other external factors that influence the actual use such as subjective norms and system accessibility (Al-Emran et al., 2020; Moritani et al., 2001). From the analysis, we realize that there is a strong correlation between content quality and perceived usefulness as (Rittgen, 2010; Willemsen et al., 2011) found. Still the strongest correlation is between perceived usefulness and behavior intention, relation that has been investigated and supported by other studies (Abioye & Adelakun-Odewale, 2017; Ong et al., 2004).

In order to investigate the hypothesis raised in this research paper, a linear regression analysis is conducted to identify unstandardized and standardized coefficients. As for the first research question to understand whether perceived usefulness could be predicted by multimedia instruction, content quality, and LMS self-efficacy, the regression analysis indicates as follows:

As results shown in Table 5, for new users, multimedia instructions, content quality of learning material in the online platform, and LMS self-efficiency positively influenced the scores regarding perceived usefulness on online learning (Lau & Woods, 2009a; Tutty & Martin, 2008; Wu et al., 2010). For experienced users, as shown in Table 5, content quality and LMS self-efficiency seem to have a positive effect on scores related to perceived usefulness (Landrum, 2020; Palmer & Holt, 2010a).

Table 1 Distribution of survey items to factor structure

Factor	Item	New users		Experienced user	
		Mean	SD	Mean	SD
Multimedia instruction		**4**	**1.06**	**4.49**	**0.72**
UM1	I like to use video media instructions in the LMS	3.76	1.10	4.35	0.68
UM2	I like to use multimedia instruction in the LMS	3.74	1.05	4.45	0.70
UM3	I like to use presentations/slides in the LMS	4.53	0.80	4.68	0.75
Content quality		**4.1**	**1.08**	**4.47**	**0.69**
SO21	Online lessons were interesting and engaging	4.11	0.99	4.63	0.55
SO22	Online lessons were divided into manageable segments	4.19	0.96	4.50	0.65
SO23	Online lessons contained the right number of visualizations, animations, and/or charts	3.94	1.23	4.33	0.77
SO24	Online lessons contained the right amount of text	4.08	1.12	4.43	0.74
LMS self-efficacy		**4.36**	**0.86**	**4.66**	**0.61**
LM1	I feel confident using the LMS	4.34	0.92	4.62	0.74
LM2	I feel confident operating functions of the LMS	4.21	0.86	4.67	0.57
LM3	I feel confident using content of the LMS	4.54	0.79	4.72	0.52
Perceived usefulness		**4.12**	**0.98**	**4.47**	**0.7**
DM1	Using the LMS gives me greater control over my work	3.95	0.95	4.30	0.77
DM2	Using the LMS improves my performance	3.97	1.06	4.35	0.73
DM3	Using the LMS makes it easier to do my job	4.15	0.95	4.53	0.68
DM4	I believe the LMS content is useful	4.41	0.87	4.68	0.57
Perceived ease of use		**4.24**	**0.97**	**4.56**	**0.72**
LP1	Learning to operate the LMS system would be easy for me	4.17	1.01	4.58	0.70
LP2	I would find it easy to get the LMS to do what I want it to do	4.27	0.94	4.55	0.72
LP3	I would find the system easy to use	4.29	0.97	4.53	0.75
Behavior intention		**4.14**	**0.99**	**4.47**	**0.78**
SP1	I intend to use the LMS to assist my learning	4.28	0.92	4.65	0.52
SP2	I intend to use functions of the LMS to assist my learning	4.21	0.95	4.52	0.62
SP3	I intend to use the LMS as an autonomous learning tool	3.89	1.00	4.20	0.95
SP4	I would like to see the LMS functions implemented further in departmental modules	4.20	1.06	4.52	0.91
Actual use		**3.69**	**1.13**	**4.1**	**0.99**
SO1	How many of the online lessons did you complete?	4.34	0.91	4.73	0.52
LD1	How many of the background readings did you complete?	3.24	1.04	3.58	0.93
LD2	How many of the readings for in-class discussion did you complete?	3.50	1.13	3.95	1.10

SD = standard deviation of the mean

Regarding to the second research question whether multimedia instruction, content quality, and LMS self-efficacy affect perceived ease of use, only LMS self-efficacy exerted positively significant effects on the scores pertaining to perceived ease of use for the new users and multimedia instruction and content quality to have a positive effect on

Table 2 Results of reliability analyses

Subscale	Items	New user	Experienced user
		Cronbach's alpha	Cronbach's alpha
Multimedia instruction	3	0.77	0.70
Content quality	4	0.84	0.77
LMS self-efficacy	3	0.84	0.73
Perceived usefulness	4	0.85	0.81
Perceived ease of use	3	0.92	0.90
Behavior intention	4	0.80	0.72
Actual use	3	0.59	0.54
Total	**24**	**0.83**	**0.79**

the perceived ease of use variable for the experienced users (Davis, 1989; Park et al., 2019) as shown in Table 6.

The third research question analyses whether perceived usefulness and perceived ease of use play a positive effect on students' behavior intention (Baki et al., 2018; Lee & Hung, 2015; Park, 2009). For both novice and experienced users, perceived usefulness and perceived ease of use seemed to have a significant effect on behavior intention to use the online platform as results shown in Table 7.

The last research question investigates whether behavior intention to use online platform predicts the actual use (Yi & Hwang, 2003). Analysis shows that behavior intention of new users has served as predictor of the actual use, and

Table 3 Pearson correlations among the factors of "new users" model

Factors	1	2	3	4	5	6	7
1. Multimedia instructions		0.362**	0.429**	0.416**	0.256**	0.471**	0.250**
2. Content quality			0.485**	0.546**	0.374**	0.490**	0.238**
3. LMS self-efficacy				0.601**	0.552**	0.537**	0.185**
4. Perceived usefulness					0.541**	0.748**	0.287**
5. Perceived ease of use						0.539**	0.209**
6. Behavioral intention							0.230**
7. Actual use							

Note **Correlation is significant at the 0.01 level (two-tailed)

Table 4 Pearson correlations among the factors of "experienced user" model

Factors	1	2	3	4	5	6	7
1. Multimedia instruction		0.435**	0.278*	0.332**	0.376**	0.441**	– 0.015
2. Content quality			0.324*	0.673**	0.405**	0.588**	0.042
3. LMS self-efficacy				0.460**	0.295*	0.445**	0.268*
4. Perceived usefulness					0.478**	0.831**	0.067
5. Perceived ease of use						0.535**	– 0.122
6. Behavioral intention							0.045
7. Actual use							

Note *Correlation is significant at the 0.01 level (two-tailed)

Table 5 Multimedia instruction, content quality, and LMS self-efficacy as predictors of perceived usefulness

	Variables	*B*	Std. error	Beta	*t*	Sig
New users	Multimedia instruction	0.175	0.076	0.137	2.317	0.022
	Content quality	0.274	0.055	0.305	5.014	0.000
	LMS self-efficacy	0.564	0.090	0.394	6.268	0.000
Experienced users	Multimedia instruction	0.003	0.135	0.002	0.021	0.983
	Content quality	0.613	0.111	0.584	5.512	0.000
	LMS self-efficacy	0.398	0.146	0.271	2.726	0.009

Table 6 Multimedia instruction, content quality, and LMS self-efficacy as predictors of perceived ease of use

	Variables	*B*	Std. error	Beta	*t*	Sig
New users	Multimedia instruction	– 0.003	0.072	– 0.002	–0.038	0.970
	Content quality	0.106	0.052	0.139	2.032	0.043
	LMS self-efficacy	0.589	0.086	0.486	6.,888	0.000
Experienced users	Multimedia instruction	0.260	0.154	0.222	1.683	0.098
	Content quality	0.246	0.127	0.260	1.945	0.057
	LMS self-efficacy	0.199	0.166	0.149	1.193	0.238

Table 7 Perceived usefulness and perceived ease of use as predictors of behavior intention

	Variables	*B*	Std. error	Beta	*t*	Sig
New users	Perceived usefulness	0.627	0.053	0.645	11.881	0.000
	Perceived ease	0.218	0.062	0.190	3.499	0.001
Experienced Users	Perceived usefulness	0.744	0.080	0.745	9.253	0.000
	Perceived ease	0.198	0.089	0.179	2.223	0.030

Table 8 Behavior intention as predictors of actual use

	Variables	*B*	Std. error	Beta	*t*	Sig
New users	Behavior intention	0.171	0.051	0.230	3.347	0.001
Experienced users	Behavior intention	0.039	0.113	0.045	0.343	0.733

Fig. 2 Derived model new users

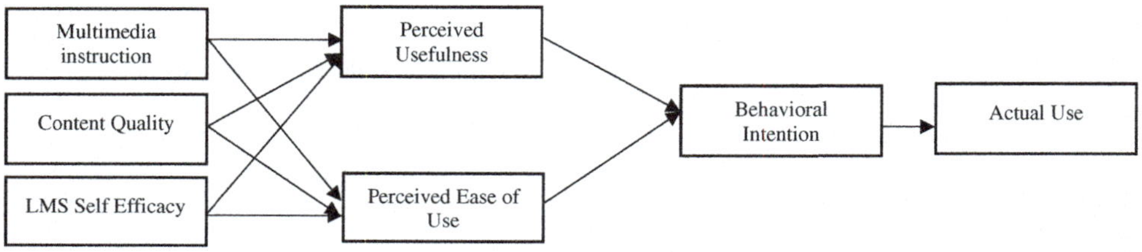

Fig. 3 Derived model experienced users

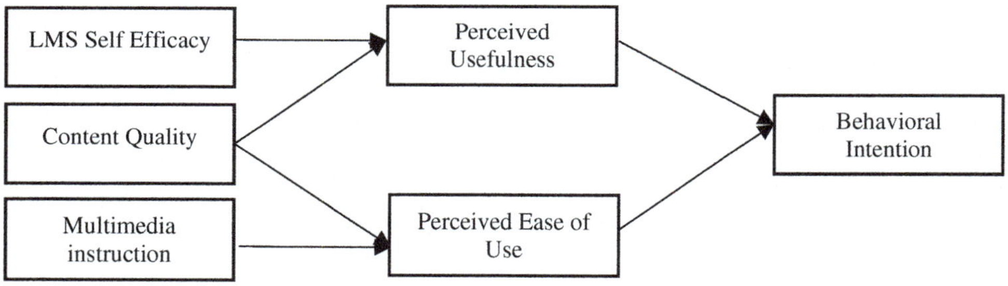

the quite opposite has happened for experienced time users (Yeop et al., 2019). Results in Table 8, are shown in detail.

On the basis of findings resulting from this analysis, an explanatory model for each user level was obtained. Arrows in Figs. 2 and 3 explain the direct influence among constructs. The three external variables multimedia instruction, content quality, and LMS self-efficacy seem to have a positive effect on perceived usefulness for new users, *indicating that an online platform that uses multimedia instruction seems to be more useful for new users*. Meanwhile, *for experienced users, perceived usefulness is affected only by content quality and LMS self-efficacy*; therefore, multimedia instructions do not play an important variable for them in using the online platform, and this result is also supported by correlation analysis that indicate that multimedia instruction and actual use have a negative correlation between them and it may come by the fact that once students are familiar with the online platform more they care for the quality of the learning materials and the efficacy of the learning management system. Related to the perceived ease of use for the new users, variables that matter most are related to the content quality and LMS self-efficacy, and the quality of online learning materials has been evaluated as the most important factor that influence both perceived usefulness and ease of use for the new and experienced users.

An important point was about behavioral intention that both perceived usefulness and ease of use for new and experienced users have a positive effect on that. Accordantly, when users believe that a system is useful and easy to use, they are more likely to use this platform only for learning purposes. This is consistent with the point proposed by Ong et al. (2004), Park (2009), Revythi and Tselios (2019), Yi and Hwang (2003), yet inconsistent with Cigdem and Topcu (2015) and Liaw (2008) that argues that only perceived usefulness affects behavior intention of the users. *A relevant result of this study is the difference that exists between new and experienced users related to the intention to use the online platform. Based on our analysis, it seems that for new users, the intention to use the e-learning platform is related to behavior intention to use the platform. Quite opposite happens with experienced users, and their behavior intention to use the platform does not have any impact on the actual use.*

The last interpretation is related with differences that exist between new and experienced users. What both have in common is the importance that users pay to the quality of learning materials. If we compare the means of all constructs, it seems to have an improvement on the evaluation of all the variables from new to experienced users. The main difference is between their perception related to the usefulness of the online learning platform and its easiness. For the new users, usefulness is related to the instructions that instructors give in the online platform, while for the experienced users it depends by the effectiveness of the learning management system. Further, with regard to perceived easiness for new users, LMS self-efficacy is the important determinant that has considerable impact on the variable, and for experienced users, multimedia instructions help them to perceive the easiness of the online learning platform.

4 Discussion

Our results show that new users' perception toward usefulness and easiness of the online learning platform affects their behavior intention and actual use of this approach. First-time users tend to utilize in a short amount of time; thus, they initially look for ease of use while getting the narrow objective done. On the contrary, experienced users' perception toward usefulness and easiness of the online learning platform affects their behavior intention to use this approach but do not affect the actual use. The science behind it tells us that the experienced user uses the instrument to get to the objective; however, its intention changes based on the content and many elements of the learning instrument.

Whether there are any differences in perception and attitudes between new users and experienced users results show that the first ones evaluate more LMS self-efficiency and ease of use. For them, LMS self-efficiency plays an important role in perceived ease of use of the platform (Aman et al., 2020). As new users, taking in consideration latest generation focus to technology and focus time span, it is arguably intuitive to follow for the self-efficacy of the learning instrument that derives from its ease of use. Quite the opposite is for experienced users; for them, content quality and multimedia instruction play an important role on perceived ease of use of the platform (Landrum, 2020; Palmer & Holt, 2010a, 2010b). This might be a result of lack of context from new users with the online platform. However, experienced users since not having the initial contextual lack of knowledge observe that in such learning environments, the two main elements that fulfill their objectives are content quality and multimedia. Research shows that these two are fundamental in the online learning environment and the findings of the study confirm it.

The similarity between new and experienced users is that both perceived usefulness and ease to use affect student's behavior intention. Thus, the more usefulness and the easier the platform and methodology stands for users, the more propensity they have to use the methodology, even when external factors that affect student's perception and attitudes of new and experienced users are different. We consider it to be of considerable interest to investigate which external factors directly influence new and experienced users' perceived usefulness and ease of use toward blended learning methodology.

5 Limitation and Future Directions

Drawing conclusions, we must consider the following limitations to this study. Students needed more explaining regarding blended learning as the notion is new for them. However, the most important limitation related with methodology and overall implications on the study is the fact that the sample of new and experienced users is not equal and also the sample size is small, thus making the recommendations cautious.

6 Conclusions

This study addressed increasingly important issues of students' perception and behavioral changes in blended learning environment. It tries to analyze student' perception and attitudes toward learning methodology but also tracking their previous experience with blended learning approach and learning management systems. Arguably, these findings will help to make more effective decisions about how to implement the right blended learning methodology. The results reported are as follows:

- Findings indicate that there are significant differences between new and experienced users on what they evaluate as important when using learning management systems. In addition, new users in terms of usefulness evaluate more content quality, multimedia instructions, and LMS self-efficiency, and on the other hand, experienced users do not find relevance in multimedia instructions.
- In terms of easiness of the online learning platform, new users evaluate as important determinant only LMS self-efficiency.
- Repetitively, we find that multimedia instruction and content quality are determinants that affect the ease of use only in case of experienced users. Behavior intention of both, new and experienced users, is affected by perceived usefulness and perceived ease of use.
- There are various differences and similarities between new and experienced users on perception and behavior toward blended learning methodology that instructors need to understand to be able to create an appropriate and effective learning environment.

Conflict of Interest StatementThe authors have no conflicts of interest to declare.All co-authors have seen and agree with the contents of the manuscript, and there is no financial interest to report.

We certify that the submission is original work and is not under review at any other publication.

Ethical StatementWe hereby, Rezart Prifti and Ana Shkreta, consciously assure that for the manuscript "Student Perception and Behavioral Changes in Blended Learning Implementation," the following is fulfilled:

1. This material is the authors' own original work, which has not been previously published elsewhere.
2. The paper is not currently being considered for publication elsewhere.
3. The paper reflects the authors' own research and analysis in a truthful and complete manner.
4. The paper properly credits the meaningful contributions of co-authors and co-researchers.
5. The results are appropriately placed in the context of prior ad existing research.
6. All sources used are properly disclosed (correct citation). Literally copying of text must be indicated as such by using quotation marks and giving proper reference.
7. All authors have been personally and actively involved in substantial work leading to the paper and will take public responsibility for its content.

The violation of the Ethical Statement rules may result in severe consequences.

We agree with the above statements and declare that this submission follows the policies of Springer as outlined in the Guide for Authors and in the Ethical Statement.

References

Abioye, A., & Adelakun-Odewale, O. S. (2017). Perceived usefulness and behavioural intention to use electronic law information resources by postgraduate law students in selected universities in South-west, Nigeria. *Ibadan Journal of Educational Studies, 14*(January/June), 6–19.

Al-Busaidi, K. A. (2012). Learners' perspective on critical factors to LMS success in blended learning: An empirical investigation. *Communications of the Association for Information Systems, 30*(1), 11–34. https://doi.org/10.17705/1cais.03002

Al-Emran, M., Arpaci, I., & Salloum, S. A. (2020). An empirical examination of continuous intention to use m-learning: An integrated model. *Education and Information Technologies, 25*(4), 2899–2918. https://doi.org/10.1007/s10639-019-10094-2

Aman, L. D. P., Sofwan, M., Mukminin, A., Habibi, A., & Yaqin, L. N. (2020). Factors affecting Indonesian pre-service teachers' use of m-LMS: A mix method study. *International Journal of Interactive Mobile Technologies, 14*(6), 137–147. https://doi.org/10.3991/ijim.v14i06.12035

Aristovnik, A., Keržič, D., Tomaževič, N., & Umek, L. (2016). Determining factors of students' perceived usefulness of e-learning in higher education. In *Proceedings of the International Conference on E-Learning, EL 2016—Part of the Multi Conference on Computer Science and Information Systems 2016* (pp. 3–10).

Azizan, F. Z. (2010). Blended learning in higher education institution in Malaysia. In *Proceedings of Regional Conference on Knowledge Integration in ICT* (Vol. 10, pp. 454–466).

Baki, R., Birgoren, B., & Aktepe, A. (2018). A meta analysis of factors affecting perceived usefulness and perceived ease of use in the adoption of E-Learning systems. *Turkish Online Journal of Distance Education, 19*(4), 4–42. https://doi.org/10.17718/tojde.471649

Boelens, R., Voet, M., & Wever, B. D. (2018). The design of blended learning in response to student diversity in higher education: Instructors' views and use of differentiated instruction in blended learning. *Computers & Education, 120*(1), 197–212.

Broadbent, J. (2017). Comparing online and blended learner's self-regulated learning strategies and academic performance. *Internet and Higher Education, 33*(September), 24–32. https://doi.org/10.1016/j.iheduc.2017.01.004

Cahya, D. K., Ciptayani, P. I., Surjono, H. D., & Priyanto. (2018). Study of instructional model on blended learning in polytechnic. *Cakrawala Pendidikan, 2*, 101–102.

Campbell, C. R. (2006). Perceptions of compressed video distance learning (DL) across location and levels of instruction in business courses. *Journal of Education for Business, 81*(3), 170–174. https://doi.org/10.3200/JOEB.81.3.170-174

Cigdem, H., & Ozturk, M. (2016a). Factors affecting student s ' behavioral intention to use LMS at a Turkish post-secondary vocational school 1 sis (SNA) in OnlineCourses. *17*(3). https://doi.org/10.19173/irrodl.v17i3.2253

Cigdem, H., & Ozturk, M. (2016b). Factors affecting students' behavioral intention to use LMS at a Turkish post-secondary vocational school 1 sis (SNA) in OnlineCourses Factors affecting students' behavioral intention to use LMS at a Turkish post-secondary vocational school Cigdem and Ozt. *International Review of Research in Open and Distributed Learning*, *17*(3). https://doi.org/10.19173/irrodl.v17i3.2253

Cigdem, H., & Topcu, A. (2015). Predictors of instructors' behavioral intention to use learning management system: A Turkish vocational college example. *Computers in Human Behavior, 52*, 22–28.

Damnjanovic, V., Jednak, S., & Mijatovic, I. (2015). Factors affecting the effectiveness and use of Moodle: Students' perception. *Interactive Learning Environments, 23*(4), 496–514.

Davis, F. D. (1989). Perceived usefulness, perceived ease of use, and user acceptance of information technology. https://doi.org/10.2307/249008

Davis, F. D., Bagozzi, R. P., & Warshaw, P. R. (1989). User acceptance of computer technology: A comparison of two theoretical models. *Management Science, 35*(8), 982–1003. https://doi.org/10.1287/mnsc.35.8.982

Dorobat, I. (2014). Models for measuring E-Learning success in universities: A literature review. *Informatica Economica, 18*(3/2014), 77–90. https://doi.org/10.12948/issn14531305/18.3.2014.07

Dziuban, C., Graham, C. R., Moskal, P. D., Norberg, A., & Sicilia, N. (2018). Blended learning: The new normal and emerging technologies. *International Journal of Educational Technology in Higher Education, 15*(1), 1–16. https://doi.org/10.1186/s41239-017-0087-5

Fishbein, M., & Ajzen, I. (1980). *Understanding attitudes and predicting social behavior*.

Ghirardini, B. (2011). *E-learning methodologies: A guide for designing and developing e-learning courses*. Food and Agriculture Organization of the United Nations.

Gómez-Rey, P., Barbera, E., & Fernández-Navarro, F. (2016). Measuring teachers and learners' perceptions of the quality of their online learning experience. *Distance Education, 37*(2), 146–163.

Hameed, S., Badii, A., & Cullen, A. J. (2008). Effective e-learning integration with traditional learning in a blended learning environment. In *European and Mediterranean Conference on Information Systems* (pp. 25–26).

Hsu, H., & Chang, Y. (2013). Extended TAM model: Impacts of convenience on acceptance and use of Moodle. *Online Submission, 3*(4), 211–218.

Kearsley, G. (2000). *Online education: Learning and teaching in cyberspace*. Wadsworth Publishing Company.

Kent, C., Laslo, E., & Rafaeli, S. (2016). Interactivity in online discussions and learning outcomes. *Computers and Education, 97*, 116–128. https://doi.org/10.1016/j.compedu.2016.03.002

Landrum, B. (2020). Examining students' confidence to learn online, self-regulation skills and perceptions of satisfaction and usefulness of online classes. *Online Learning Journal*, *24*(3), 128–146. https://doi.org/10.24059/olj.v24i3.2066

Lau, S. H., & Woods, P. C. (2009a). Understanding learner acceptance of learning objects: The roles of learning object characteristics and individual differences. *British Journal of Educational Technology, 40*(6), 1059–1075. https://doi.org/10.1111/j.1467-8535.2008.00893.x

Lau, S. H., & Woods, P. C. (2009b). Understanding the behavior changes in belief and attitude among experienced and inexperienced learning object users. *Computers and Education, 52*(2), 333–342. https://doi.org/10.1016/j.compedu.2008.09.002

Lear, J., Ansourge, C., & Steckelberg, A. (2010). Interactivity/community process model for the online education environment. ... *Online Learning and Teaching*, *6*(1), 71–77.

Lee, L. T., & Hung, J. C. (2015). Effects of blended e-Learning: A case study in higher education tax learning setting. *Human-Centric Computing and Information Sciences*, *5*(1). https://doi.org/10.1186/s13673-015-0024-3

Lee, Y.-H., Hsieh, Y.-C., & Hsu, C.-N. (2011). Adding innovation diffusion theory to the technology acceptance model: Supporting employees' intentions to use e-learning systems. *American Economic Journal: Applied Economics, 14*(4), 124–137. https://doi.org/10.1257/app.3.4.i

Liaw, S. (2004). Considerations for developing constructivist web-based learning. *International Journal of Instructional Media, 31*(3), 309–321.

Liaw, S., & Huang, H. (2011). A study of investigating learners attitudes toward e-learning. *12*, 28–32.

Liaw, S.-S. (2008). Investigating students' perceived satisfaction, behavioral intention, and effectiveness of e-learning: A case study of the Blackboard system. *Computers & Education, 51*(2), 864–873.

Liaw, S.-S., & Huang, H.-M. (2013). Perceived satisfaction, perceived usefulness and interactive learning environments as predictors to self-regulation in e-learning environments. *Computers & Education, 60*(1), 14–24.

Martín-García, A. V., Martínez-Abad, F., & Reyes-González, D. (2019). TAM and stages of adoption of blended learning in higher education by application of data mining techniques. *British Journal of Educational Technology, 50*(5), 2484–2500.

Mbengo, P. (2014). E-learning adoption by lecturers in selected zimbabwe state universities: An application of technology acceptance model. *Journal of Business Administration and Education, 6*(1), 15–33.

Moritani, M., Fujimitsu, Y., Shinohara, M., Niido, T., Tsuchida, A., Aoki, T., & Koyanagi, Y. (2001). A case of AFP producing early gastric cancer successfully treated with small dose CDDP and 5-FU (PF) therapy. *Gan to kagaku ryoho. Cancer & Chemotherapy, 28*(13).

Moskal, P., Dziuban, C., & Hartman, J. (2013). Blended learning: A dangerous idea? *Internet and Higher Education, 18*, 15–23. https://doi.org/10.1016/j.iheduc.2012.12.001

Norberg, A., Dziuban, C. D., & Moskal, P. D. (2011). A time-based blended learning model. *On the Horizon, 19*(3), 207–216. https://doi.org/10.1108/10748121111163913

Oh, E., & Park, S. (2009). How are universities involved in blended instruction? *Educational Technology and Society, 12*(3), 327–342.

Ong, C., Lai, J., & Wang, Y. (2004). Factors affecting engineers ' acceptance of asynchronous e-learning systems in high-tech companies. *41*, 795–804.https://doi.org/10.1016/j.im.2003.08.012

Ouadoud, M., Nejjari, A., Chkouri, M. Y., & El Kadiri, K. E. (2018). Educational modeling of a learning management system. In *Proceedings of 2017 International Conference on Electrical and Information Technologies, ICEIT 2017, 2018-Janua* (pp. 1–6). https://doi.org/10.1109/EITech.2017.8255247

Padilla-meléndez, A., Aguila-obra, A. R., & Garrido-moreno, A. (2013). Perceived playfulness, gender differences and technology acceptance model in a blended learning scenario. *Computers & Education, 63*, 306–317.

Palmer, S., & Holt, D. (2010a). Students' perceptions of the value of the elements of an online learning environment: Looking back in moving forward. *World Academy of Science, Engineering and Technology (WASET)*, 18(2), 135–151.

Park, C. W., Kim, D. G, Cho, S., & Han, H. J. (2019). Adoption of multimedia technology for learning and gender difference. *Computers in Human Behavior*, *92*(December), 288–296.https://doi.org/10.1016/j.chb.2018.11.029

Park, S. Y. (2009). An analysis of the technology acceptance model in understanding university students' behavioral intention to use e-learning. *Journal of Educational Technology & Society, 12*(3), 150–162.

Pituch, K. A., & Lee, Y. (2006). The influence of system characteristics on e-learning use. *47*(March 2002), 222–244. https://doi.org/10.1016/j.compedu.2004.10.007

Revythi, A., & Tselios, N. (2019). Extension of technology acceptance model by using system usability scale to assess behavioral intention to use e-learning. *Education and Information Technologies, 24*(4), 2341–2355. https://doi.org/10.1007/s10639-019-09869-4

Rittgen, P. (2010). Quality and perceived usefulness of process models. In *Proceedings of the ACM Symposium on Applied Computing* (pp. 65–72).https://doi.org/10.1145/1774088.1774105

Ross, B., & Gage, K. (2006). Global perspectives on blending learning. In J. C. Bonk, R. C. Graham (Eds.), *The Handbook of Blended Learning* (pp. 155–168).

Sethy, S. S. (2008). Distance education in the age of globalization: An overwhelming desire towards blended learning. *Turkish Online Journal of Distance Education, 9*(3), 29–44. https://doi.org/10.17718/tojde.04460

Staker, H., & Horn, M. B. (2012). Classifying K—12 blended learning. *Innosight Institute*. https://doi.org/10.1007/s10639-007-9037-5

Thai, N. T. T., De Wever, B., & Valcke, M. (2017). The impact of a flipped classroom design on learning performance in higher education: Looking for the best "blend" of lectures and guiding questions with feedback. *Computers and Education, 107*, 113–126. https://doi.org/10.1016/j.compedu.2017.01.003

Tselios, N., Daskalakis, S., & Papadopoulou, M. (2011). Assessing the acceptance of a blended learning university course. *1996*.

Tutty, J. I., & Martin, F. (2008). Learning management system self-efficacy of online and hybrid learners in instructional technology. *Citeseer*.

Vonderwell, S., & Zachariah, S. (2005). Factors that influence participation in online learning. *Journal of Research on Technology in Education, 38*(2), 213–230. https://doi.org/10.1080/15391523.2005.10782457

Willemsen, L. M., Neijens, P. C., Bronner, F., & de Ridder, J. A. (2011). "Highly recommended!" The content characteristics and perceived usefulness of online consumer reviews. *Journal of Computer-Mediated Communication, 17*(1), 19–38. https://doi.org/10.1111/j.1083-6101.2011.01551.x

Wong, K. T., Hamzah, M. S. G., Goh, P. S. C., & Yeop, M. A. B. (2017). Blended E-learning acceptance as smart pedagogical tools: An initial study in Malaysia. *Turkish Online Journal of Educational Technology, 2017*(November Special Issue IETC), 115–122.

Wu, J. H., Tennyson, R. D., & Hsia, T. L. (2010). A study of student satisfaction in a blended e-learning system environment. *Computers and Education, 55*(1), 155–164. https://doi.org/10.1016/j.compedu.2009.12.012

Yeop, M. A., Yaakob, M. F. M., Wong, K. T., Don, Y., & Zain, F. M. (2019). Implementation of ICT policy (blended learning approach): Investigating factors of behavioural intention and use behaviour. *International Journal of Instruction, 12*(1), 767–782. https://doi.org/10.29333/iji.2019.12149a

Yi, M. Y., & Hwang, Y. (2003). Predicting the use of web-based information systems: Self-efficacy, enjoyment, learning goal orientation, and the technology acceptance model. *59*, 431–449. https://doi.org/10.1016/S1071-5819(03)00114-9

Zain, F. M., Sahimi, S. M., & Hana, E. (2016). *Envisioning the future of online learning, October 2019s*. https://doi.org/10.1007/978-981-10-0954-9

Technological Advancements and Sustainability-Based Innovations in Creative Industries

The Collaboration Between Academic and Industry in Creative Industry and Sustainability-Based Programs: The Academic Perspective

Astrid Kusumowidagdo and Marini Yunita Tanzil

Abstract

This research aimed to examine various aspects affecting the collaboration between academics and industries, specifically creative industries as well as sustainability-based programs in Indonesia. It explored collaboration motives, activities, processes, main actors, issues, benefits, and expected results. The research also primarily emphasized sustainability-based activities, that is, how to be financially sustainable and efficient for the environment as well as the community, specifically in the realm of creative industries. Data were collected through various qualitative research methods, such as the literature, observation, and focus group. The research participants were a total of 11 deans who led creative industries-based learning programs and owned sustainability-based programs, as well as seven industry partners in Indonesia. The results showed that collaboration motives, such as academic development, students and teaching competence, as well as the awareness of industries, were the focus and existence of the participants. The projects or activities performed included design workshops and competitions, employment, on-site opportunities for students and faculty, corporate research, professional conferences, corporate grants, philanthropic donation, as well as an advisory board. The collaboration process generally involved face-to-face dialog, trust building, commitment, shared understanding, as well as seeking the public's target and local actor. This relationship could provide great benefits for both academics and industries. The urgency of this research was to develop Indonesian education based on real experience and creative industry actors.

Keywords

Academic collaboration · Industry support · Sustainability · Creative industries

1 Introduction

The General Assembly of the United Nations declared 2021 as the International Year of Creative Economy for Sustainable Development (Cruz & Teixeira, 2021). This emphasizes the need to foster the internal aspects, promotion plans, and culture preservation. The United Nations explained that sustainable development goals included zero poverty and hunger, good health and well-being, quality education, gender equality, clean water and sanitation, affordable and clean energy, climate action among others (Fonseca et al., 2020; Harper, 2021). Also, based on the 2030 agenda, they believed that creative industries, in collaboration with other economic sectors, can play a significant role in achieving the mission and plan of "Transforming Our World" (Wiktor-Mach, 2020).

To confront external challenges, education needs to be strongly related to industries and strive for a better world. Academics are relevant to the public's needs as they produce graduates who can work optimally. Creative industries are faced with several challenges, such as the effort to apply sustainability, remote work, and technology, among others. Therefore, collaboration has emerged as one of the answers to becoming creative in the working world (Lucietto et al., 2021; Sjöö & Hellström, 2019).

Education for the creative industries is expected to observe the external challenges and strive for an answer through the learning process offered in universities.

A. Kusumowidagdo (✉)
Architecture Department, School of Creative Industry, Universitas Ciputra Surabaya, Surabaya, Indonesia
e-mail: astrid@ciputra.ac.id

M. Y. Tanzil
Fashion Product Design and Business, School of Creative Industry, Universitas Ciputra Surabaya, Surabaya, Indonesia
e-mail: Marini.yunita@ciputra.ac.id

M. Nawaz Tunio et al. (eds.), *Sustainability in Creative Industries*, Advances in Science, Technology & Innovation, https://doi.org/10.1007/978-3-031-48453-7_14

Collaboration is one of the alternatives used for solving this problem. Camacho and Alexandre (2019); Jones and Coates (2020); Kuys et al. (2021) investigated industries and their relationship with academics. Furthermore, Arshed et al. (2022), Fernandes and O'Sullivan (2021), Fernandes et al. (2020), Marinho et al. (2020) discussed crucial indicators regarding project management, obtaining benefits, measuring collaboration, as well as highlighting the cause of the failure in collaboration between the industry and academic sectors.

This research focuses on creative industries based on successful spelling in the indicator process and aims to explore all aspects related to the optimization of successful project work between academics and industries. This success emanates from various parties due to the nature of multi-stakeholders' collaboration. Collaboration enables the accessibility and affordability of technology, institution, education, research qualities, as well as financial service and production cost because it tends to satisfy all parties when the process is well executed (Arshed et al., 2022).

To ensure the smooth operation of the process, several aspects need to be considered to create a bond regarding the benefits and motives between the industries and academics. A good understanding among the involved parties can lead to a successful process and optimized results. Chung et al. (2021) stated that several aspects were needed to examine a perfect partner. Issues, such as initiation and planning in the internal, project, execution, closing, monitoring, and control phase, also need to be considered (Bertello et al., 2022). According to Roberts (2007), relationship activities between academics and industries include classroom learning, design workshop, design competition and expos, site visits, internship and working opportunities, research, community service, a new invention for research methods and titles, industry grant, as well as advisory teams.

In the implementation phase, collaboration usually involves initiators, such as lecturers and students from the academic world, as well as owners, directors, management, and staff from the industry sector. Design teams and other experts play a crucial strengthening the creative industries (Belfiore, 2020; Roberts, 2007). Collaboration promotes the growth of knowledge, technology, and skill as well as facilitates industry transformation to survive competition (Fernandes et al., 2020). It can also promote customer retention and profits while ensuring the industries achieve innovative ideas (Arshed et al., 2022). On the academic side, collaboration serves as a bridge, connecting academics with practitioners. This can promote research quality and linkages with the real world (Austin et al., 2021; Burova et al., 2022).

According to Roberts (2007), the expected outcome (achievement) encompasses several aspects. Firstly, it involves widening collaboration on a local and global scale. Secondly, it involves establishing a designer framework for lecturers, practitioners, and professional associations in order to obtain proportional achievement befitting their roles. Thirdly, it aims to promote diversity in the design community. Students and academic teams need to train their thinking with societal problems and be well connected. Lastly, new models and creative synergy are achieved to produce joint activities based on sustainability principles. This research further explores the collaboration between academics and industries as well as focuses on creative industries and universities based on the sub-sector competency. It also aims to identify the collaboration model between them by focusing on sustainability-based activities. Therefore, the question is how to find the right collaboration model with detailed parameters including collaboration motives, processes, activities, acting actors, issues, benefits, and results.

The novelty of this research lies in the context of creative industries in Indonesia and the scope of design education. Moreover, the value is to provide input for driving the collaboration of academics and industries, which will be especially important for direct interaction in finding opportunities and problem-solving in the field.

2 Research Methods

This research was conducted on the structural academics in Java and Bali, Indonesia, as well as industry partners from May to October 2022, using both online and off-line models. Data were collected using various qualitative research methods, such as the literature research, focus group discussion, and observation.

According to Akyıldız and Ahmed (2021), the use of focus groups offers several benefits, such as the collection of data from various participants. A total of 11 people were selected from universities, and seven people were selected from industries in the national level for this discussion (Akyıldız & Ahmed, 2021; Barrett & Twycross, 2018; Sim & Waterfield, 2019). The results of the focus group were documented in the discussion forum.

The submitted questions focused on the collaboration process, activities, acting actors, constraints, benefits, and results as shown in Table 1. This was to address research questions that aim to identify the right collaboration model, with detailed parameters, such as collaboration motives, processes, activities, acting actors, issues, benefits, and results as stated in the introduction. Generally, the questions submitted are shown in Table 1.

The results of the focus group showed triangulated sources from experts in social research as seen in Figs. 1 and 2. These include the literature research, data sources from participants, experts, and field observation.

Table 1 Questions of focus group discussion

No	Topic	Questions
1	Collaboration motives	What is the motive of collaboration? How does collaboration affect industries? How does collaboration affect academics?
2	Collaboration process	How to carry out a good collaboration process? What are the factors that encourage collaboration processes for both academics and industries in terms of creative industries?
3	Collaboration activities	What sort of activities are usually carried out with the industry partners? What sort of activities are usually carried out with academic partners?
4	Acting actors	Who plays an important role in the success of this project from the academic partners' side? Who plays an important role in the success of this project from the industry partners' side?
5	Issues	Mention and explain the issues that arise when implementing collaboration projects from the industry side Mention and explain the issues that arise when implementing collaboration projects from the academic side
6	Benefit	What are the benefits of this collaboration?
7	Results	What is the outcome of this collaboration?

Fig. 1 Focus group discussion through zoom meeting. *Source* private data

In detail, research informants from universities and industries are presented as follows: The 11 deans, as research informants from universities, engaged in Indonesian creative industries' education as seen in Table 2.

Table 3 shows the seven informants from industries, and they also gave their opinions on the focus group discussion event.

3 Results and Discussions

3.1 Results

This section presents the focus group discussion between academic participants and creative industries. The discoveries displayed include the collaboration motives, processes, activities, acting actors, issues, benefits, and results. The discussion aspect is explained below.

3.1.1 Collaboration Motives

Collaboration motives, such as academic development, students and teaching competence, as well as the awareness of the industries, were the focus of the participants in this research. Some opinions in the focus group discussion process are as follows:

When seeking for collaboration, there should be efforts to advance both parties, specifically to improve the industries and academics. The main goal is to improve course materials, student and faculty experiences, as well as provide support for the industries (IVG, F, Maranatha Christian University).

We, from the industry welcome this collaboration by providing innovative ideas and contributions to education. Some activities are indeed part of corporate social responsibility, aimed at improving the conditions of Indonesian society (AA. PT Gistex).

Fig. 2 Field observation in Jakarta (Trisakti University) and fashion hub supported by industries. *Source* private data

Table 2 Informants from the faculty of design and creative industries

No	Universities	Informants
1	Trisakti University	Dr. Sangayu Ketut Laksemi, M.Ds
2	Indonesia Institute of Arts, Denpasar	Dr. AA Gede Bagus Udayana
3	Telkom University, Bandung	Dr. Roro Retno Wulan, S.Sos, M.Pd
4	Maranatha Christian University	Irena V Gunawan, ST., M.Com
5	Indonesia Institute of Arts, Yogyakarta	Dr. Timbul Rahardjo, M.Hum
6	Bandung Institute of Technology	Dr. Agus Sachari, M.Sn
7	Indonesian Institute of Informatics	Eddy Triswanto ST, M.Kom
8	Jakarta Institute of Arts	Dr. Indah Tjahjawulan
9	Pelita Harapan University	Dr. Martin Katoppo, ST, MT
10	Sepuluh Nopember Institute of Technology	Imam Baihaqi, ST, M.Sc. Ph.D
11	Tarumanagara University	Dr. Kurnia Setiawan, S.Sn, M.Hum

Table 3 Informants from industries

No	Industries	Business fields
1	PT Gistex, Mr. Anto Adiguna	Textile industry
2	Nippon Paint, Mrs. Alun Hutoyo	Residential materials/paint business
3	Ciputra World Hotel Surabaya, Jeffry Ferbrianto	Hotel business
4	Ciputra Group, Mrs. Mieke Wibisono	Property business
5	Teh Villa, Mr. Ronald Sitolang	Packaged food/tea business
6	PT Karyamitra Budi Sentosa, Mrs. Agnes Widjaja	Garment and retail business
7	Blum, Mr. Bambang Jatmanto	Interiors business

The motives concluded from the discussion are as follows:

- Improvement in Education and Industries
  The parties involved in the group discussion believe education is a great way to develop public welfare. However, the improvement in output and graduates' quality, as well as resources, are usually based on corporate achievement, profits, and the superiority of long-term competition.
- Teaching Resource Competency
  People need to consider returning to the industries in the form of internships and sabbaticals. In the academic world, fresh ideas contribute to the development of industries.
- Awareness Toward Industries
  This awareness enables students to properly adapt to challenges in the fields and also allows industries to gain exposure from their stakeholders in future.
- Contribution Toward the Sustainability Aspect
  Participants are required to consider sustainability principles and their three pillars, namely integrated manner, economic development, and social and environment. These pillars should be considered and emphasized as the main mission of collaboration projects.
- Community Transformation
  Penta helix activities only contribute to community development, while projects promote community development, engagement, confidence, empowerment, a sense of responsibility, belonging, as well as social cohesion. The growth in physical, social, and economic tends to have a huge impact on life quality.

3.1.2 Collaboration Process

Education sectors tend to have a good collaboration that includes immediate outcome, face-to-face dialogue, trust building, commitment, joint understanding, as well as the pursuit of public target and local actors (Austin et al., 2021).

Some opinions from informants are provided below:

This collaboration requires a strong commitment from both parties to run smoothly. Clear milestones and detailed work can promote collaborative work (AH, F, NP).

Finding the target community and recognizing local actors are important, hence our contribution will truly benefit the surrounding environment (MK, M, Pelita Harapan University).

The summary of the focus group discussion is as follows:

- Immediate Outcome
  A good collaboration accompanied by a successful job tends to produce great acceptable outcomes which satisfy the project actors and their stakeholders. In a project, it is important to assess how the outcome manages to fulfill the performance indicator.
- Face-to-face Dialogue
  Face-to-face dialogue is necessary because it helps to build joint agreements. It can be effectively run by conducting on-site, off-line, and scheduled meetings. Moreover, online tools can be used to record its progress and outcomes.
- Trust Building
  Trust building is necessary when initiating a Penta helix project. However, a written agreement significantly aids all involved parties. Trust building is based on the equal profit that tends to be achieved by every actor of Penta helix.
- Commitment Toward Collaboration Process
  According to each of the job's scope with the agreed milestone, commitment toward the collaboration process is the will and capability in carrying out the activity. There are intense and routine involvements to optimally run the projects. Goal milestone, resources, and sustainable funds are the main factors that promote commitment toward collaboration.
- Joint Understanding
  This includes having a vision and mission, targets and common goals, sustainable milestones, avoiding conflict of interest, as well as orienting on common goals. The understanding process is usually implemented before the collaboration project. This helps the people to become aware of external ownership and consider intellectual work as registered.
- Finding Target Community and Local Actor Participants
  It is appropriate to select a target community befitting the project's goal. Local actors significantly contribute to the success of the activity. However, a sense of belonging to the project provides an opportunity for other actors to integrate effectively.

3.1.3 Collaboration Activities

In order to achieve successful collaboration, lecturers work on classroom learning, short-term workshops, design competition as well as provide on-site opportunities for students and faculty. This is explained below.

Our collaboration usually involves classes that have been given joint project briefs. This is followed by a visit to the industry to gain product knowledge. Together, we determine the objectives and activities to be carried out (ET, M, I).

We can also carry out commercial projects together, create mockups of some of our products, and develop them into new innovations. Therefore, we have performed several corporate sponsored research projects (MW, F, CG).

Various opinions in the focus group discussion are outlined in the conclusions below:

- Classroom Learning
  In collaboration with industries, experts or guest lecturers are invited for additional class activities. This helps industries and academics to gain an audience regarding product knowledge update.
- Sustainability-Based Short-Term Workshops
  Applicable practical workshops can be conducted in partnership with the students. The industries and experts are the strength of this activity as they help to understand the process of real problem-solving.
- Sustainable Design Competition
  Design competitions with interesting prizes are one of the activities performed in terms of processing industry materials.
- Student and Faculty On-site Opportunity
  A short-term internship and permanent visit to the plants, workshops, showrooms, and offices make the academics understand the real world. Industries need to obtain proper manpower, while academics have to review the input processes.
- Corporate-Sponsored Research Projects
  Industries are required to fund sustainable or outsource research projects to the academic world. Fresh ideas are appreciated and realized practically because they serve as the best solutions for the development of new products and business processes. This sustainable direction needs to be in line with the vision and mission of industries.
- Professional Conferences and Community Organization
  Industries and academics can cultivate conferences or initiate other activities for the benefit of the community. Industries normally gain promotion, while academics actualize science in the real world.
- Advisory Board for Academic
  Industries serve as academic boards that donate corporate grants for university development. Their involvement is in the form of corporate social responsibility and philanthropic activities.
- Sustainability Aspect Conducted in a Collaboration Project
  The project process involves sustainability aspects in planning, selecting materials, working, and maintenance.

3.1.4 Issues

Common issues identified during the implementation of collaboration include program continuity, innovations, subpar, lack of communication, and change of plans. Some opinions regarding the focus group discussion process are as follows:

The continuation of programs is a crucial aspect to be considered as many good programs often fail to continue. Therefore, a long and clear roadmap should be immediately considered, at least for the next 10–20 years, to ensure the collaboration becomes more meaningful (MK, M, Pelita Harapan University).

While changes in a project are sometimes normal, caution should be exercised due to force majeure problems, such as natural disasters or even COVID-19 disease (RS, M, TV).

Based on the focus group discussion, the following issues are identified during the collaboration:

- Program Continuity
  Projects typically have a specific time frame; hence, it is necessary to establish a long-term milestone in maintaining sustainability, and provide a target for each milestone.
- Innovations
  A long-term project needs to always innovate at every step of the milestone. The orientation toward sustainability also needs to be maintained. In order to address relevant challenges, scheduled monitoring and evaluation are intended to favor well-implemented innovations on both products and business processes.
- Subpar Implementation
  Field situations, such as catastrophes and Covid-19, often pose obstacles for optimum realization and force majeure. Therefore, additional efforts are required to achieve optimum outcomes.
- Lack of Communication
  Lack of communication usually arises where personnel are tightly occupied and areas without direct access to a network. The use of integrated communication services needs to be considered from the outset to ensure the project runs smoothly.
- Change of Plans
  Collaboration actors need to be flexible and plan the adjustment in achieving the maximum outputs. This is because there tends to be a change of plans, which is addressed through joint understanding and communication.
- Lack of Sustainability Aspect
  Sustainability-based planning encounters numerous challenges that hinder its effective operation. Therefore, strong commitment is required to discover the solution-based problem.

3.1.5 Benefits of Collaboration for Industries

Industries aid credibility, the planning process, and the creative players in developing their knowledge and practices. Some opinions in the focus group discussion are as follows:

We have received ample input and feedback from our joint training and programs, and each program has provided valuable insights. Therefore, the academic team has had the opportunity to learn about the industry sector to a greater extent (AAGBU, M, ID).

Based on detailed discussion, the following conclusions can be drawn:

- Improving Credibility by Collaborating with the Academics
  Industries provide sustainability and access to new knowledge, research, and unique insights, while collaborating with academics.
- Aiding the Planning Process and the Activity Implementation
  The academics assist in the planning and activity development as well as monitoring of the application.
- Aiding the Creative Players in Developing Knowledge and Practices
  The academics extensively explore trends and provide new knowledge aside the products and old patterns. They also help to recommend new working procedures in practice.

3.1.6 Collaboration Benefits for Academics

Based on the discussion of the focus group below, collaboration with the industry team provides valuable insights for the academic team:

As a team from the university, we have experienced several benefits from collaborating with industries. Through these collaborations, we can have access to real-life cases as well as provide new solutions and innovations. Students not only receive royalties but are fully financially supported throughout the working process. There are also other results in the form of intellectual property rights (AAGBU, M, ID).

Of course, we are delighted to collaborate with industries. There is funding for research and projects aimed at producing joint innovations (RRW, F, Telkom University, Bandung).

Based on various opinions in the focus group discussion, some benefits for the universities are concluded as follows:

- Bringing real experiences to universities cases.
  Industries bring field experience, issues, and new trends to enrich cases that need to be solved. This enables students to face similar cases and turn them into relevant learning.
  In the industry sector, challenges help to explore new research topics or ideas, thereby enriching the existing conditions in universities. New research has the potential to result in innovations.
- Receiving funding from industries
  Financial support from industries helps academics to creatively develop learning products and processes. This funding is often combined with corporate social responsibility.

3.1.7 Academic Actors

Academic actors can come from various parties as discussed in the forum:

The initiator can be bottom-up or top-down, as long as there is a project problem that can be worked on, specifically depending on the sustainability aspect in the creative industry sector. For example, bottom-up can come from the initiative of lecturers or students, while top-down can come from university officials, such as rectors or deans (IVG, F, Maranatha Christian University).

The following conclusions can be drawn from various discussions:

The main actors include rectors, vice-rectors, deans, and heads of departments. Usually, lecturers, students, and alumni with specific expertise are involved. These findings are new and more detailed. Previous research generally stated that academic actors are academic leaders.

3.1.8 Industry Actors

Some opinions from the discussion are as follows:

In the industry sector, the key actors are usually from the managerial team and above because they have a clear understanding of the direction, goals, and milestones. However, the project can be accepted and carried out by anyone. This workflow is common in the industries (AW, F, KMBS).

It can be concluded from various discussions that collaboration in the industries is initiated by owners, managers, advisory boards, as well as other investors (BJ, M, B).

The previous process usually involves a Memorandum of Understanding and a Memorandum of Agreement (BJ, M, B).

3.1.9 Some Examples of Collaboration Between Indonesian Universities and Industries

Examples of collaboration and processes carried out to illustrate the collaboration of academics and industries are as follows:

- **Industry collaboration in managing waste between Ciputra University and PT Karya Mitra Bersama Sentosa**
  PT Karya Mitra Bersama Sentosa (KMBS) is a shoe and bag manufacturing industry that supplies well-known brands worldwide and generates a significant amount of waste from leather materials. In order to improve both the industry and academia, PT KMBS entrusted leather waste management to Ciputra University. The initiators from the university were lecturers and heads of the study program, while the initiator from the industry was the management team. The project was carried out with a clear road map and involved several community groups. As a result, students produced several prototypes and

received royalties as well as intellectual property, which were marketed by PT KMBS. Students and the industry developed creative industry products based on sustainability concepts for the environment, community, and business.

- **Collaboration to increase knowledge between Maranatha Christian University (Bandung) and suppliers of building materials**
  In the design world, knowledge about materials is essential, specifically for environmentally friendly and adaptable materials. Through collaboration with industries, the Interior Design Study Program at Maranatha Christian University organizes a product knowledge program for students. However, students can visit material industries to gain real knowledge about sustainable materials for use in the field.
- **Collaboration in an art exhibition between the education world, artists, and creative industries, initiated by the Indonesia Institute of Arts, Yogyakarta**
  Pakuwon Mall, a property industry, initiated a joint exhibition program to accommodate artists and creative industry actors. The event was expected to accommodate creativity as well as mobilize the potential of Indonesia's creative industries post-prolonged Covid-19. It was jointly initiated by academic parties as mediators and well received by artists as well as creative industry figures. The exhibition went smoothly with the title Visual Arts and Creative Industry Collaboration.
- **Collaboration between the Sepuluh Nopember Institute of Technology, Surabaya, for the application of 3D printer-based digital methods in pre- and post-surgery planning**
  Collaboration has been established for the development of medical research applications between a laboratory and the industry product design study program at the Faculty of Design and Creative Business. The collaboration is carried out with Airlangga Hospital to leverage the educational potential in the design field, specifically in the realm of Technology 4.0. A 3D printer can be used to facilitate pre- and post-surgery processes for modeling.

3.2 Discussion

3.2.1 Model of Collaboration

Based on the various findings, focus group discussions, and activities carried out, the model is summarized in Fig. 3:

It is necessary to consider the implications of inventions by highlighting the linkages and contributions to the progress of creative industries as seen in Fig. 3. In order to strengthen collaboration, various methods are used to accelerate the process and solve existing issues, widen the activity range, promote industry and academic actors, as well as initiate projects.

3.2.2 Discussion on the Model Findings

The findings on the theme of collaboration motives and processes are in line with Kinsella (2018) and Snepvangers et al. (2018). A creative partnership with other sectors can lead to creative teaching and learning. In terms of collaboration activities, the findings support previous research, such as Roberts (2007), Fernandes and O'Sullivan (2021). In collaboration issues, several findings, such as innovation, program continuity, lack of communication, and change of plan, are in line with Muscio and Vallanti (2014), Marinho et al. (2020), Bertello et al. (2022). Meanwhile, on the theme of benefits for industries and academics, the findings are in line with Roberts (2007), Fernandes and O'Sullivan (2021). The overall findings support previous research as shown in Table 4.

Some of the new findings in this research, as seen in Table 4, showed that the model discovered was very specific to Indonesia, specifically in the context of sustainability and creative industries. On the theme of issues, there were other notable findings, namely subpar implementation and lack of sustainability. Focusing on both industry and academic actors is a new phenomenon in research of this kind.

Important factors, such as outputs of good activities, monitored process and collaboration with the actors, paying attention to sustainability aspect and its process, as well as clear milestones to achieve desired goals, are observed from the motivation side. It is crucial to exhibit empathy for local conditions and the participation of the local actors.

The expansion of collaboration needs to be explored with technology-based innovation output (Austin et al., 2021; Jones & Coates, 2020; Rybnicek & Königsgruber, 2019), intellectual property right arrangement, and collaboration experience (Sjöö & Hellström, 2019), as well as local cultural aspects (Arshed et al., 2022).

New tools are employed with the development of technology in hardware, software, virtual reality, augmented reality, and suitable applications. Digital collaboration is one of the best ways for efficient and optimized uplift, as it enables the union of several industries and academics. This concept not only solves communication issues but also enables the platform to produce new products, systems, and complementary products.

Intellectual property rights are also an aspect that should be further discussed, specifically for findings with commercial potential. It is crucial to explore collaboration experiences and identify suitable instruments as good parameters to meet the needs of academics and industries. Lastly, culture is also a consideration due to the different

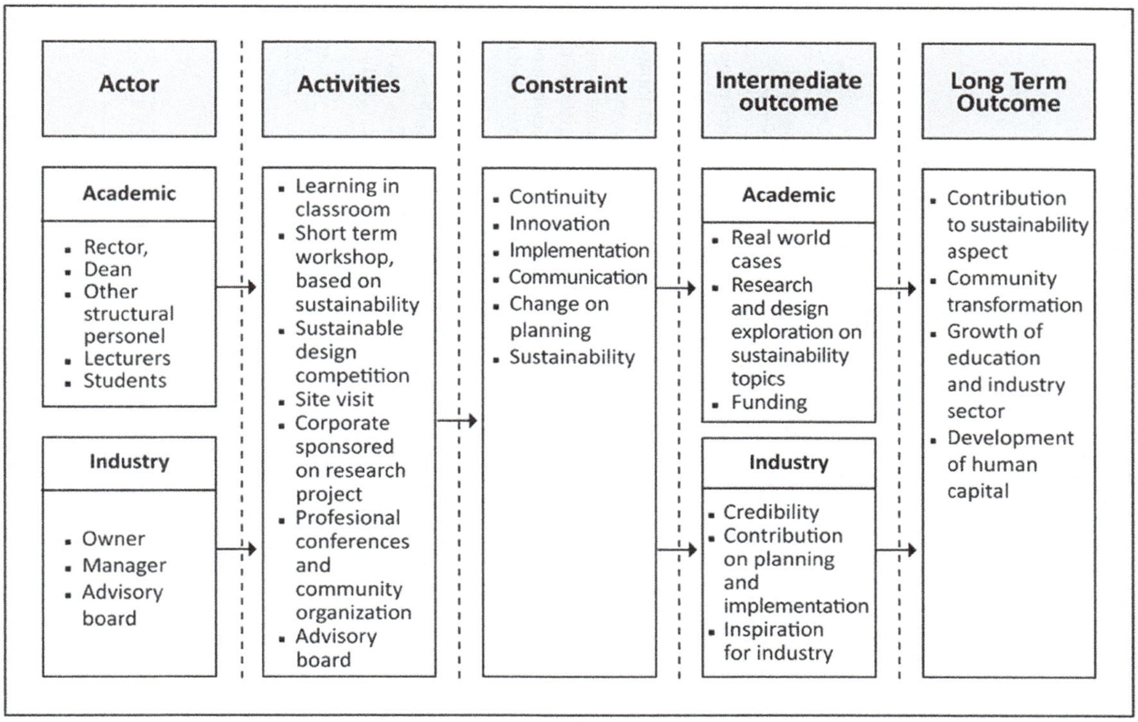

Fig. 3 A systematic explanation of the collaboration. *Source* private data

nature between education and industries as well as the existence of local and spatial characters that might influence collaboration.

Several issues need to be addressed. Firstly, continuity should be handled with great planning and collaboration project management. The continuity of innovation is vital to ensure the projects remain relevant and offer additional values. Secondly, changes in plans should be based on sustainability work to ensure the outcome remains accountable. Underwhelming implementations can be achieved by giving additional activities and corrective processes in the works. Currently, issues are solved through technologies used in various departments. Thirdly, a structured reward system tends to promote good collaboration and improve desired outcomes. Lastly, the lack of flexible sustainability and process needs to be considered during the planning.

4 Conclusion

This research addressed several crucial points, including collaboration processes, activities, actors, issues, benefits, and outcomes, with a focus on managing collaboration programs between industries and academics in Indonesia. In answering the research question, each aspect has been clearly described with the appropriate parameters. The parameters of collaboration activities include learning in the classroom, sustainability, sustainable design competition, site visit, sponsored research project, conference and community organization, advisory board. The industries' actors that played a major role were owner, managers, and advisory board; while the academic actors were rectors, dean, other structural personal, lecturers, and dean. The most frequently encountered issues are the continuity, innovation, implementation, communication, change on planning, and sustainability, and the benefits often found include intermediate and long-term outcome.

This mapping is the first related sustainability-based programs and Indonesian creative industries. These aspects can serve as parameters in future, enabling smoother collaboration management and a clearer road map. The value of this research lies in providing information about important aspects of joint collaboration that can support various programs from academics, industries, and even government institutions, such as the Ministry of Education and Culture as well as the Ministry of Tourism and Creative Economy teams.

This research shows novelty as its implementation focuses on the relationship between academics and industries. Further literature is recommended to employ different methods, such as quantitative surveys and a significant number of participants. The method implemented in this research is directed to different perspectives, such as academics and governments.

Table 4 Alignment between previous findings, theories, and new findings

Points	Actors from academics and industries	Activities	Constraints/issues	Intermediate outcome	Long-term outcome
Result	Rector, deans, other structural personnel, lecturers, students Owner, manager, advisory board	Learning in the classroom Short-term workshop Sustainable competition Site visit Corporate research project Professional conference and community organization Advisory board	Continuity Innovation Communication Change on planning	Academic: Real-world cases Research and design exploration Funding Industry: Credibility Contribution to planning and implementation	Contribution to sustainability Community Transformation Growth of education and industry sector Development of human capital
Supporting literature	–	Kinsella (2018), Snepvangers et al. (2018)	Muscio and Vallanti (2014), Marinho et al. (2020), Bertello et al. (2022)	Roberts (2007), Fernandes and O'Sullivan (2021), Lucietto et al. (2021)	Roberts (2007), Fernandes and O'Sullivan (2021), Lucietto et al. (2021)
New findings	Actors from Indonesian industries and academics	–	Subpar implementation, lack of sustainability	–	–

Acknowledgements The authors are grateful to the Ministry of Education and Culture for the The Kedaireka Program (Directorate General of Higher Education) as well as the Ministry of Tourism and Creative Economy for the internal funding.

References

Akyıldız, S. T., & Ahmed, K. H. (2021). An overview of qualitative research and focus group discussion. *International Journal of Academic Research in Education, 7*(1), 1–15. https://doi.org/10.17985/ijare.866762

Arshed, N., Ahmad, W., & Hanif, U. (2022). A spatial temporal exploration of factors motivating academia-industry collaboration. *Journal of the Knowledge Economy, 13*(1), 521–540. https://doi.org/10.1007/s13132-021-00729-6

Austin, D., May, J., Andrade, J., & Jones, R. (2021). Delivering digital health: The barriers and facilitators to university-industry collaboration. *Health Policy and Technology, 10*(1), 104–110. https://doi.org/10.1016/j.hlpt.2020.10.003

Barrett, D., & Twycross, A. (2018). Data collection in qualitative research. *Evidence-Based Nursing, 21*(3), 63–64. https://doi.org/10.1136/eb-2018-102939

Belfiore, E. (2020). Whose cultural value? Representation, power, and creative industries. *International Journal of Cultural Policy, 26*(3), 383–397. https://doi.org/10.1080/10286632.2018.1495713

Bertello, A., Ferraris, A., De Bernardi, P., & Bertoldi, B. (2022). Challenges to open innovation in traditional SMEs: An analysis of pre-competitive projects in university-industry-government collaboration. *International Entrepreneurship and Management Journal, 18*(1), 89–104. https://doi.org/10.1007/s11365-020-00727-1

Burova, A., Hanna, H., Palma, P. B., Keskinen, T., Hakulinen, J., Opas, V., Makela, J., Ronkainen, K., Siltanen, S., Raisamo, R., & Turunen, M. (2022). Towards measuring the aggregated debt of trustworthiness level. *International Conference on Technical Debt*, 1–11.https://doi.org/10.1145/1234567890

Camacho, B., & Alexandre, R. (2019). Design education. University-industry collaboration, a case study. *The Design Journal, 22*(1), 1317–1332. https://doi.org/10.1080/14606925.2019.1594958

Chung, J., Ko, N., & Yoon, J. (2021). Inventor group identification approach for selecting university-industry collaboration partners. *Technological Forecasting and Social Change, 171*, 120988. https://doi.org/10.1016/j.techfore.2021.120988

Cruz, S. C. S., & Teixeira, A. A. C. (2021). Spatial analysis of new firm formation in creative industries before and during the world economic crisis. *Annals of Regional Science, 67*(October), 385–413. https://doi.org/10.1007/s00168-021-01052-3

Fernandes, G., & O'Sullivan, D. (2021). Benefits management in university-industry collaboration programs. *International Journal of Project Management, 39*(1), 71–84. https://doi.org/10.1016/j.ijproman.2020.10.002

Fernandes, G., O' Sullivan, D., Pinto, E. B., Araújo, M., & Machado, R. J. (2020). Value of project management in university-industry R&D collaborations. *International Journal of Managing Projects in Business, 13*(4), 819–843.https://doi.org/10.1108/IJMPB-08-2019-0191

Fonseca, L. M., Domingues, J. P., & Dima, A. M. (2020). Mapping the sustainable development goals relationships. *Sustainability, 12*(8), 3359. https://doi.org/10.3390/SU12083359

Harper, G. (2021). Sustainable development and the creative economy. *Creative Industries Journal, 14*(2), 107–108. https://doi.org/10.1080/17510694.2021.1952735

Jones, S. E., & Coates, N. (2020). A micro-level view on knowledge co-creation through university-industry collaboration in a multinational corporation. *Journal of Management Development, 39*(5), 723–738. https://doi.org/10.1108/JMD-08-2019-0365

Kinsella, V. (2018). Transforming creative classroom contradictions through activity theory analysis. In K. Snepvangers, P. Thomson, & A. Harris (Eds.), *Creativity policy, partnerships, and practice in education* (1st ed., Vol. 1, pp. 45–64). Palgrave Macmillan.

Kuys, J., Al Mahmud, A., & Kuys, B. (2021). A case study of university-industry collaboration for sustainable furniture design. *Sustainability, 13*, 10915. https://doi.org/10.3390/su131910915

Lucietto, A. M., Peters, D. L., Rusi Taleyarkhan, M., & Tan, S. (2021). Academic and industry collaboration: A literature review. In *ASEE Annual Conference* (pp. 1–22). https://docs.lib.purdue.edu/enepubs/74

Marinho, A., Silva, R. G., & Santos, G. (2020). Why most university-industry partnerships fail to endure and how to create value and gain competitive advantage through collaboration—A systematic review. *Quality Innovation Prosperity, 24*(2), 34–50. https://doi.org/10.12776/QIP.V24I2.1389

Muscio, A., & Vallanti, G. (2014). Perceived obstacles to university-industry collaboration: Results from a qualitative survey of Italian academic departments. *Industry and Innovation, 21*(5), 410–429. https://doi.org/10.1080/13662716.2014.969935

Roberts, J. L. (2007). The Future of Academic-Industry Collaboration. *International Association of Societies of Design Research*, 1–17.

Rybnicek, R., & Königsgruber, R. (2019). What makes industry-university collaboration succeed? A systematic review of the literature. *Journal of Business Economics, 89*(2), 221–250. https://doi.org/10.1007/s11573-018-0916-6

Sim, J., & Waterfield, J. (2019). Focus group methodology: some ethical challenges. *Quality and Quantity, 53*, 3003–3022. https://doi.org/10.1007/s11135-019-00914-5

Sjöö, K., & Hellström, T. (2019). University-industry collaboration: A literature review and synthesis. *Industry and Higher Education, 33*(4), 1–11. https://doi.org/10.1177/0950422219829697

Snepvangers, K., Harris, A., & Thomson, P. (2018). Propositions for creativity policy, partnerships and practice in educational creative futures. In *Creativity policy, partnerships and practice in education* (pp. 353–358). Springer International Publishing. https://doi.org/10.1007/978-3-319-96725-7_16

Wiktor-Mach, D. (2020). What role for culture in the age of sustainable development? UNESCO's advocacy in the 2030 agenda negotiations. *International Journal of Cultural Policy, 26*(3), 312–327. https://doi.org/10.1080/10286632.2018.1534841

Sound Visualization Based on Font Modification Using the Sound of Angklung

Citra Fadillah, Annisa Luthfiasari, and Laurensius William Julio Hartono

Abstract

This paper describes how typography is explored as a medium to adapt the intangible musical characteristic of angklung into font modifications to preserve the value of angklung as the cultural identity of Sundanese people in Indonesia. Capturing the intangible form of sound of angklung and translated it using the theory of association and semiotics as a catalyst to depict a visual form of sound visualization, the sound of angklung can be developed as base for elements of design. Typography is one of the core elements of visual communication design, which can be interpreted verbally, visually, to be heard, and to be deciphered by a sound. Font as a product of typography is widely used as the visual supporting format in computer system and text and comes as a suitable instrument in design sector to develop alternative context and design element based on themes which are required by designers or end users. This method is expected to open the possibilities to translate any other traditional musical instruments into visual forms and embody the intangible form of culture into the daily lives of people.

Keywords

Font modification · Traditional music · Angklung · Sound visualization · Typography · Semiotics

C. Fadillah (✉) · A. Luthfiasari · L. W. J. Hartono
Bina Nusantara University, West Jakarta, Indonesia
e-mail: citra.fadillah@binus.edu

A. Luthfiasari
e-mail: annisa.luthfiasari@binus.ac.id

L. W. J. Hartono
e-mail: laurensius.hartono@binus.ac.id

1 Introduction

Typography is very close to our everyday lives. As the main visual communication tool, it is apparent that typography is ubiquitous and inescapable. Typography not only carried meanings textually, but as a part of visual communication, typography is a creation that can conceive many kinds of value that are expressed by its visuals. Typography can be such a powerful tool that functions to inform, advocate, play, give caution, entertain, express, educate, and transform things (Heller & Ilic, 2012). As a very expressive media, there is no doubt that typography has a huge potential to carry very complex entities such as the value of a culture.

On the other side, Indonesia as a country which own multicultural elements either in tangible or intangible forms has endeavor to preserve it over in a long period of time with various ways. Many efforts have been made to continue the reviving of these cultural values, such as altering cultural values and translating them as an inspiration for new design products. However, the exploration of cultural artifacts was completed mostly in tangible forms. Whereas, Indonesia also has a lot of intangible traditional culture such as traditional music, performance art, songs, and many more. Furthermore, these intangible arts are becoming farther by their people's lives since they cannot be seen or touched directly.

Moreover, the development of the local industry is raising together with the needs of design elements that embody the characteristic of Indonesian. Typography is perfect to fill the gap, because it is applicable dan highly functioning. The exploration in typography will be beneficiary, not only to preserve traditional artifacts but also to enrich the graphic design industry with design products and support Indonesian values.

In this research, typography exploration used as a medium to adapt the intangible musical characteristic of angklung into font modifications to preserve the value of

M. Nawaz Tunio et al. (eds.), *Sustainability in Creative Industries*,
Advances in Science, Technology & Innovation, https://doi.org/10.1007/978-3-031-48453-7_15

angklung as the cultural identity of Sundanese people in Indonesia. Angklung was chosen as the source of inspiration since it is one of the indigenous traditional instruments in Indonesia that has strong and unique sound as they characteristic. This research result as a font modification has strong cultural value that embodied angklung's sound characteristics. The font is adequate to be applied in various design products, either printed or in digital media.

2 Method

To create font modification based on the sound characteristic of angklung, a few steps should be developed as a process. These steps as described in Fig. 1 are required to orderly translate the angklung's sound characteristics. This is a linear step that is projected as a recommended method to be applied to any other musical instruments. This method has involved some theories, especially the relationship between music dan visuals, and also the theory of adaptation.

2.1 Relationship Between Music and Visual

The main basis of the relationship between music and visuals is communication. In communication theory, there are at least three different lenses with which to view how this process works: communication as a one-way process of meaning construction, in which the sender attempts to construct or reconstruct the meaning developed by the receiver; communication as a two-way process of meaning construction, in which two or more people construct new meanings together; and communication as an omnidirectional diachronic process of meaning construction, in which the focus is on the continuous development of meaning itself (Ruler, 2018). Language is a communication tool used by everyone in their daily life as a means to convey information and arguments to others. In this case, language cannot be separated from culture because language represents its nation and has a close relation to the attitude or behavior of groups of speakers of the languages (Rabiah, 2012).

The representation of language comes in diversity. Music and visuals are only some of the types of it. Putting the context of the music as the basis before putting the sound as

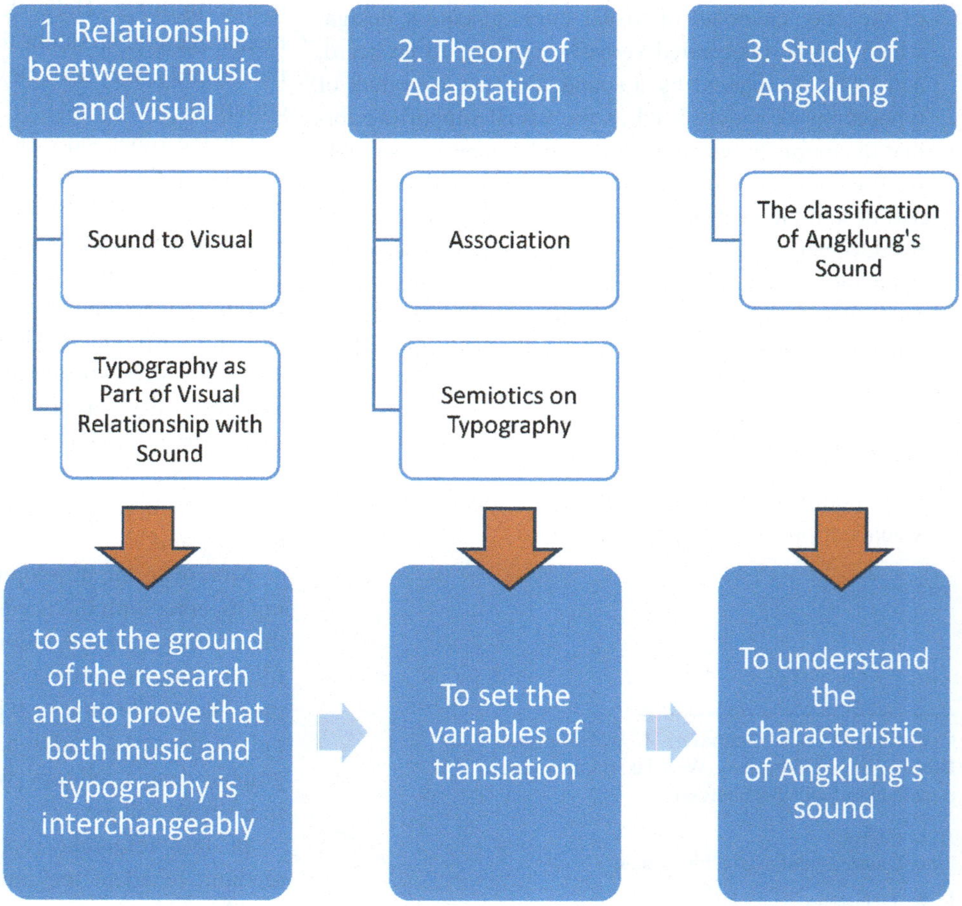

Fig. 1 Theories that underly the methods

the main character to be discussed is also important, to let the view and existence of the relationship visible. What we hear from music is also reflected in our brains as imagination; visuals. Küssner et al. (2019) mentioned that visual imagery is a potential mechanism of emotion induction during listening as well as several studies dealing with music, emotion, and visual imagery from several exciting and novel perspectives. Therefore, conducting the relationship as communication and also with emotion is concluded as an important basis to begin with before we continue to the part of music as sound. Thus in music, there are parts of writing in it. The writing itself connects us to language as visual and later on may definite into typography. The writing has a message and meaning. The study of music can help to access this aware, inventive side that can enhance anyone's writing (Kennedy, 2016).

2.1.1 Sound to Visual

Representing a message from one form to another different kind of form is a way to manifest our communication and limitation. The mixture or collaboration between two or more types of message messenger would overcome some innovations and new ideas. Moreover, the ideas that come up might become the answers to some situations that happened in some sectors or conditions. Sound is a part of music. Musical attributes, based most probably on implicit and/or explicit learning, affect the way sound information is encoded by the human brain. It should be kept in mind; however, that in those studies the sound stimulation was somewhat repetitive (Tervaniemi, 2004).

Sound visualization is an implementation of form representation. Some messages need to be decoded into another part, for specific audiences or events. Not to order or replace some parts of another, but to fulfill the needs of our senses. Exclusively in design, many parts of the senses might complete the clarity of design. Cross-media content and products represented by audio visualization have exploded in terms of quantity, enabling people to gradually get adapted to getting information through audio and video and also transforming modern society to a post-text society. Audio visualization plays an irreplaceable role in current society (Zhang et al., 2018).

2.1.2 Typography as Part of Visual Relationship with Sound

Typography is a major function to design. It is the art or process of setting and arranging typefaces to stylize the appearance (Yadav et al., 2014). As part of visual communication design, typography plays an important role as an element of design. The contribution of typography which contains letters that have meaning and messages gives the idea to designers to develop many kinds of objects from many spaces and form, tangible and intangible. Sound is an intangible version of things that can be sensed by us are one of the options that can be maintained and collaborated with typography, not to mention the typeface of the letters that consist of the drawing of the letter itself. Sayegh (2018) explained that typography is a subset of lettering because it is the study of letters applied to typefaces. Typography does have similarities to lettering. It is still dealing with letters but within the context of typefaces and their proper use. Therefore, it is not a good idea to refer to typography as lettering, since they have different connotations.

The system that typography has that also has a relation to sound in communication and can be preserved also in the example of typography kinetic. This kind of artwork can show us the movement or any design principles. Kinetic typography is that it improves the emotional presentation and delivery of information through the text. Visual representation form combined with the concept of time, space, and acoustic will me more attractive and engage attention. Therefore, the formal language is to maximize information transfer efficiency (Lim, 2016). In kinetic typography, letters have dynamic structures that can enter from the different aspects of the display and change in style and color according to the word's meaning, music/sound rhythm. In this sense, typography has great importance being a visual language, style, and identity as much as its dynamic communication tool with a multi-faceted structure (Turgut, 2012).

2.2 Theory of Adaptation

Regardless of its differences and similarities, the conversion of sound and typography is an interesting subject with very high potential. The adaptation of sound into a visual form, especially typography, can produce a new approach to typeface design. There are several ways to adapt sound into visual form; however, in this research, there will be two main theories that are going to be used as the basic methodology to adapt angklung's musical characteristics into a font modification. The first theory is called the theory of association which shows that the quality of a sound can be associated directly with a form of visual shapes like letterforms through an association library that we have in our minds. Secondly, the theory of semiotics also going to be used in this research. This theory will be aligned both angklung's sound with typography as a form of language that has the same ground interchangeably.

2.2.1 Typography as Part of Visual Relationship with Sound

Everyone has a set of experiences in their brain that have accumulated over a lifetime that act as an association library. The result is everybody unconsciously connect things that have similar feel, appearance, emotion, context, history, and many more. In the term of typography, this happens when we see a typeface in the context of how it is used. For example, the black letter font used by Hitler's campaign gives it a strong 'Nazi' feel. In addition, the letter Gotham used in Barack Obama's campaign makes the typeface imply inspiration and optimism.

Furthermore, letterforms association also continues in the form of intangible products like music. Letters can represent the human voice and convey the tone and quality of that sound visually. Hyndman (2016) revealed that sound can be linked directly to a typeface. One example is by associating vocal ranges with the size and thickness of the letters and how the letters are arranged on the page. Even simple text can 'sound' slower, scream, and have different characters and intonations as we can see in Fig. 2.

Lewis and Walker (1989) consider the visual appearance of letters as a 'visual code of speech' which represent the meaning and the sound produced when the word is spoken. Furthermore, Hyndman (2016) suggests a term called visual onomatopoeia, which is how a visual form is associated with sound. Instinctively, it is known that letters reflect the quality of physical objects around us with a common understanding. Bold letters with large sizes are agreed to 'sound' louder than light letters which will sound softer. Hyndman also studied the relationship between the visual form of letters and sounds which was then concluded in Fig. 3.

From Hyndman's research (2016), it is concluded that volume can be represented by letter size, sound characters such as timbre and bass can be represented by letter thickness and width, while tempo can be represented by letter kerning. In another study, Hyndman (2016) tried to associate the shape of letters with a sound and how these shapes affect human feelings. According to her, the shape of the letter is related to emotion, while the axis (slope) is related to energy. Rounded letters are considered calming, while angular letters seem to be anxious. Balanced (upright) letters 'sound' static, while italics 'sound' dynamic, as we can see the difference in Fig. 4.

The study of letterform and sound association mentioned above shows that the quality of a sound can be associated directly with a form of visual shapes like letterforms through an association library that we have.

3 Semiotics in Typography

A lecturer and graphic designer, Armstrong (2003), through his research entitled 'Hearing Type' found a relationship between the aural field (something related to the sense of hearing) and the visual field. In his research, Armstrong (2003) described musical variables in a visual form, which is in the form of typographic compositions. According to Armstrong (2003), there is a parallel relationship between the interaction of visual forms (including typographical forms and other elements) and the interaction of sound in a spatial area.

Both music and typography came from the same origin, which is verbal language. Derived from the rhythm in speech, music is a form of storytelling whose structure consists of the expression of information in the form of audio. On the other hand, typography is a visualization of spoken language, where the notion of time and grammatical syntax is embedded in the reading process of a composition of typographic elements.

In addition, Armstrong also argues that musical notation can be directly compared with typography since both of them are systems of symbols that represent elements of each language. On that basis, typography and music are considered to have their semiotic code and can be linked through it. In semiotic theory, the concept of code is called a semiotic resource which can act as a media or a mode.

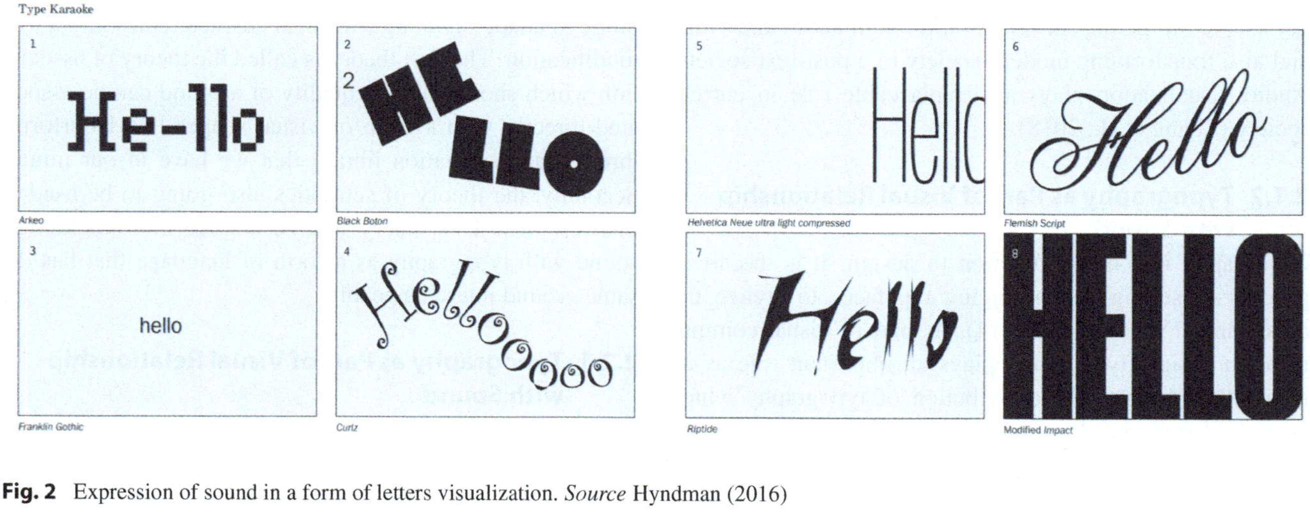

Fig. 2 Expression of sound in a form of letters visualization. *Source* Hyndman (2016)

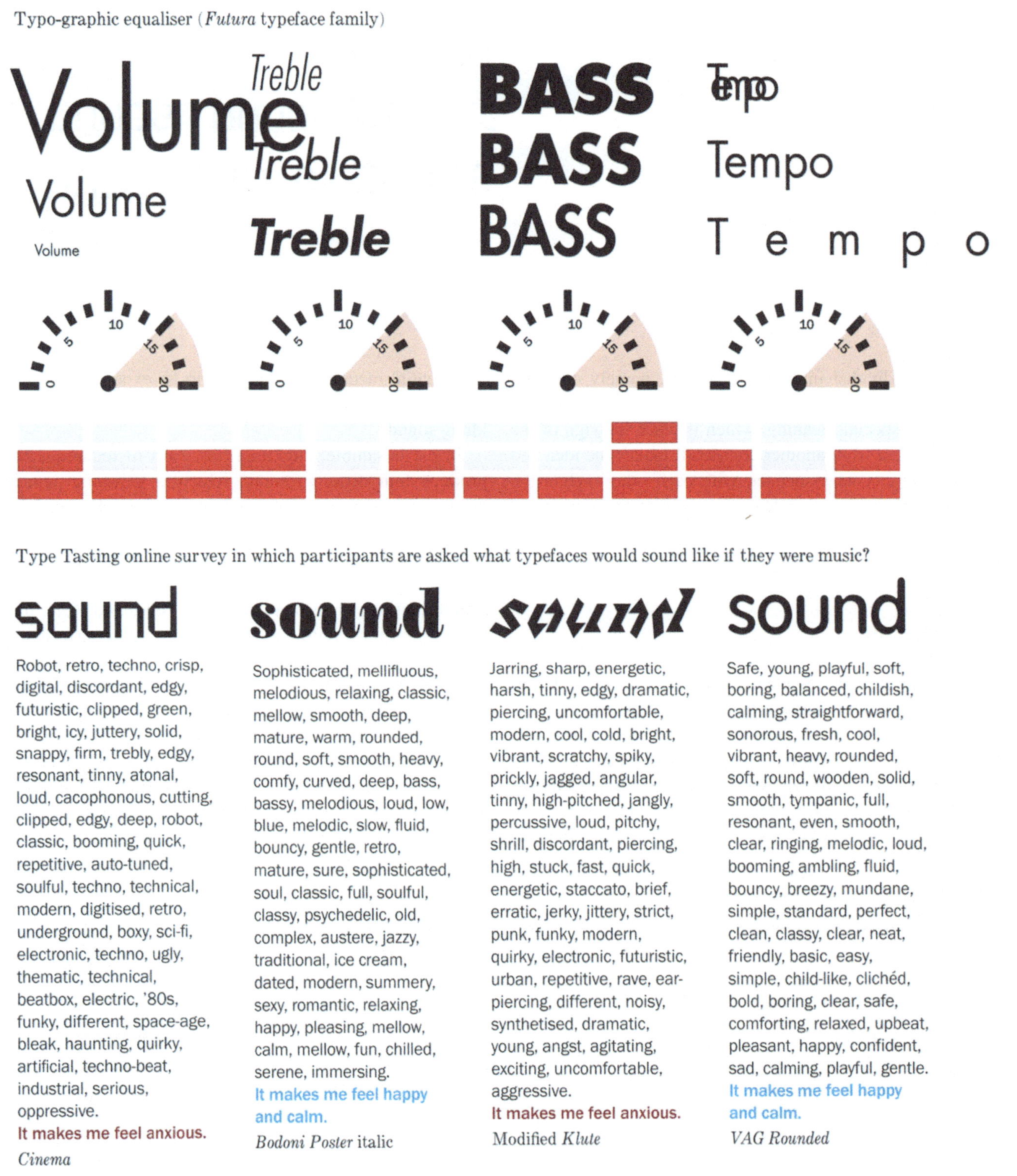

Fig. 3 Relationship between sound quality and the visual appearance of letters. *Source* Hyndman (2016)

If it is considered a mode, it consists of grammar and lexis. Meanwhile, if it is considered a media, it only consists of lexis.

Typography is often seen as a media, which is a collection of typefaces that have certain qualities, not seen as a system. When the semiotic reference comes from the

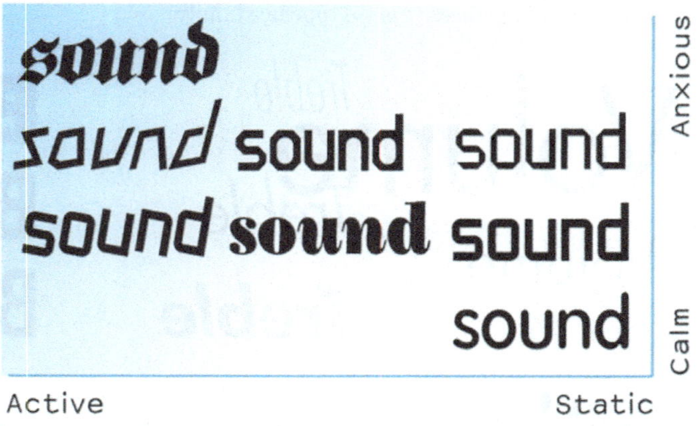

Fig. 4 Relationship between type, sound, and human emotions. *Source* Hyndman (2016)

media, it is divided into two principles, namely connotation and experiential metaphor. The word connotation here is used in a special meaning, which is when the sign of the idea is adapted from another context. It makes the idea and its meaning can be associated with each other. While the experiential metaphor is where the signifier has a meaning that comes from personal experience which is then developed in a metaphor (Van Leeuwen, 2005).

4 Result

As stated in the theories mentioned above, it is possible to conjoin the musical characteristic of angklung into a typographic form using the theory of association. Moreover, the quality of the angklung sounds can be equated more precisely by bridging it with semiotics. According to Van Leuween's theory, typography will be seen as a medium that took angklung's sound as its connotation and experiential metaphor. It resulted in an experiment, which has four groups of letters exploration that resembles of angklung sound.

In the ensamble, angklung has ways of technique that produce four different sound which is kurulung pendek, kurulung panjang, centok, and tengkep. Each sound has its characteristic that resulted in four different visualizations. In this adaptation, the first visual came in the form of line association. It is intended to give a basic understanding of how the sound should be visualized. After that, the exploration continues in the form of letterforms. It shows several letters in the making. After that, the adaptation process continues in the search for the basic typeface for font modification. These exploration as experiment described in Table 1. Lines association in Table 1 Figures a, d, g, and j describe the angklung sound. The combination to font was also made from letterforms exploration in Table 1 Figures b, e, h, and k, together with base fonts in Table 1 Figure c, f, i, and l.

Table 1 Eexploration of modification to font

Angklung sound	Sound characteristic	Lines association	Letterforms exploration	Explanation	Base font
Kurulung pendek	High amplitude, standard duration, high vibration, and high pitch		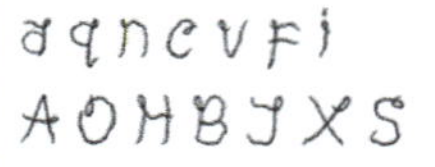	Dynamic lines, slightly round edges, and organic shapes	
Kurulung panjang	High amplitude, long duration, high vibration, and high pitch		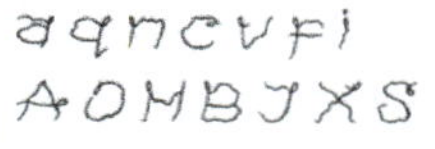	Dynamic and wavy lines, round edges, expanded length, and organic shapes	
Centok	High amplitude, short duration, low vibration, and high pitch	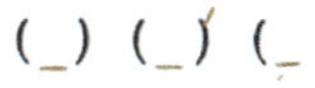	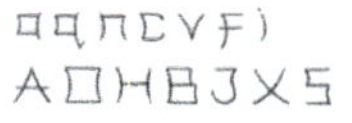	Sturdy and angular structure, concave shapes, sharp edges	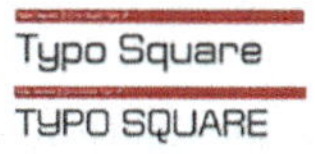
Tengkep	Low amplitude, low duration, high vibration, and low pitch		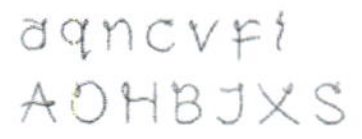	Dynamic lines, slightly round edges, and organic shapes	

5 Discussion

Typography is a very versatile medium that can convey meaning and represent an identity. This research proves that typography can represent not only a tangible form of inspiration but also intangible ones. According to the experimentation, it is very possible to translate any other traditional musical instruments into visual forms. It can produce new typefaces that embody the identity of a culture and brings values into the daily lives of the people. On the other hand, with its nature as a communication medium, typography always existed in everyday life. It makes typography such an ideal medium. The adaptation of angklung sound into a font modification gives the potential to as an example of how to enjoy music in a different form. This method is expected to open the possibilities to translate any other traditional musical instruments into visual forms and embody the intangible form of culture into the daily lives of the people.

References

Armstrong, F. (2003). Hearing type. *The Education of a Typographer Magazine*. issue (#42).

Blakey, N., Guinea, S., & Saghafi, F. (2017). Transforming undergraduate nursing curriculum by aligning models of clinical reasoning through simulation. In R. Walker & S. Bedford (Eds.), *Research and development in higher education: Vol. 40. Curriculum transformation* (pp. 25–37). Higher Education Research and Development Society of Australasia. http://www.herdsa.org.au/research-and-development-higher-education-vol-40-25

Hyndman, S. (2016). *Why font matters*. Virgin Books.

Kennedy, E. P., & Santos, R. T. (2016). *Music and words: Connecting the love of music with language*. Kingsborough Community College.

Küssner, M. B., Eerola, T., & Fujioka, T. (2019). Music, emotion, and visual imagery: Where are we now? *Psychomusicology: Music Mind, and Brain, 29*(2–3), 59–61. https://doi.org/10.1037/pmu0000245

Lewis, D. (2013). *The brain sell: When science meets shopping*. Nicholas Brealey Publishing.

Lewis, C., & Walker, P. (1989). Typographic influences on reading. *British Journal of Psychology, 80*, 241–257. https://doi.org/10.1111/j.2044-8295.1989.tb02317.x

Lim, S. (2016). Design of kinetic typography interaction based on the structural characteristics of Hangul. *International Journal of Contents, 12*(3), 42–46. https://doi.org/10.5392/IJoC.2016.12.3.042

Mari Tervaniemi, E. B. (2004). From sounds to music towards understanding the neurocognition of musical sound perception. *Journal of Consciousness Studies*.

Rabiah, S. (2012). Language as a tool for communication and cultural reality discloser. In *International Conference on Media, Communication and Culture "Rethinking Multiculturalism: Media in Multicultural Society*.

Sayegh, A. M. E. (2018). The aesthetics impact of the typographic on the logo advertising and meaning (analytical research). *OSR Journal of Research & Method in Education*. https://doi.org/10.9790/7388-0802051828

Turgut, O. P. (2012). Kinetic typography in movie title sequences. *Procedia—Social and Behavioral Sciences*. https://doi.org/10.1016/j.sbspro.2012.08.209

Van Leeuwen, T. (2005). Typographic meaning. *Visual Communication, 4*(2), 137–143. https://doi.org/10.1177/1470357205053749

Van Ruler, B. (2018). Communication theory: an underrated pillar on which strategic communication rests. *International Journal of Strategic Communication, 12*(4), 367–381. https://doi.org/10.1080/1553118X.2018.1452240

Yadav, P., Chakrabarti, D., Bisoyi, D. (2014). Typography as a statement of design. In *International Ergonomics Conference HWWE*.

Zhang, Y., Pan, Y., & Zhou, J. (2018). Study on application of audio visualization in new media art. *Journal of Physics: Conference Series*.https://doi.org/10.1088/1742-6596/1098/1/012003

The Implementation of Deep Learning Technique in Mobile Application as a Preservation and Learning Media of Javanese Letter

Raymond Sutjiadi, Timothy John Pattiasina, and Peter Santoso

Abstract

Indonesia as the largest archipelagic country in the world has many local heritages. One of the valuable heritages is local language, which is 718 local languages in total from 37 provinces in Indonesia. Javanese language is a local language used by many people in Java Island, the most crowded mainland in Indonesia. This language has specific form of letter, differ from Latin alphabet. Because of that, nowadays people rarely learn how to write Javanese letter because of its difficulty. In this research, mobile application is developed using deep learning technique with VGG-16 convolutional neural network (CNN) architecture, as the cutting-edge method of artificial intelligence, to recognize Javanese letter and convert it into Latin alphabet. By using this application, younger generation can learn Javanese letter easily and in attractive way. Three main features are provided, i.e., scanning the letter, writing the letter, and converting the letter. Users are able to learn how to read and write Javanese letter, also receive feedback score from the system. This application is aimed as a preservation and learning media for those who want to learn Javanese letter using information technology. Based on testing result, this application has 89% of accuracy to recognize Javanese letter.

Keywords

Deep learning · Convolutional neural network · Artificial intelligence · Mobile application · Javanese letter

R. Sutjiadi (✉) · P. Santoso
Informatics Department, Faculty of Information Technology, Institut Informatika Indonesia Surabaya, Surabaya, Indonesia
e-mail: raymond@ikado.ac.id

T. J. Pattiasina
Information System Department, Faculty of Information Technology, Institut Informatika Indonesia Surabaya, Surabaya, Indonesia
e-mail: temmy@ikado.ac.id

1 Introduction

Indonesia is the largest archipelagic country located in South East Asia with more than 272 million population and inhabit more than 1.9 million square kilometer area (BPS, 2022). Based on the census data in 2010 held by Center of Statistics Bureau, Indonesia comprises more than 300 groups of ethnic (Sari et al., 2018). This factor makes Indonesia rich of traditional cultures as the invaluable heritage that need to be preserved in this modern era. Every ethnic possesses specific traditional culture in form of traditional apparel, house, song, and language. Some of them have already been recognized as world heritage by the United Nations Educational, Scientific, and Cultural Organization (UNESCO). Indonesia owns 718 local languages in total (Kwary, 2019) from 37 provinces in Indonesia.

One of the biggest ethnics in Indonesia is Javanese, which settles in most part of Java island (Central Java, Special District of Yogyakarta, East Java), and some of them also migrate in other parts of Indonesia island (Muttaqien et al., 2018). This ethnic has many specific traditional cultures, and most of them still exist until now. Javanese has their own specific letter, differs from Latin alphabet, as a written communication medium since the mid-fifteenth century called Javanese letter or sometimes also called as Hanacaraka, Carakan, and Dentawyanjana (Suria, 2018). This letter consists of 20 basic characters represent vowel sound called Aksara Nglegena as seen in Fig. 1. Besides Aksara Nglegena, there are additional characters called Aksara Murda (Fig. 2), Aksara Mahaprana (Fig. 3), Aksara Swara (Fig. 4), and Aksara Ganten (Fig. 5). Javanese letter has similar format with the letter of Balinese and Sundanese because those letters are descended from Aksara Kawi, the older letter used in the era of Sriwijaya Kingdom (Zurbuchen, 1976).

On the other side, nowadays younger generation is exposed by many modern cultures from other countries via

M. Nawaz Tunio et al. (eds.), *Sustainability in Creative Industries*,
Advances in Science, Technology & Innovation, https://doi.org/10.1007/978-3-031-48453-7_16

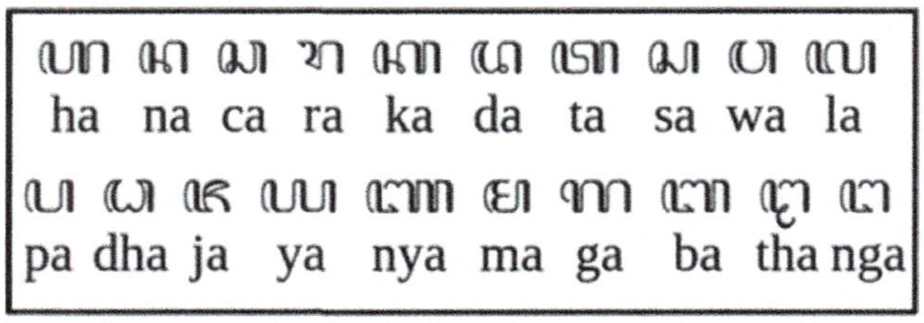

Fig. 1 Twenty characters of Aksara Nglegena

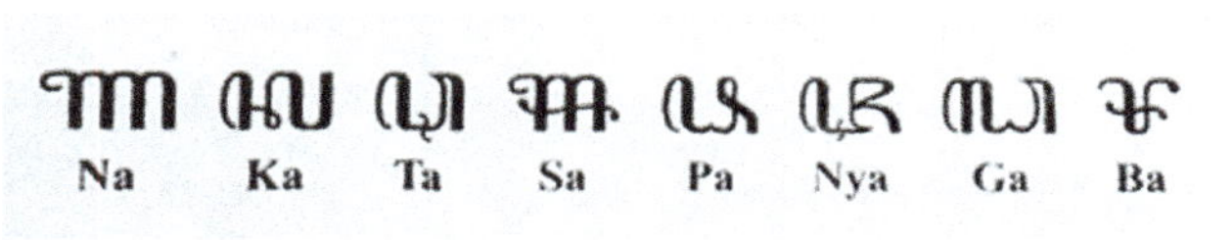

Fig. 2 Eight characters of Aksara Murda

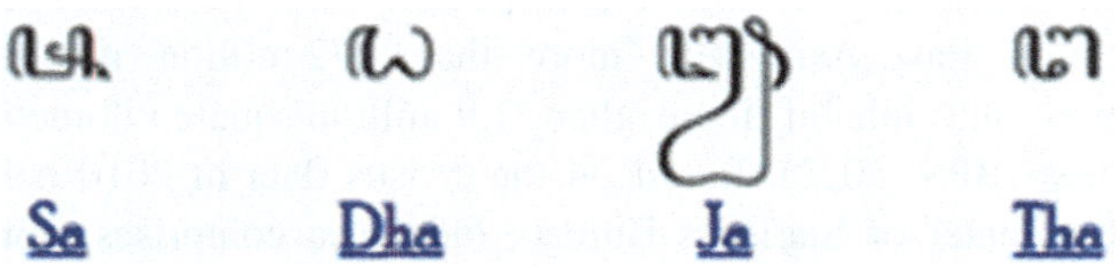

Fig. 3 Four characters of Aksara Mahaprana

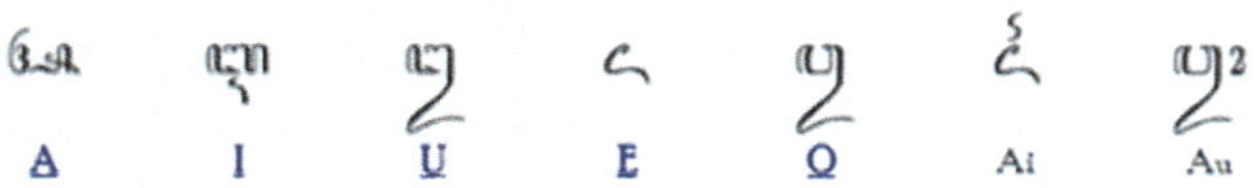

Fig. 4 Seven characters of Aksara Swara

Fig. 5 Two characters of Aksara Ganten

television program, music, and other media. Foreign culture is more appreciated, instead of Indonesian local culture itself, because it can be accessed easily via electronic device, such as computer, smartphone, and tablet, which in general already owned by younger generation. The content is more enjoyable, so many people idolize them, for example, Anime, Manga, and J-pop from Japan. It is spreading quickly online, and as more people are exposed to and aware of this Japanese popular culture, the influence will be even more profound (Bangsa & Sihombing, 2022). This factor becomes an impediment to introducing local traditional culture to the younger generation and further, it will eliminate the actors who preserve the Indonesian culture.

The preservation of Javanese letter becomes a tough challenge, not only for government institutions, but also for educators. The Indonesia government via the Ministry of Education, Culture, Research, and Technology issues a policy where Javanese language and literature become a local course curriculum in school located in Central and East Java (Rahadini & Nurhayati, 2022). This effort is to increase the number of people who understand the Javanese language, including how to write Javanese letter. But in practice, this subject is often considered less important than other courses, and students' interest in learning Javanese language is still low comparable to the other languages (Nurmasari et al., 2017), like English and Mandarin.

Based on the above situation, it needs innovative way how to introduce Javanese letter to younger generation. In this research, mobile application based on Android is developed as a media to learn Javanese letter. Not only for educational purpose, this application is also implemented the artificial intelligence to recognize Javanese letter and convert it into Latin alphabet characters. So by using this application, younger generation is able to learn Javanese letter in more fun and enjoyable way.

2 Research Methodology

This research uses some methodologies to build the application. Rapid application development (RAD) is a robust methodology for developing mobile applications rapidly. Convolutional neural network (CNN) as a deep learning method is implemented for the intelligent system to support the recognition features. This research also adopted VGG-16 as a CNN model that has proven to be used as an image recognition engine.

2.1 Rapid Application Development (RAD)

To develop this application, it will be used rapid application development (RAD) methodology as a software development life cycle (SDLC). RAD model is a prototyping-based method and develop iteratively without specific planning involved (Fatima et al., 2018). RAD also focuses on users requirements by building prototyping model, which developed iteratively based on suggestions and testing from users. Different from waterfall methodology, which states that software developer teams should have a detailed roadmap for the project, RAD does not have complicated planning. Basically, RAD methodology aims to develop the applications faster. Continuous feedback analysis and frequent interactions are needed to achieve good result. RAD supports quick program design progression and makes without question transport on adjusted time (Yashod, 2021). Figure 6 shows RAD processes, where every phase will be continued by the next phase until finish. This continuous phase will be repeated until prototypes fulfill all system objectives and requirements.

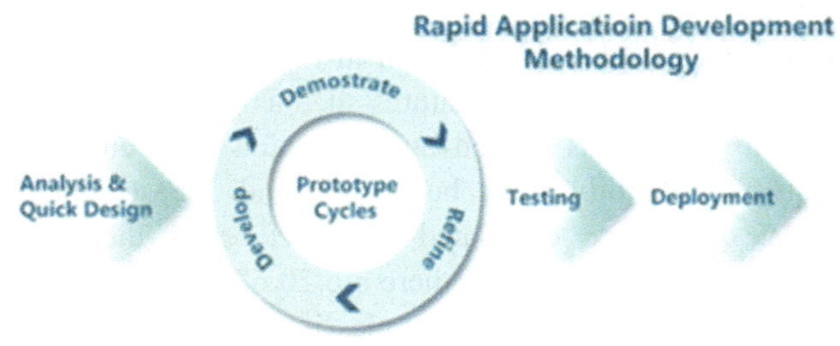

Fig. 6 Rapid application development (RAD) methodology

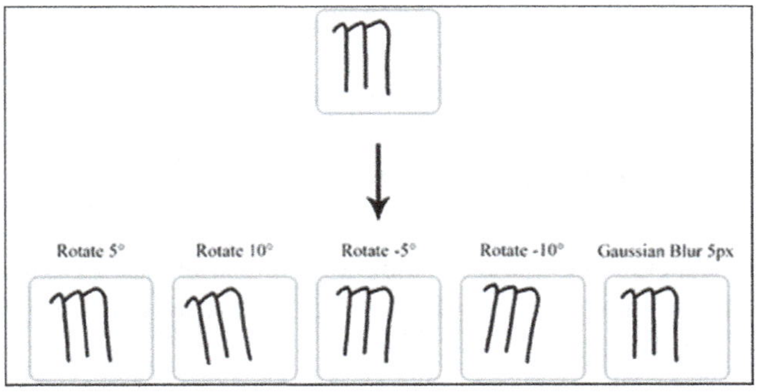

Fig. 7 Dataset image preprocessing

The first phase of the RAD is to gather all the information related to the system requirements. In this phase, developer should do system requirement analysis by interviewing the users, distributing questionnaire, studying the literatures (books or journals), or observing the business process. Based on the result of system requirement analysis, developer then creates quick design of the application. The second phase is the repetitive stage for prototyping development phase. During this phase, developer will implement quick design into product prototype. This product prototype should be expanded by accommodating user needs. Developer should demonstrate this prototype to the users and refine it based on user feedback. The third phase is testing the prototype. In this phase, product prototype is already mature and ready to be tested for final deployment. The last phase is deployment in the actual environment after all system functionalities have been validated.

This application comprises into two modules, and each module is separated into five stages of development (Sasmito et al., 2020), i.e.,

(a) Module for dataset training and testing, written using Python programming language, TensorFlow framework, and Keras library. This module results data model which will be used by the Android application and becomes an important part to recognize Javanese letter accurately based on artificial intelligence training process.

- Business modeling: In this first stage, dataset is collected by gather hand written samples of Javanese letter from some people. Dataset used to build a data model consists of 250 vector images per 20 characters of Aksara Nglegena only. So, the total dataset is 5000 vector images.
- Data modeling: In this stage, collected sample data is preprocessed to be fitted to the required specification to optimize the data training results. To generate dataset and maximize the variant of the image, every image data is entering preprocessing stage where there are image binarization process and other five image processing methods (apply image rotating and Gaussian filter) as shown in Fig. 7. The objective of this process is to minimize the overfitted data model.
- Process modeling: In this stage, final dataset is uploaded to the Kaggle platform and openly published to be used by other developers.
- Application generation: This is the main stage, where this module is developed using Python programming language in Kaggle environment system.
- Testing: the last stage where resulted data model is optimized by tuning hyperparameter to search the most optimum accuracy value. The final data model then will be hosted on the cloud server as a source of knowledge for Android application to recognize Javanese letter and convert it into particular Latin alphabet characters.

(b) Module for Android application, written using Java programming language. This application will be used by the users to scan Javanese letter using smartphone camera and convert it into Latin alphabet characters, based on data model created by first module.

- Business modeling: In this first stage, literature study is conducted to search similar research and find state-of-the-art to be developed.
- Data modeling: In this stage, prototyping model is developed by designing user interface to accommodate every feature on application.
- Process modeling: In this stage, final prototyping model is developed by arranging some improvements based on user needs.
- Application generation: This is the main stage, where this module is developed using Java programming language in Android Studio IDE.
- Testing: the last stage where mobile application is ready to be used and tested by the users to find some bugs and optimize its accuracy.

2.2 Convolutional Neural Network (CNN)

Convolutional neural network (CCN) is a popular deep learning algorithm that processes certain input digital images to provide the knowledge to detect the item categories by giving some important variables (biases and weights) to various characteristics and objects in the image. Convolutional neural network (CNN) has shown excellent performance in many computer vision and machine learning problems (Khan et al., 2020). The development of CNN has had a tremendous influence in the field of computer vision in recent years and is responsible for a big jump in the ability to recognize objects (Voulodimos et al., 2018). This fast achievement has been enabled by increasing the amount of data that available for data training and also supported by the computing power enhancement.

The CNN design has kernels or filters that can find edges or features throughout a digital image. Every digital image on CNN is represented by an array of pixel values for each color channel, which is often an RGB or 3-channel image. More precisely convolutional layers are able to detect patterns on images. A kernels is a collection of weight values in form of matrix which is trained to detect specific features on an image. The fundamental principle of CNN is to spatially convolve the kernel on an input image and determine whether any features have been identified. The spatial translation of the input from a characteristic detection layer will be conveyed to the output unchanged since each filter or kernel seeks to identify a specific characteristic at each input point. A convolution process is performed by computing the dot product of the kernel and the input area where the kernel is overlapped to obtain a value expressing how sure it is that a certain feature is present (Kumar & Kumar, 2018). Figure 8 depicts the fundamental CNN structure (You et al., 2017).

2.3 VGG-16 Architecture Model

VGG-16 is a CNN model developed by Simonyan and Zisserman (2015) from University of Oxford, to classify large-scale image and maintain its peak performance to result good accuracy. This model won as runner-up position in ILSVRC-2014 challenge with less than 7.5% test error. VGG-16 model is good to be used as image classification algorithm because of its robustness.

This model comprises 13 convolutional layers and three fully connected layers as shown in Fig. 9. Max Pooling is used to extract low-level features such as edges and points. Softmax is used as an activation function to calculate the relative probability value between 0 and 1 based on the model confidence correlated to which class the image belong to. In this research, there are 20 classes of Javanese letter (Aksara Nglegena) to be classified as an output. The highest probability value will be taken as a final recognition result. Detailed per layer VGG-16 architecture process is shown in Table 1.

3 Result and Discussion

3.1 Use-Case Diagram

In general, this Andorid mobile application has four features or menus that can be accessed by the users (as shown in Fig. 10), i.e.,

- Scanning: Users can access this menu to scan a letter of Javanese letter (Aksara Nglegena) using smartphone camera, and application will recognize that letter utilizes VGG-16 (CNN) method. The recognition result in form of particular Latin alphabet characters will be shown on screen.
- Learning: Users can learn how to write Javanese letter (Aksara Nglegena) by writing the letter directly via smartphone touch screen. Application will recognize the result and give a score based on artificial intelligence.
- Converting: Users can type the Latin alphabets, and application will convert it into respective Javanese letters (Aksara Nglegena). Also, users are able to copy the characters or share it to another application.
- Setting: Allows users to change the functionality and behavior of an application. Users may access this menu to change some tuning parameters related to the function of the application.

Fig. 8 Basic structure of CNN

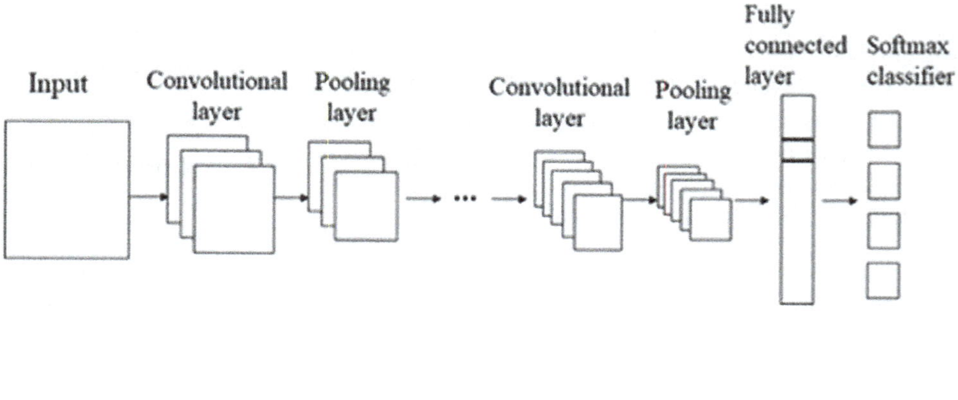

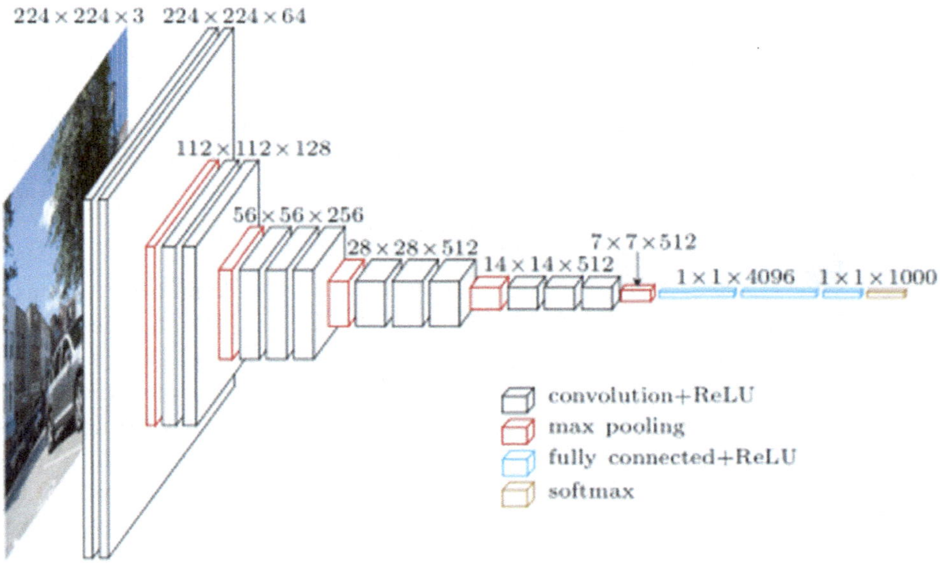

Fig. 9 Architecture of VGG-16

Table 1 List of VGG-16 per layer process

No	Layer	Description
1	Conv3-64	64 channel convolution layer, 3 × 3 matrix, padding = 'same', ReLU activation rectified
2	Conv3-64	64 channel convolution layer, 3 × 3 matrix, padding = 'same', ReLU activation rectified
3	Maxpool	Max pooling process, matrix 2 × 2, stride 2 × 2
4	Conv3-128	128 channel convolution layer, 3 × 3 matrix, padding = 'same', ReLU activation rectified
5	Conv3-128	128 channel convolution layer, 3 × 3 matrix, padding = 'same', ReLU activation rectified
6	Maxpool	Max pooling process, matrix 2 × 2, stride 2 × 2
7	Conv3-256	256 channel convolution layer, 3 × 3 matrix, padding = 'same', ReLU activation rectified
8	Conv3-256	256 channel convolution layer, 3 × 3 matrix, padding = 'same', ReLU activation rectified
9	Conv3-256	256 channel convolution layer, 3 × 3 matrix, padding = 'same', ReLU activation rectified
10	Maxpool	Max pooling process, matrix 2 × 2, stride 2 × 2
11	Conv3-512	512 channel convolution layer, 3 × 3 matrix, padding = 'same', ReLU activation rectified
12	Conv3-512	512 channel convolution layer, 3 × 3 matrix, padding = 'same', ReLU activation rectified
13	Conv3-512	512 channel convolution layer, 3 × 3 matrix, padding = 'same', ReLU activation rectified
14	Maxpool	Max pooling process, matrix 2 × 2, stride 2 × 2
15	Conv3-512	512 channel convolution layer, 3 × 3 matrix, padding = 'same', ReLU activation rectified
16	Conv3-512	512 channel convolution layer, 3 × 3 matrix, padding = 'same', ReLU activation rectified
17	Conv3-512	512 channel convolution layer, 3 × 3 matrix, padding = 'same', ReLU activation rectified
18	FC-4096	Fully connected dense layer of 4096 units, ReLU activation
19	FC-4096	Fully connected dense layer of 4096 units, ReLU activation
20	FC-20	Fully connected dense layer of 20 units, Softmax activation

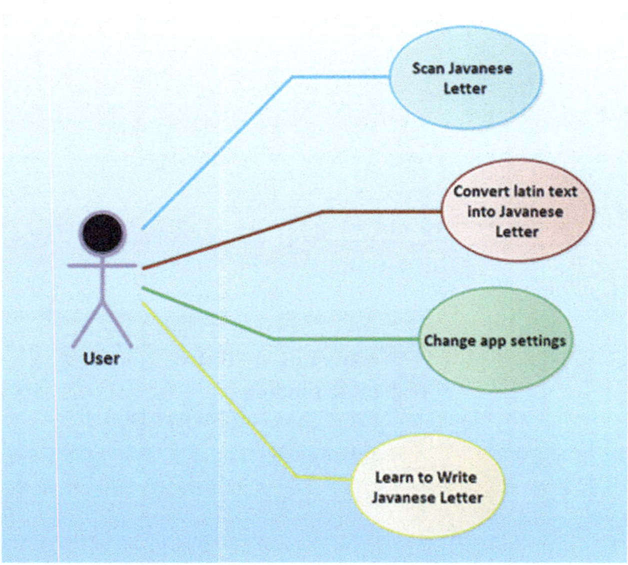

Fig. 10 Use-case diagram

3.2 Architectural Design

Architectural design depicts the relationship between the user and many subsystems related to the application. Based on Fig. 11, it can be seen that user will interact directly to application using Android smartphone device. This application has to access the cloud backend server via Internet connection, where data model hosted and run a Python web service. This web service stand by to receive a sample image from smartphone and process the recognition. In this case, a neural network model that has been trained is needed to be used in the prediction process of scanning a Javanese letter and learning to write in Javanese. The recognition result then will be sent back to the application via Internet connection and inform to the user on display screen.

3.3 User Interface Design

Designing good user interface plays an important part because user will have experience using this application via all features provided in user interface. All menus must be stated clearly about the function to help user operate all

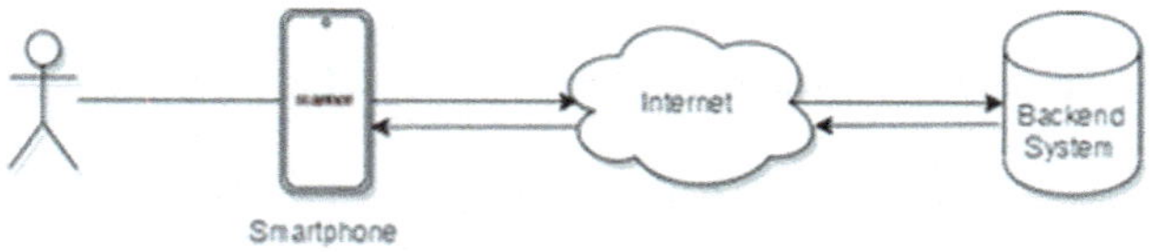

Fig. 11 Architectural design

features smoothly. Also, user interface design must represent the theme of the application. Because this application is about Javanese culture, user interface uses brown as the main color combined with white color, inspiring from the color of Batik—traditional textile design from Java. There are five pages in this application that can be accessed by the user, i.e.,

- Main page: On this page, user will find four main buttons to access four features provided by the application. Scan menu to access the scanning page, convert menu to access the conversion feature, exercise menu to access the learning feature, and setting menu to access setting page. At the top, there is information of application name written using typography similar to Javanese letter, version number, and the photo of Borobudur temple—historical temple from Java—as the background image. The user interface design for main page is shown in Fig. 12.
- Scanning page: This is the main feature of this application. By accessing this page, application will activate smartphone camera automatically, and user just aims the Javanese letter as the camera focuses. If system recognizes the character, there will be pop up menu to inform the respective Latin alphabet. The user interface design for scanning page is shown in Fig. 13a and b.
- Learning page: By accessing this page, application will show 20 buttons, one button for each Javanese letter (Aksara Nglegena). User has to select one of them, and system will open white drawing space, where user can practice how to write selected Javanese letter using touch screen. After submit it, system will show the result and the similarity score based on artificial intelligence. The user interface design for learning page is shown in Fig. 14a, b, and c.

Fig. 12 Main page user interface design

- Converting page: On this page, there are two text boxes, the top for typing Latin alphabet character and the bottom to show the Javanese letter. Also there is copy button, where the user can copy the text and paste it to another application. The user interface design for converting page is shown in Fig. 15.
- Setting page: This is the special page to adjust the application setting. Also in this page user can access the tutorial page, asking question via email, and read the developers identity. The user interface design for setting page is shown in Fig. 16.

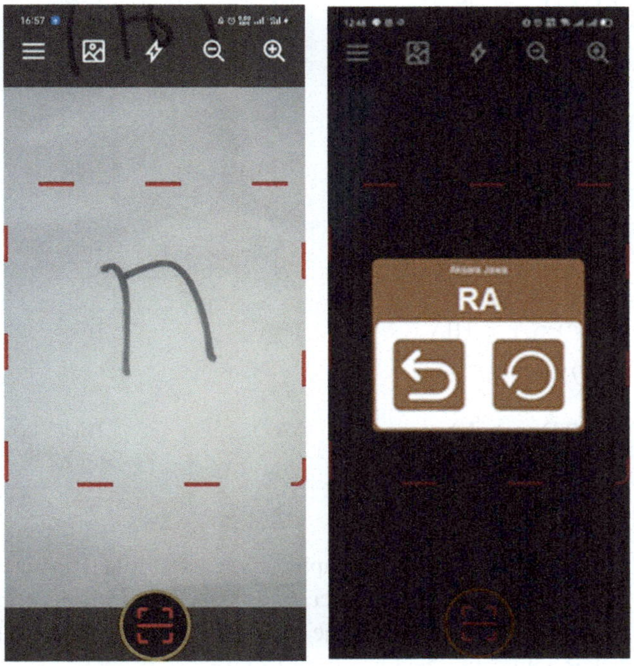

Fig. 13 Scanning page user interface design

3.4 Application Testing

Developing artificial intelligence application must be tested its accuracy to improve user experience. To test the accuracy of recognition feature, the application will be used to scan several handwritten Javanese letters (Aksara Nglegena) as shown in Table 2 Figures a–i. Javanese letter is written on a white paper sheet using black permanent ink with proper thickness. An Android smartphone equipped with back camera is used as a device to install and test the application. The testing conducted in well-lit environment to maximize the image quality. From the result of testing as shown in Table 2, it is found that the system accuracy is 89%.

4 Conclusion

The development of Android application to recognize Javanese letter using deep learning method is able to run all features well. The recognition system also works reliably with 89% accuracy to predict handwritten Javanese letter. By using this application, user can learn how to write Javanese letter as well and get a feedback score using artificial intelligence. This application also has a feature to make it easy to type Javanese letter by using Latin alphabet characters.

Fig. 14 Learning page user interface design

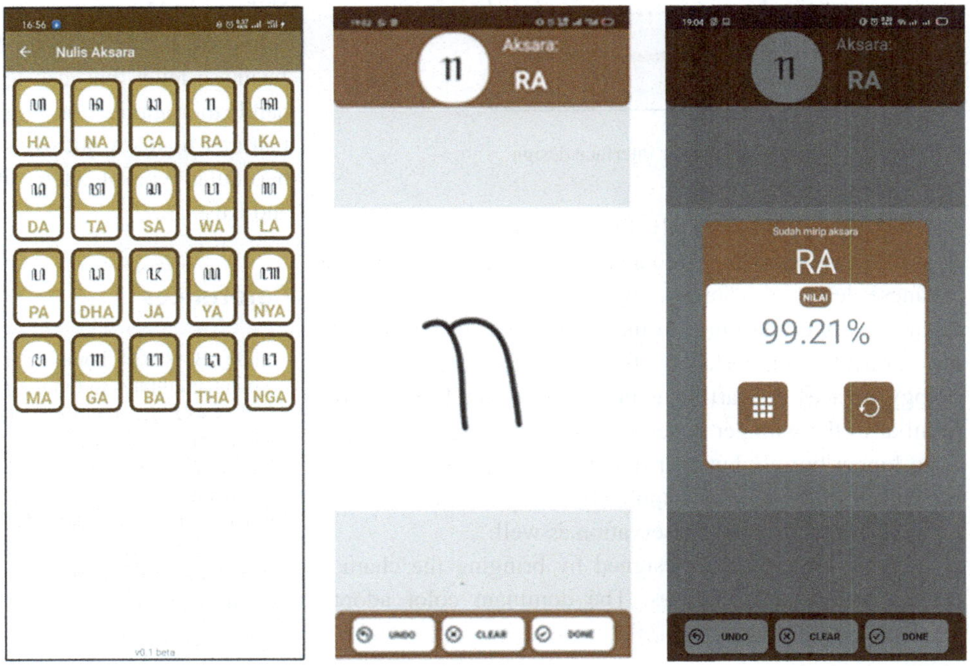

Fig. 15 Converting page user interface design

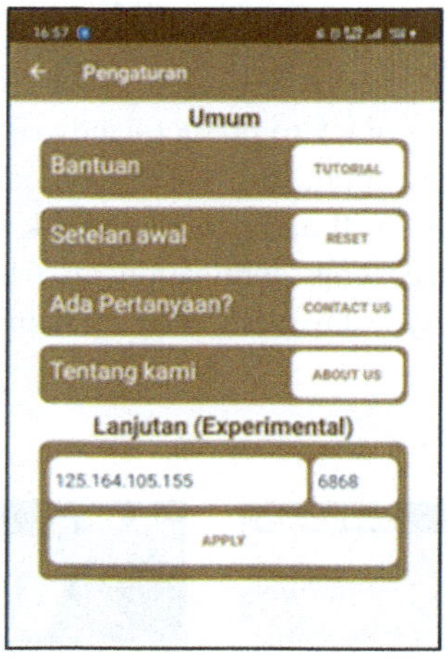

Fig. 16 Configuration page user interface design

Table 2 Application testing

Input image	Expected result	Recognition result	Conclusion
	DHA	THA	False
	GA	GA	True
	JA	JA	True
	KA	KA	True
	LA	LA	True
	NGA	NGA	True
	RA	RA	True
	THA	THA	True
	YA	YA	True

This application not only functions as a tool, but also works as an innovative educational media to introduce Javanese letter to someone who wants to learn writing Javanese letter. Learning Javanese letter becomes more fun and enjoyable supported by the use of information technology devices and artificial intelligence which is already familiar to the younger generation. Learning Javanese letter is no longer boring, but user may use this application interactively. Furthermore, this application may contribute to the effort of Javanese letter preservation as well.

The user interface is designed by bringing the characteristic of Javanese cultures. The dominant color adopts the color of Batik, typography style uses font character that resembles Javanese letter, and background image uses Borobudur temple, one of the Javanese pride cultures. This makes application feeling closer to Javanese people.

There are recommended suggestions for application improvement in the future. This application can be leveraged to recognize Javanese letter other than Aksara Nglegena. Also this application can be added the spelling sound of Javanese letter. So users not only learn writing Javanese letter, but also learn how to spell it. On the other hand, this application could be improved by adding some features to learn and recognize other languages that have particular type of letter, like Mandarin Chinese, Japanese, and Russian.

References

Bangsa, B. P. & Sihombing, L. H. (2022). The impact of Japanese popular cultures to Indonesian younger generation: A case study of PUNICO. *Humaniora, 13*(3), 241–246. https://doi.org/10.21512/humaniora.v13i3.8131

BPS. (2022). *Statistical yearbook of Indonesia 2022*. Badan Pusat Statistik.

Fatima, F., Javed, M., Amjad, F., & Khan, U. G. (2018). An approach to enhance quality of the RAD model using agents. *Journal of American Science, 14*(9), 47–55. https://doi.org/10.7537/marsjas140918.08

Khan, A., Sohail, A., Zahoora, U., & Qureshi, A. S. (2020). A survey of the recent architectures of deep convolutional neural networks. *Artificial Intelligence Review, 53*, 5455–5516. https://doi.org/10.1007/s10462-020-09825-6

Kumar, S., & Kumar, M. (2018). A study on the image detection using convolution neural networks and tenser-flow. In *International Conference on Inventive Research in Computing Applications (ICIRCA)* (Vol. 2, pp. 1080–1083).

Kwary, D.A. (2019). A corpus platform of Indonesian academic language. *SoftwareX*, 202–106.https://doi.org/10.1016/j.softx.2019.01.011

Muttaqien, A., Sologon, D., & O'Donoghue, C. (2018). Earnings polarization, ethnicity, and regional perspective in Indonesia. In *WIDER Working Paper 2018/106 United Nations University* (pp. 1–23).

Nurmasari, L., Subiyantoro, S., & Fadhilah, S. S. (2017). Primary school students' barriers on learning Javanese language: A case study in central Java, Indonesia. In *International Conference on Teacher Training and Education 2017 (ICTTE 2017)* (Vol. 158, pp. 436–444).

Rahadini, A. A., Nurhayati, E., & Suwarna. (2022). Teachers' beliefs about Javanese learning materials on local wisdom curriculum in Indonesia. *Journal of Language and Linguistic Studies, 18*(Special Issue 1), 348–358.

Sari, B. T., Vijver, F. J. R., Chasiotis, A., & Bender, M. (2018). Contextualized bilingualism among adolescents from four different ethnic groups in Indonesia. *International Journal of Bilingualism, 23*(6), 1469–1482. https://doi.org/10.1177/1367006918803678

Sasmito, G. W., Wibowo, D. S., & Dairoh, D. (2020). Implementation of rapid application development method in the development of geographic information systems of industrial centers. *Journal of Information and Communication Convergence Engineering, 18*(3), 194–200. https://doi.org/10.6109/jicce.2020.18.3.194

Simonyan, K., & Zisserman, A. (2015). Very deep convolutional networks for large-scale image recognition. In *The 3rd International Conference on Learning Representations (ICLR2015)*. https://doi.org/10.48550/arXiv.1409.1556

Suria, O. (2018). Development of word game algorithm for learning Javanese script. *CommIT (Communication and Information Technology) Journal, 12*(2), 73–86. https://doi.org/10.21512/commit.v12i2.4792

Voulodimos, A., Doulamis, N., Doulamis, A., & Protopapadakis, E. (2018). Deep learning for computer vision: A brief review. *Computational Intelligence and Neuroscience, 2018*, 1–13. https://doi.org/10.1155/2018/7068349

You, W., Shen, C., Guo, X., Jiang, X., Shi, J., & Zhu, Z. (2017). A hybrid technique based on convolutional neural network and support vector regression for intelligent diagnosis of rotating machinery. *Advances in Mechanical Engineering, 9*(6), 168–176.

Yashod, R. (2021). Overview of System Development Life Cycle Models. *Journal of Management and Science, 11*(1), 12–22. https://doi.org/10.26524/jms.11.3

Zurbuchen, M. S. (1976). *Introduction to old Javanese language and literature: A Kawi prose anthology*. University of Michigan Center for South and Southeast Asian Studies.

Web-Based Human Resource Information System Design AT PT. Cakra Mandala Sakti Surabaya

Alexander Wirapraja, Rizky Basatha, and Viky Yoviantono

Abstract

Human resources are defined as people who are ready, willing, and able to contribute to the achievement of organizational goals. Every organization has different goals; therefore, the required human resource capabilities will be different for each company. The recording and archiving process at this company is still carried out semi-manually using spreadsheet processing applications such as Excel so that errors are very susceptible to errors both during input and calculations. Judging from the level of urgency of these problems, the company will develop a Web-based human resource information system (HRIS) to optimize the operational management process, especially the management of records and personnel data. This Web site was developed using HTML and PHP scripting with a database using MySQL 5.7, while the development method will use the waterfall method. The results obtained are that by using a Web-based HRIS, employees at the company are very helpful and easy in managing employee data such as the recruitment process, permits, attendance, and KPI assessments.

Keywords

Website · HRIS · Management · Operational

A. Wirapraja (✉) · V. Yoviantono
Information Systems Department, Institut Informatika Indonesia, Surabaya, Indonesia
e-mail: alex@ikado.ac.id

R. Basatha
Informatics Management Department, Institut Informatika Indonesia, Surabaya, Indonesia
e-mail: rizky@ikado.ac.id

1 Introduction

Indonesia is currently in the Industrial era 4.0, an era where the role of technology plays an important role in managing information so that it becomes useful information for its users. The process of managing data into accurate, effective, and efficient information is an important thing needed by every company or organization in increasing work productivity, time, and costs.

However, it should also be noted that technology alone is not enough for the role of humans, in this case, HR is a critical point that needs to be considered in ensuring the success of the company's business processes. HR personnel owned by the company must have the ability and competitive power to be able to compete, especially in free competition in the current Industrial 4.0 era. The current condition of Indonesia's HR competitiveness based on a report from the World Economic Forum states that the global competitiveness index of Indonesian human resources in 2019 was ranked 50 out of 141 countries or a decline from the previous position which was at 45. Companies should not only treat HR as employees but also need to think about how they can develop into a workforce that is productive, skilled, and has good performance.

Based on the problems above, for companies, especially PT. Cakra Mandala Sakti needs to change from a manual process to digital-based data management, specifically in the management aspect of human resource management in an effort to increase the effectiveness of data and information as a source of managerial decision making and assist the company's operational management process.

PT Cakra Mandala Sakti itself is a company located on Jalan Romokalisari, Bumi Maspion warehouse block A-3 Gresik East Java Indonesia. This company is engaged in the sale of snack products. The initial conditions that exist in this company are the increasing number of employees, incomplete recording of information about employees, the recruitment process, and HR management that has not

M. Nawaz Tunio et al. (eds.), *Sustainability in Creative Industries*,
Advances in Science, Technology & Innovation, https://doi.org/10.1007/978-3-031-48453-7_17

been maximized. This has resulted in the HR department being overwhelmed and struggling to organize and manage human resource management within the company.

2 Theoritical Background

2.1 Human Resources Management

The definition of the term human resource management (HRM) refers to the field of science or art in managing potential, roles, and labor relations in realizing efforts to achieve company goals through the implementation of management policies in the company both from the side of internal employees and external factors such as the community, including function of planning, organizing, effective, and controlling effectively and efficiently (Afifah & Sary, 2020). The quality factor and development of HR management in corporate organizations is a great potential for organization, of course it must be supported by the implementation of good HR management strategies. This is because human resources are increasingly seen as a strategic asset in improving organizational performance to achieve an organization's competitive advantage (Kavanagh & Johnson, 2018). Based on this definition according to Nurjaman et al. (2020), it can be concluded that human resource management (HRM) has the following characteristics:

1. HRM system consisting of various HRM elements such as practices and subsystems of the management system.
2. Employees are considered as a source of sustainable competitive advantage; in other words, they are a strategic resource.
3. From an analytical point of view, all of the above definitions identify subsystems as part of the system in the organization as a whole.
4. The next characteristic is to focus on the fit between HRM and management strategy.
5. Focus on the effects that affect organizational performance.

Menggison et al., in Maghfiroh (2021) provide a definition as the core in the process of developing human resources, namely:

1. **Training:** is a systematic effort to transfer knowledge or expertise from someone who knows and understands to people who do not know or do not understand.
2. **Development:** is a long process carried out by an individual to increase his potential and effectiveness.
3. **Learning:** is a process that aims to change oneself from the original state.
4. **Education:** is seen as a planned learning structure with the aim of training the mind.
5. **Human resource development:** is a term used to describe an integrated and holistic approach to changing work behavior by using learning techniques and strategies as an effort to maintain and improve the company's existence or as an effort to anticipate future business demands.

2.2 The Role of Information Systems in Organizations

The role of information systems in organizations as systems that use the exchange of data and information with the aim of meeting the needs of daily transaction processing. Its purpose is to support the operational management and strategic activities of an organization by providing relevant parties with the necessary reports (Hussein & Kertahadi, 2014). In information management systems, data is an important element including the following definitions: (1) Data is an important element or "lifeblood" of an organization. (2) The quality of data production and data maintenance are important elements that determine the smooth operation of every part of the organization. (3) Data represents the "facts" of transactions that occur especially in daily management operations. (4) A transaction can be considered a consequential event in a procedure such as the process of recruiting and managing new employees for certain positions with certain salaries. (5) The organization will definitely try to capture data (facts) related to each of these transactions such as employee profiles, work dates, names of employees who are hired, positions, departments of new employees will work, and so on, until the data can be stored in the organization's archives so that can be used in the future as part of the decision-making process (Kavanagh & Johnson, 2018).

Application and implementation of information systems is a big investment for every business company today. But poor IS selection and implementation can be an obstacle to achieving business goals. When IS does not have the capability and capacity needed to collect, store and transfer important information for business organizations, it can affect decision making such as limited decision-making options, taking wrong steps to lead to losses for the company causing loss of customers (Pearlson et al., 2016).

2.3 Human Resource Information Systems

Literature from Tannenbaum in 1990 cited by Majeed and Özyer (2016) that the human resource information system is a system used to obtain employee data, carry out the process of storing and sorting data, manipulating the data if necessary, analyzing data, carrying out the process of changing data and information and distributing-related information about the organization's human resources to users. In addition, Kovach and Cathcart in 1999, cited also in Majeed and Özyer (2016) also has the same opinion in defining HRIS as a systematic procedure in data processing such as collecting, storing, maintaining, data retrieval process, and data validation required by an organization regarding human resource activities including employee activities within the scope of work and characteristics of organizational units.

In Fig. 1, it is shown that in activities in various business companies, the use of human resources information systems (HRIS) has been proven to be able to increase productivity, employee performance, and business competitiveness against competitors (Rusjiana, 2016). Understanding the term relationship between the term employee performance and the use of HRIS emphasizes the demands of mastery of information technology by employees in an effort to increase productivity and quality of their work.

HRIS in practice is also seen as an integrated system designed to improve the efficiency and effectiveness of data processing of company employees by collecting HR data and producing documentation in the form of HR reports that can be useful as a source of information in decision making (Susanto & Andriana, 2019). Especially when companies need information that is informative, relevant, fast, accurate, and available for the decision-making process, this can be seen in most companies that have implemented management information systems in the company's operational activities. Based on this explanation, the thing that needs to be considered for the company is increasing the capability and ability of all company employees to be able to master the concept of information systems and use them according to the needs of the work in their respective departments (Rosadi & Purnomo, 2020).

According to Veithzal Rivai (2009) in (Rusjiana, 2016) the HRIS is formed from various elements. Every element must function properly for this system to benefit the company. The three components that are the main functions of the HRIS are: (1) Input function: Entering information about employees into HRIS, (2) The data maintenance function: is a function where after the data is entered into the application, the maintenance function will update and add new data to the existing database, (3) Output function: Function that produces high value output to the user. Meanwhile, according to Wibawa et al. (2018), in the scope of operational management, HRIS is designed to achieve the following objectives: (1) Offer a reliable, comprehensive, and sustainable information system for users to help complete their work, (2) Provide up-to-date information with optimization of cost feasibility, and (3) To provide data security and privacy functions.

The development of the HRIS in this study will be carried out on a Web-based basis, this is done because the

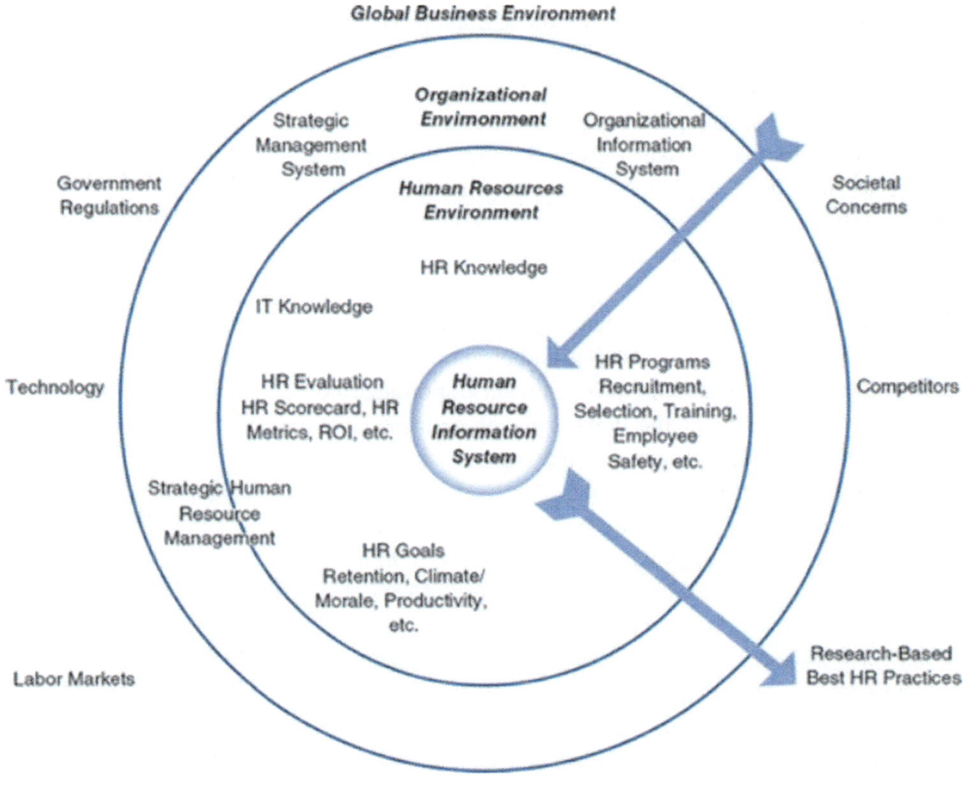

Fig. 1 HRIS on organizational and global business environments (Kavanagh & Johnson, 2018)

Web site is also a form of digital platform that presents a number of characteristics that explain its appeal as an organizing model. (1) digital platforms contribute to a significant reduction in transaction costs, including distribution, search, contracting and monitoring costs. (2) digital platforms help in organizing and coordinating the development of complementary product technologies through modularity and proper governance structures and (3) Generativity is defined as the technological capability for organizational management to generate new results driven by the presence of a large and heterogeneous volume of users (Asadullah et al., 2018).

2.4 Key Performance Indicator

Good organizational performance refers to good employee performance as well. Good employee performance does not just happen automatically but must be formed. Aspects such as managerial standards, knowledge and skills, commitment, and performance appraisal can affect employee performance. One method of performance appraisal is to use key performance indicators (KPI). KPI is a formal system used by organizations within a certain period of time to assess the work performance of an employee. KPI also has a function to identify, observe, measure, collect data, and see the strengths and weaknesses of employees in carrying out their work. The purpose of this KPI is in order to improve the work performance of employees, and then, the performance appraisal will be designed in such a way according to the circumstances in the company to help the company achieve organizational goals and motivate employee performance (Setiobudi, 2017).

Based on the explanation above, it is known that KPI is important for companies that need to conduct performance appraisals for their employees. According to Dessler in 2008 (Evita et al., 2017), the company's reasons for evaluating employee performance are as follows:

1. In practice, the KPI process can help management make decisions about employee salaries and promotions.
2. Employee performance appraisal has an important role in the performance of the company's process management. KPI assessment can link the company's strategic goals into employee specific goals.
3. Assessment provides input to leaders and their subordinates to develop change plans through efforts to improve deficiencies and strengthen systems or procedures that have been carried out correctly.
4. KPI assessment can be input in doing career planning. The assessment provides an opportunity to review the employee's career plan by taking into account the scope of the employee's strengths and weaknesses.

3 Research Methodology

3.1 Waterfall Method

The application development method used is the waterfall method. The waterfall method is a system life cycle development method which is also known as a linear sequential model or a classical system life cycle. The waterfall model starts from the stages of analysis, design, coding, testing, and maintenance given a sequential approach to the software life cycle (Christian, 2020). The essence of the waterfall method is to work on a system that runs sequentially and linearly. Therefore, each step must be completed before moving on to the next step to avoid repeating the step (Lumbangaol & Ridho, 2020). The following are the stages of developing an HR information system at PT. Cakra Mandala Sakti using the waterfall development model:

1. **Analysis:** is the stage where the activities in it include collecting requirements that are carried out intensively to identify the initial requirements of the software such as what is needed by the user. The specification of software requirements is carried out by distributing questionnaires of initial user requirements and equipped with interviews, and the results of questionnaire transcripts and interviews have also been documented in the company archives.
2. **Design:** This is the stage where the activities are carried out after the analysis stage is complete. At this stage, there is a merging process that focuses on the design of making software including data structures, software architecture, interface representation, and coding phases. This stage changes the results of the needs analysis and displays it in the form of a design so that it can be implemented by the developer in the form of software at a later stage. At this stage, it is also necessary to document the design of the software to be developed.
3. **Writing program code:** Pada tahap ini desain yang telah direncanakan pada tahap sebelumnya akan diterjemahkan ke dalam baris kode pada pembuatan perangkat lunak. Hasil dari tahap ini adalah program komputer yang sesuai dengan desain yang telah dibuat. At the coding stage, the developer continues to coordinate and is supervised directly by representatives from PT. The Mandala Sakti Chakra.
4. **Testing:** Pengujian berfokus pada pengujian perangkat lunak yang telah di baut secara logis dan fungsional pada semua bagian dan fitur dalam aplikasi. Pengujian dilakukan untuk meminimalisir kesalahan dan memastikan output yang dihasilkan sesuai dengan yang diinginkan oleh organisasi. Testing of this system is carried out on system users, namely HRD managers and a sample of users, namely employees of PT. The Mandala Sakti Chakra.

5. **Maintenance:** The condition of a software can change when it is implemented by the user. Changes can occur due to errors that were not detected during the testing phase, and this can be caused when the software has to adapt to a new environment. In the maintenance phase, the stages of the development process from specification analysis can be repeated so as to get the changes needed to the existing software. At the maintenance stage, periodic inspections and updates have been scheduled according to the policy of PT. The Mandala Sakti Chakra.

3.2 Context Diagram

In Fig. 2, it is explained what data flow is carried out by each entity (actors or Web site system users) into a Web site system process of PT. The Mandala Sakti Chakra. In the context diagram of the website PT. The Mandala Sakti Chakra there are 5 entities, namely (1) HR managers, (2) employees, (3) prospective employees, (4) accounting, and (5) division heads. The five entities interact directly or indirectly with the website system of PT. The Mandala Sakti Chakra.

In the HR manager entity, the data provided into the system is the main information data provided by the system in the form of employee information, employee assessment results, approval of permits and leave, and the recruitment process. While in the employee role, users will receive notifications on transaction requests and status information including: employee status, attendance information, and recapitulation of absenteeism such as leave or permission that was previously submitted as well as employee assessments that have not been carried out.

This application also provides features for prospective employees or job applicants, where they can view information on job vacancies or the new recruitment process for vacant job positions, so that prospective employees can send personal data such as a biodata form, cover letter, and curriculum vitae which will be assessed later and interviewed by HR manager. Prospective employees will also receive notification whether or not they are accepted at the company.

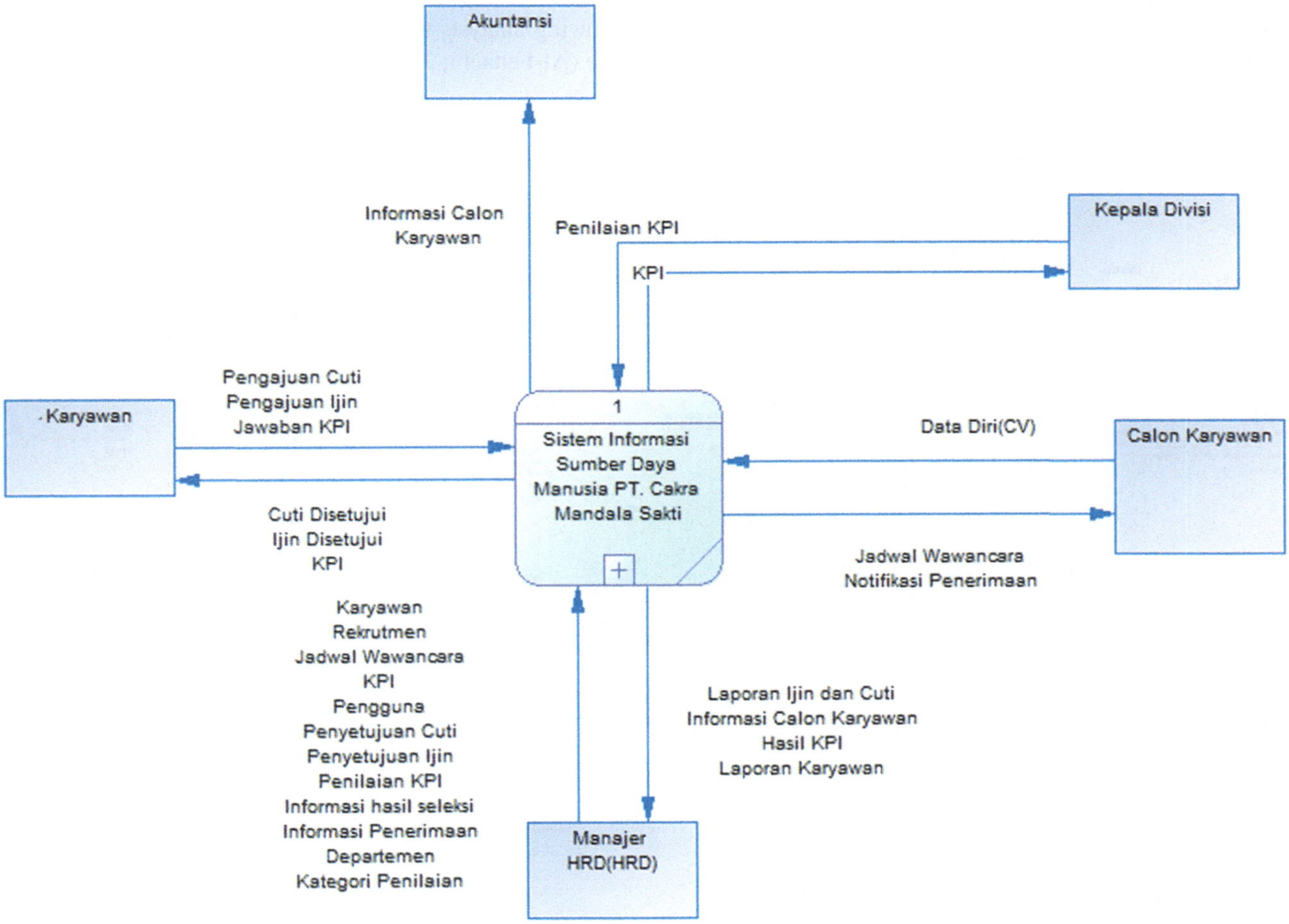

Fig. 2 Context diagram PT. Cakra Mandala Sakti

3.3 Conceptual Data Model (CDM)

In software engineering, software development models often represent sets of activities, objects, and network transformations using a particular notation, syntax, or semantics suitable for computational processing. Application development models are used with the aim of providing visualization of the design process, to assist in idea generation, problem solving, and evaluation and can be used to facilitate interaction and communication (Al-Fedaghi, 2021). Conceptual data model (CDM) is a main concept design of the main database which includes the relationships between tables in the database management system (Irianti et al., 2021).

Modeling with CDM is also one of the most powerful and effective analytical techniques for understanding the information needed to support any organization (Al-Fedaghi, 2021). In general, in the implementation of system development, model design with CDM is the initial stage of database modeling before the original database concept is created. The following is the CDM of the developed system:

Based on Fig. 3, it is known the tables will be used for the needs of the new system. Some of the main tables above require detail tables, which will later be used to record important details. In this conceptual data model, as shown in Fig. 3, there are 12 entities including employees, job titles, KPIs, KPI categories, users, departments, off work, off work categories, work permits, recruitment, attachment candidates, and employee candidates. To create the detailed table, the conceptual data model needs to be converted into a physical data model. If the conceptual data model describes the main tables used, the physical data model describes the detailed tables formed from the main tables.

3.4 Physical Data Model (PDM)

Physical data model is a model that displays a detailed design description of the database model design in physical form (Irianti et al., 2021). A complete PDM will cover all the parts of the database needed to describe the relationships between tables. PDM is usually used to calculate storage estimates, including specific storage allocation details for the database management system used at PT. The Cakra Mandala Sakti Indonesia.

The physical data model as in Fig. 4 provides a logical sequence in forming the basis of the physical model and elaborating on the conceptual model. Physical data models add more detail but usually remain technology neutral, allowing analysts to discuss and agree on the logical structure (Al-Fedaghi, 2021).

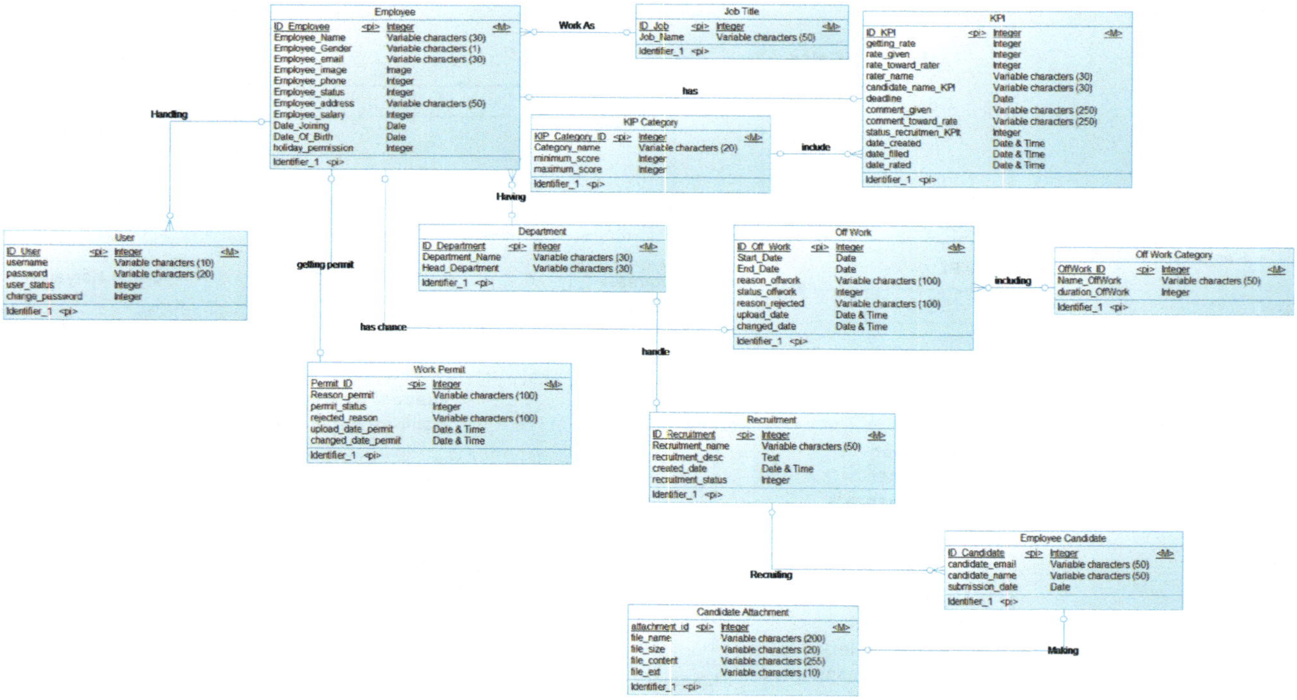

Fig. 3 Conceptual data model at PT. Cakra Mandala Sakti

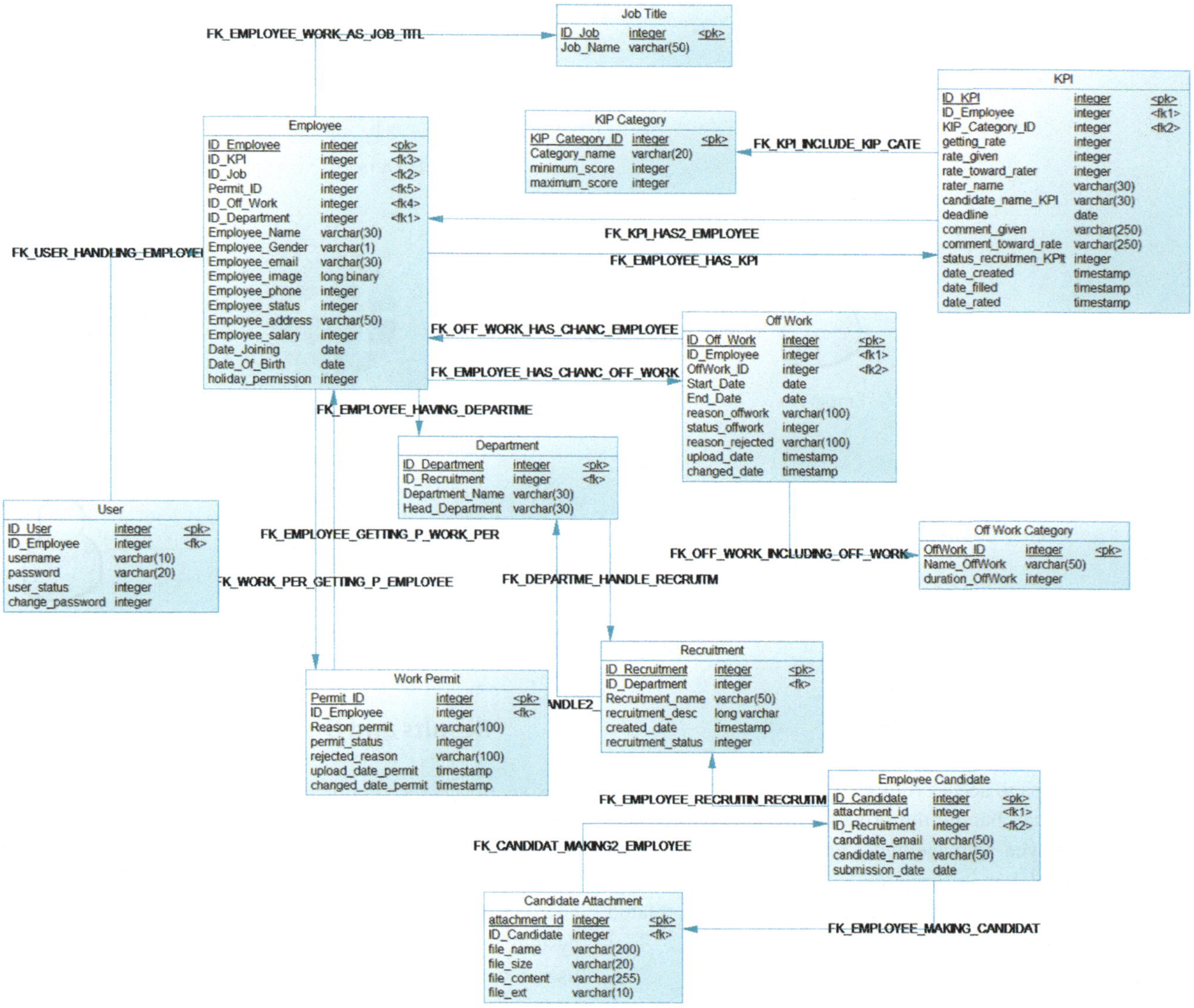

Fig. 4 Physical data model at PT. Cakra Mandala Sakti

3.5 Architectural Design

Architectural design is used to explain the client–server architecture used in building the HR information system of PT. The Mandala Sakti Chakra. Determining the right design is a consideration in building a system. Network structure and relationship between HRD manager (admin), employees, HR department, and Web server on HR information system of PT. Chakra Mandala Sakti can be seen in Fig. 5:

3.6 User and System Requirements

In this key performance indicator information system, the main access of the system is divided into two roles, namely:

- Employees: Users who use several features on the Web site include applying for permits, applying for leave, and filling out reviews that are addressed to them.
- HR Department: Includes HR managers and users in the HR department who manage the entire HRIS Web site of PT. Cakra Mandala Sakti includes inserting, updating, and deleting employee data, users, and assessment criteria data, approving requests for permission and employee leave and creating and opening recruitment processes.

HR information system at PT. This Mandala Sakti Cakra has several system features and capabilities that can be divided based on needs including functional requirements and non-functional requirements as follows:

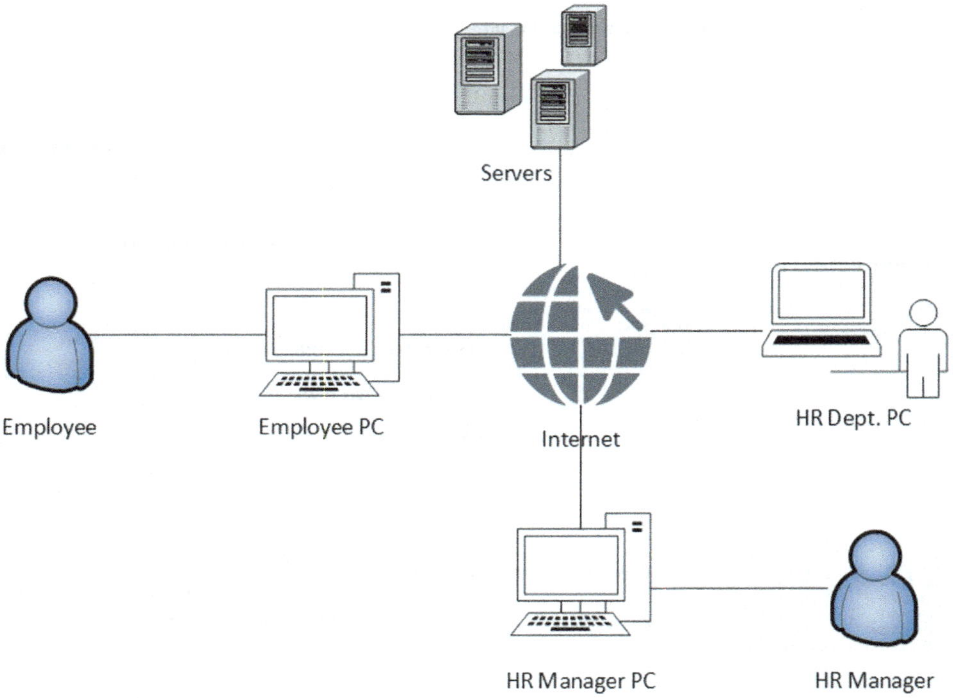

Fig. 5 Architectural design in PT. Cakra Mandala Sakti

3.6.1 Functional Requirements

1. Have admin page access rights.
2. The system can handle the process of employee recruitment, employee management, leave and permits, and personal employee information.
3. Has a confirmation feature before taking actions related to changes to the database.
4. Has a leave application feature.
5. There is a KPI assessment feature for employees.
6. The system is able to generate individual assessment reports and employee information reports.

3.6.2 Non-functional Requirements

1. Information: Provide up-to-date information about employees.
2. Economics: The system is intranet, and the data used is not too large, so the system will help companies, especially in the efficiency of managing operational costs.
3. Control: Admin, in this case the HR department, has full control over the data in the system.
4. Efficiency: Operations are more efficient, as the system has moved from semi-computerization to system integration.
5. Security: The system requires the user to login before being able to perform activities. Each page provides security conditions so that the URL cannot be accessed if the user is not logged in, even though the user knows the URL on a particular page.

4 Results and Discussion

This section is the final result of the design and implementation of HR information systems at PT. Chakra Mandala Sakti Surabaya. This section also shows the interface of the Web site that is used to interact between employees and the admin and human resources department. This system is used for internal companies, so the first time a user accesses the site, the initial display is a login form as shown in Fig. 6a and b:

On this page, the admin and employees when logging in using the default password for the first time will be asked to change the password on the change password page as shown in Fig. 6b so that for subsequent logins the user will use their own password personally to maintain data security.

After successfully logging in, the user will be faced with a dashboard page where on this page the user can perform several processes related to personal data management such as applying for permits and leave, providing KPI assessments, and maintaining personal information data.

The application display as shown in Fig. 7 is a page that can be used by employees to apply for the leave process by entering the type of leave, leave start date, and leave end date along with the reasons stated for taking leave, and after the data is input by the user, this application is made, will be done, and verified whether the requested leave is approved or disapproved by the employee's department head through the human resources department.

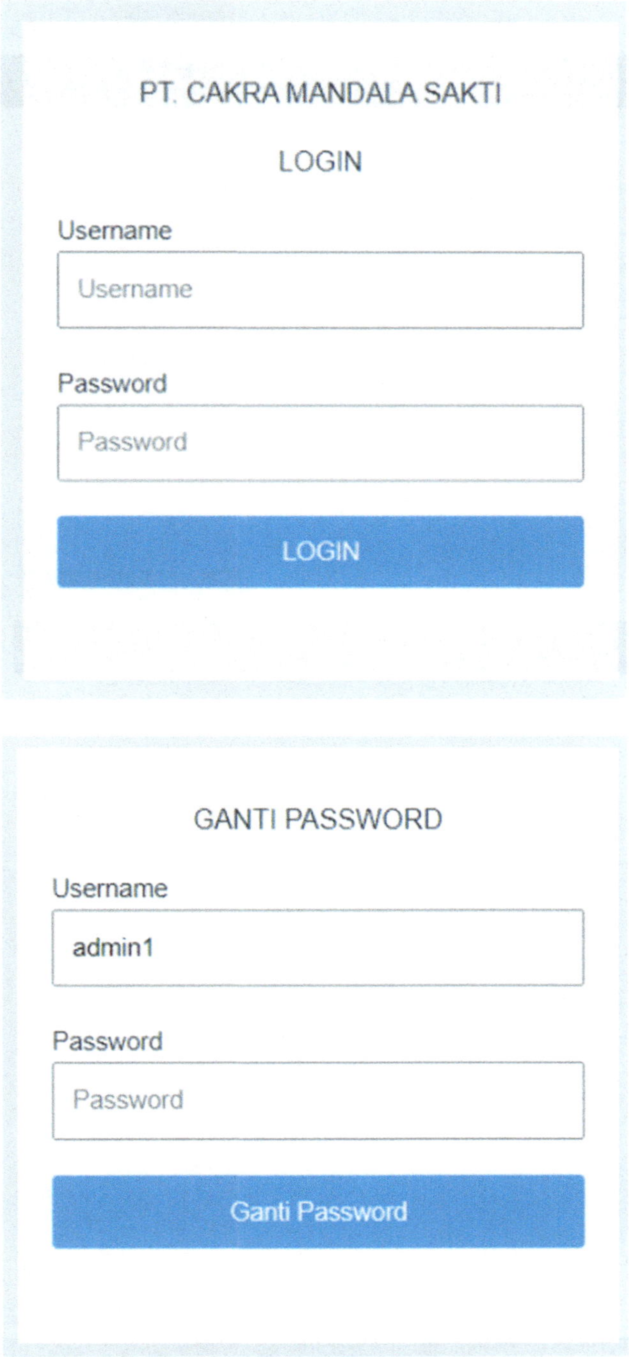

Fig. 6 **a** Login form. **b** Change password form

Figure 8 shows the display of the employee leave application page, where on this page there is also a validation process for employees who wish to apply for leave, the difference is determined based on gender, and both men and women each have a different type of leave. On this leave application page, there is also a validation of the date chosen to meet the allotted leave given or not, if the leave taken exceeds the annual leave allowance according to company policy, the user will not be able to take leave, but if the number of days selected is less or equal to the leave allowance given, the system will direct the user to choose another schedule.

The initial stage before carrying out the assessment process is usually the HR department; admin will provide a page on this application Web site that contains the criteria for the assessment to be carried out. As shown in Fig. 9, the following contains the page used by the admin to be able to add KPI categories that are used as preparation for the assessment that will be carried out on employees.

The KPI assessment process at PT. Cakra Mandala Sakti uses the point method where the HR department of the company gives points to employees who are rated between 0 and 100. The technical filling in the performance appraisal in accordance with company policy is carried out directly by the supervisor of the assessed employee, and the filling page for the assessed employee can be seen in Fig. 10.

Figure 10 shows that the form for filling out the assessment/review will be carried out by the department head from the employee being assessed, on this KPI assessment page, the department head fills in the minimum and maximum values, and the system has determined the deadline for the assessment. In this system, to maintain credibility, a two-way assessment is also carried out between the management and employees.

Figure 11 shows a page that is used by the admin in order to be able to manage the list of employees in the company either by adding, editing, or deleting employee data. This page can only be accessed by the company admin based on the approval of the company leadership.

5 Conclusion

Based on the research and implementation that has been done, several conclusions are obtained from the process of making the HRIS Web site at PT. The Mandala Sakti Chakras includes:

1. Implementation of the HRIS Web site of PT. Cakra Mandala Sakti has made it easier for management and HRD to manage employee data working in the company, both in the process, including: submitting and approving employee permits and leave, providing performance appraisals for employees, and managing the process of recruiting new employee candidates.
2. The Web site is developed using an intranet network, with the consideration that the Web site is not easily accessed by networks outside the company network.
3. This human resource information system Web site has succeeded in displaying up-to-date information. In daily operations, the admin can manage and access the Web

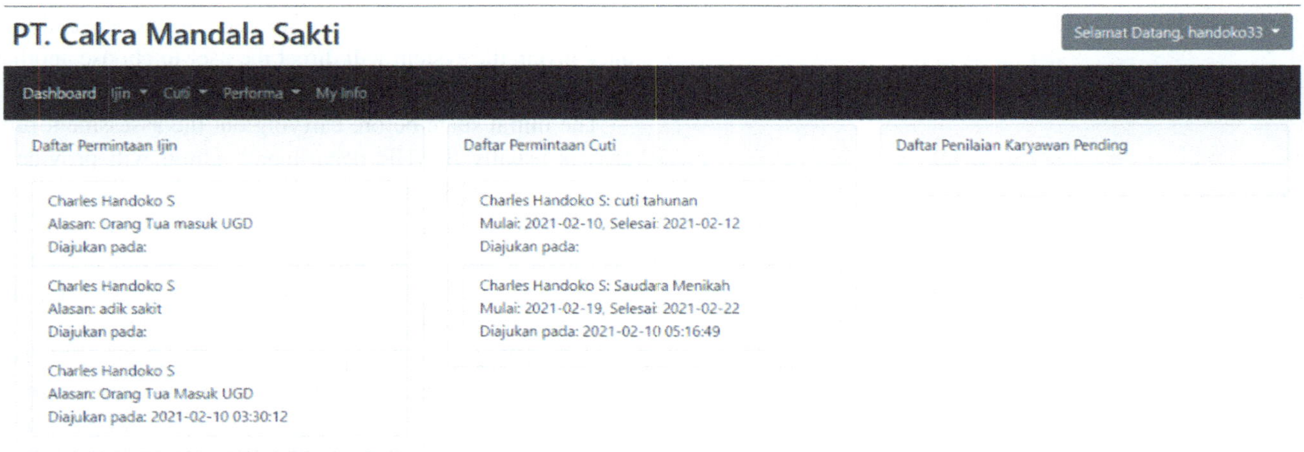

Fig. 7 Dashboard page PT. Cakra Mandala Sakti

Fig. 8 Employee leave application page

Fig. 9 Category page of employee performance assessment

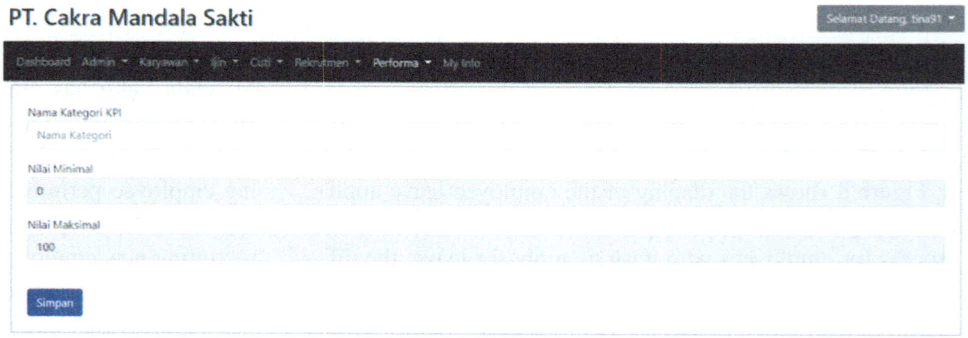

Fig. 10 Employee performance assessment page

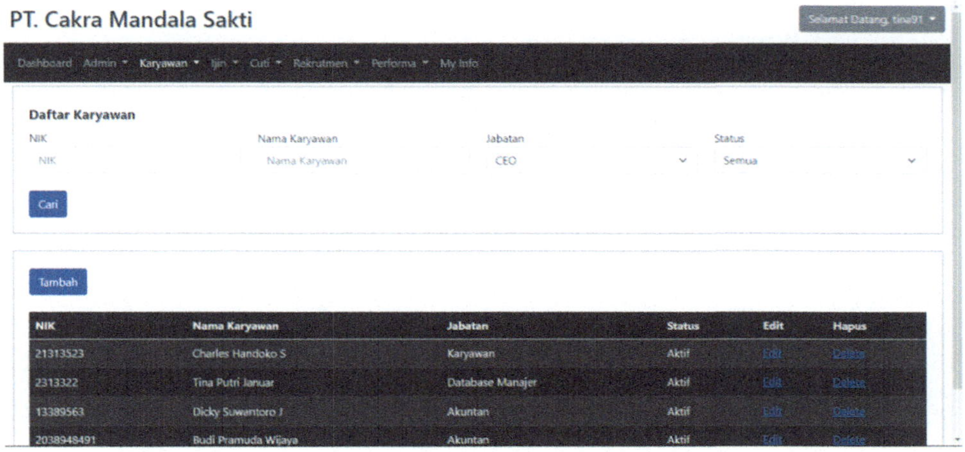

Fig. 11 Employee list page

site through various devices, either with a personal laptop or with a mobile display as long as it is connected to the company intranet.

4. The system has integrated the existing HR management system at PT. Cakra Mandala Sakti with the HRIS Web site that has been developed to achieve effectiveness and efficiency in the development of the company's business processes.

References

Afifah, I. N., & Sary, F. P. (2020). The influence of human resource information system (HRIS) effectiveness on employees' performance at brankas. *Journal of Educational Management and Leadership*, *1*(2), 61–67. https://doi.org/10.33369/jeml.1.2.61-67

Al-Fedaghi, S. (2021). Conceptual data modeling: Entity-relationship models as thinging machines. *IJCSNS International Journal of Computer Science and Network Security*, *21*(9). https://doi.org/10.22937/IJCSNS.2021.21.9.33

Asadullah, A., Faik, I., & Kankanhalli, A. (2018). Digital platforms: A review and future directions. In *Proceedings of the 22nd Pacific Asia Conference on Information Systems—Opportunities and Challenges for the Digitized Society: Are We Ready?, PACIS 2018, January 2019*.

Christian, A. (2020). Perancangan Sistem Informasi Penjualan Rumah Berbasis web. *Jurnal Manajemen Informatika, 4*(2), 61–70.

Evita, S. N., Muizu, W. O. Z., & Atmojo, R. T. W. (2017). Penilaian Kinerja Karyawan Dengan Menggunakan Metode behaviorally anchor rating scale dan management by objectives (Studi kasus pada PT qwords company international). *Pekbis Jurnal, 9*(1), 18–32.

Hussein, F., Kertahadi, & Riyadi. (2014). Implementasi Sistem Informasi Sumber Daya Manusia (Studi Kasus pada Perusahaan Jasa PT. Wiranas Laundry and Dry Cleaning Service). *Jurnal Administrasi Bisnis*, *10*(1), 1–11.

Irianti, C. N. D., Cholissodin, I., & Arwan, A. (2021). Pengembangan Sistem Informasi Manajemen Proyek Properti Berbasis Website (Studi Kasus: PT. Sona Citra Mandiri). *Jurnal Pengembangan Teknologi Informasi Dan Ilmu Komputer*, *5*(6), 2478–2485. http://j-ptiik.ub.ac.id

Kavanagh, M. J., & Johnson, R. D. (2018). *Human resource information systems* (4th ed., Issue July). Sage Publications.

Lumbangaol, M. H., & Ridho, M. R. (2020). Rancang Bangun Sistem Informasi Penjualan dan Penyewaan Properti Berbasis WEB Di Kota Batam. *Jurnal Comasie, 01*(03), 83–92.

Maghfiroh, A. (2021). Analisis Pengembangan Sumber Daya Manusia Dalam Perspektif Syariah. *Jurnal Ilmiah Ekonomi Islam*, *7*(1), 403. https://doi.org/10.29040/jiei.v7i1.2138

Majeed, Z. A., & Özyer, S. T. (2016). Implementation of the human resources information systems and comparative study of various platforms. *Universal Journal of Engineering Science*, *4*(4), 66–78. https://doi.org/10.13189/ujes.2016.040402

Nurjaman, K., Mustajam, A., Syaifuddin, S., Lubis, Y., & Abadi, Y. (2020). Meningkatkan Kinerja Perusahaan Dengan Menerapkan Konsep Manajemen Sumber Daya Manusia Strategik Dalam Menyongsong Persaingan Global. *Komitmen: Jurnal Ilmiah Manajemen*, *1*(2), 73–82.

Pearlson, K. E., Saunders, C. S., & Galletta, D. F. (2016). *Managing and using information system: A strategic approach* (6th ed.). Wiley. http://weekly.cnbnews.com/news/article.html?no=124000

Rosadi, A. A. S., & Purnomo, Y. J. (2020). Pengaruh Sistem Informasi Sumber Daya Manusia Terhadap Kinerja Pada Pegawai PT Raudah Utama Cianjur. *Jurnal Sains Sosio Humaniora*, *4*(2), 357–367. https://doi.org/10.22437/jssh.v4i2.10865

Rusjiana, J. (2016). Pengaruh Sistem Informasi Sdm Terhadap Kinerja Stuctural Equaltion Modeling SEM-PLS. *Jurnal Computech & Bisnis, 10*(1), 21–29.

Setiobudi, E. (2017). Analisis Sistem Penilaian Kinerja Karyawan Studi pada PT. Tridharma Kencana. *JABE (Journal of Applied Business and Economic)*, *3*(3), 170. https://doi.org/10.30998/jabe.v3i3.1768

Susanto, R., & Andriana, A. D. (2019). Analisis Metode 360 Derajat Untuk Penilaian Kinerja Karyawan Pada Sistem Informasi Manajemen Sumber Daya Manusia. *Majalah Ilmiah UNIKOM, 17*(1), 19–24.

Wibawa, J. C., Izza, M., & Sulaeman, A. (2018). HRIS (human resources information system) design for small for micro, small and medium enterprises. *IOP Conference Series: Materials Science and Engineering*, *407*(1). https://doi.org/10.1088/1757-899X/407/1/012134

User Experience Toward Sustainable Choice: Case Study ZALORA Indonesia

Sheena Y. Liman and Geoffrey Tjakra

Abstract

In 2020, ZALORA launched a new initiative called "ZALORA Earth Edit," as one of its sustainability strategies for the next five years. It enables customers to have choices of items that fit into several criteria as sustainable products. However, despite its good objectives, the appearance of this feature seems to be rather unnoticeable and passive, thus less encouraging for the customers to utilize. A quick survey was conducted with a few ZALORA Indonesia users, and the result shows that more than half of them were not aware of this function, while only a few of the rest have ever tried it. This could be a sign for ZALORA to make some changes in the way customers interact with the platform. By mapping the user flow and using a sustainable-oriented marketing framework, named SHIFT framework, this research aims to analyze and identify the intervention needed on the user experience aspect that could encourage a more conscious shopping through ZALORA Earth Edit. On a bigger picture, it provides a recommendation of methods to improve digital user experience that could influence sustainable consumer behaviors.

Keywords

ZALORA Indonesia · Sustainable choice · User experience · Conscious shopping

1 Introduction

In the age of ever-growing fashion products, the term "fast fashion," which used to apply to big brands that are well-known worldwide only, now has evolved to include smaller and local brands that also meet the criteria as cheap, low quality, and trendy. With the advancement of technology and the internet, consumers are exposed to this large selection of products that encourage them to continuously purchase new clothing which leads to excessive consumption. The Covid-19 pandemic also plays a big part by accelerating the digital transformation, thus affecting the e-commerce business. In Southeast Asia, nearly 20 million people became digital consumers in H1 of 2021, almost the same as the total new digital consumers of 2020 (Zalora, 2021). In a narrower scope, Digital Market Outlook reported that the number in Indonesia is expected to grow to 221 millions of consumers in 2025, and this e-commerce industry is usually dominated by products revolving around electronics, fashion, and beauty.

This rapid development of the fast fashion market has a great impact on our environment and society, where at the same time, more and more consumers have become aware of the issue and start to be more conscious of their actions. According to Think with Google, the search interest in "sustainability" is rising over 50% among Indonesian consumers, indicating that it has become an increasingly important topic. People are starting to act and holding brands accountable for their environmental impact (Maini, 2022). To keep the business running while responding to the consumers' demands, businesses now make changes in their business model. Many of them have gone on an overnight formation of a highly visible web presence under the "green" or "responsible" header, to show their commitment on the environmental crisis while sustaining their own (Kramer, 2012). Some also try to engage customers in taking sustainable action by providing green selections among their regular items. In this research, we will be discussing

S. Y. Liman (✉) · G. Tjakra
Pelita Harapan University, Tangerang, Indonesia
e-mail: sheena.liman@uph.edu

G. Tjakra
e-mail: geoffrey.tjakra@uph.edu

M. Nawaz Tunio et al. (eds.), *Sustainability in Creative Industries*, Advances in Science, Technology & Innovation,
https://doi.org/10.1007/978-3-031-48453-7_18

about improving this digital engagement on the fast fashion market through making intervention on the user experience, with a guiding framework that would encourage sustainable consumer behaviors. To give an in-depth examination and more practical results, the authors have chosen a fashion company as the case study: ZALORA Indonesia.

1.1 Background: ZALORA Indonesia

Founded in 2012, ZALORA is one of the largest and leading online fashion retailers in Asia. Although the headquarters is in Singapore, ZALORA has many localized sites in other countries, such as Indonesia, Malaysia, and Philippines. They offer a wide range of fashion products, i.e., apparels, bags, accessories, as well as home and lifestyle goods, that come from more than 900 of local and international brands. Every week, ZALORA Indonesia claims to release hundreds of new products, making it the most visited B2C e-commerce fashion Web site in the first quarter of 2021, with approximately 2.9 million monthly web visits (Wolff, 2021). Meeting the criteria that have been mentioned earlier in this section, while indeed they do work with popular labels, e.g., Mango, H&M, and Michael Kors, ZALORA can also be considered a fast fashion company. However, they have taken a major step in responding to the sustainability issue with their Earth Edit initiative.

Committed to conducting their business in a responsible way, ZALORA designed the "Sustainability Strategy 2020–2025" to create a sustainable fashion ecosystem. It is centered around four key sustainability pillars: Environmental Footprint, Ethical Sourcing, Responsible Workplace and Community Engagement, and Sustainable Consumption, in which customers are directly involved. Through this strategy, ZALORA aims to inspire the customers to make sustainable fashion choices and to participate in the circular fashion movement, by increasing their sustainable fashion assortment. By 2022, 20% of ZALORA's product assortment is targeted to meet the sustainability criteria, and it will be increased to 50% by 2025. To reach the target, ZALORA launched the "Earth Edit" feature that enables customers to choose products that fit into one or more criteria as sustainable products. Those criteria are as follows: (1) eco-production; (2) community involvement; (3) earth-friendly materials; (4) fair production processes; and (5) animal friendly. This feature gives access for customers to conscious shopping, and for suppliers to increase the visibility of their products with special tags and campaigns.

To access the Earth Edit feature, customers could start from the homepage of ZALORA Web site and click "Earth Edit" (Fig. 1), where they will be directed to the product page that contains options of the selected products. Figure 2 shows that these products have special tag that could identify them even when being listed on a regular page, together with non-Earth Edit products. Customers could also choose between one to five categories on the left side of the page to filter their options. While using this feature and applying the filters may look uncomplicated and effortless, at the same time it also seems rather passive and unappealing.

At the beginning of this paper writing, the author conducted a quick online survey on Instagram Story in less than 24 h, asking ZALORA users whether they knew about the Earth Edit filter, and 17 people joined the survey. Ten of them chose the answer A: "No, I don't" (59%), and four chose the answer B: "I know but I've never used it" (24%), while one chose the answer C: "I know, and I've tried it" (6%) and two chose the answer D: "I know, and I often use it" (12%). Considering how Instagram Story works where anyone could view the post (survey) without joining the poll, and the concept of *social desirability* that consumers endorse high-involvement sustainable options to convey social status to others (Griskevicius et al., 2010), the author assumes that if more people (ZALORA users) joined the survey, the percentage result of answer A and/or answer B is likely even higher. There might be various causes behind each answer score that need further research, and in this paper, we will be discussing how to intervene the user flows that are related to the first two answers.

2 Literature Review

2.1 User Experience for Sustainability

Our Common Future, also known as the Brundtland Report (1987), defines the sustainable development as "development that meets the needs of the present without compromising the ability of future generations to meet their own needs." It emphasizes that what we do today determines what is possible in future. In today's practice, sustainability has become an important aspect that drives key initiatives within the society, from the lowest level to industries and companies, as reported by the Global Reporting Initiative (GRI). Many companies and countries have been actively reporting on their sustainability initiatives, including the change that happens within their internal work responsibilities to advance the movement. Designers play one of the most important roles as the design of everyday products and services is often the problem that underlies the environmental issues we face today (Kramer, 2012). This said, design should be a central in finding solutions for these issues, which many have been grasping for practical tools and methods around sustainable design thinking. One thing to note, the design areas that are affected here go beyond

Fig. 1 Earth Edit on ZALORA Indonesia homepage

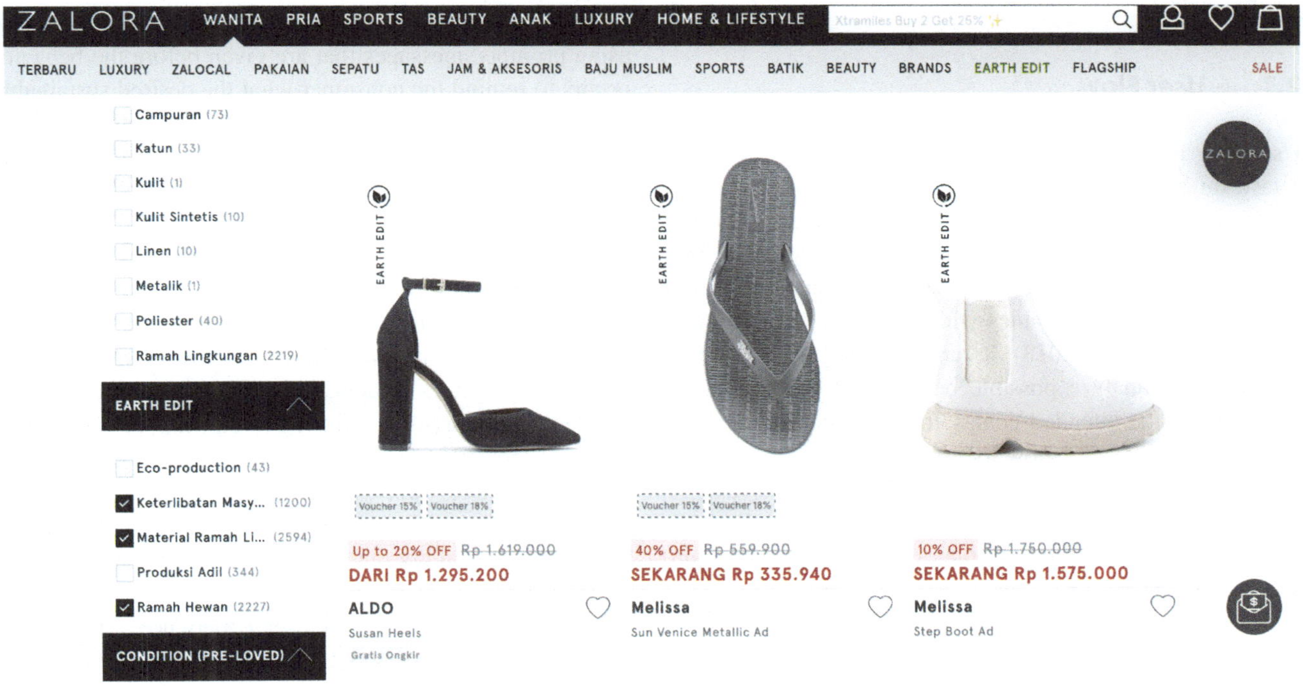

Fig. 2 Earth Edit filter and tag

physical or hard product but also include the design of user interface and user experience of digital products. For example, many waste management services have built their own smartphone application where users can register and manage their household waste collection and disposal more efficiently than making a direct appointment with the service provider. In the context of fast fashion, Carousell, an application to sell and purchase second-hand goods, allows the users to easily market their pre-loved items with clear labeling on every item's condition. Users can also list each item in a group with other sellers' products from the same brand, making it easy for buyers who want to explore various stuffs from a particular brand, similar to ZALORA. All these features are designed to give users the best experience in using the service or buying products. Research shows that user experience has a positive and significant effect on purchase intention (Vandecandelaere, 2018; Watulingas & Permana, 2020), and a customer's buying behaviors are largely dependent on factors such as usability and esthetics (Sinha & Fukey, 2020).

2.2 Persona

A persona, or sometimes called user model or user profile, is a fictional character constructed to represent the needs of real users. The character would usually have made up name and information such as age, occupation, hobby, interest, and goal or desire that is consistent with the user target from the research. Creating personas can help ensure that designers keep the users in mind during the design process.

2.3 User Flow

User flow is the process steps taken by a user to complete a task on a Web site or app. It is presented with a visual diagram or flowchart. A user flow from one user may differ from the other, or even from the one created by the designers. In that case, new user flows are often created to predict and show the possible routes with which the user interacts with a product (Xia, 2018). It can help designers, therefore the brand, to analyze possible behavior of users to better lead them to the intended goal. Often mistaken as the same with user journey, user flow focuses on the process of using the product, while the former focuses on the user experience design of the entire process. Both start with creating (one or more) persona to find the objective, the motivation, and the pain point of users.

2.4 SHIFT Framework

Taking the optimistic view of the important role of marketing in sustainable consumption, the SHIFT framework helps for conceptualizing and encouraging consumer behavior change (White et al., 2019a, 2019b). This framework is comprehensively developed through the literature review over a set of 320 articles that covers behavioral sciences, business, psychology multidisciplinary, economics, and management. The acronym SHIFT stands for the psychological factors that influence pro-environmental behaviors: Social influence, Habit formation, Individual self, Feelings and cognition, and Tangibility. This framework helps to examine the most effective ways to shift consumer behaviors to be more sustainable, including through identifying the barriers that are already existing. Considering the scope of this research, the author has identified which factors from the framework that could be applied in the context of user experience: Habit Formation, and Feelings and Cognition.

2.4.1 Habit Formation

Making It Easy. New habit formation is required when shaping many sustainable behaviors that involve repeated actions, such as taking shorter showers, carrying shopping bag as alternative to plastic bag, and choosing not to get the cutlery when ordering food delivery. To develop and strengthen the positive habits, we could design sustainable actions that encourage repetition, and one of the easiest way is to make it the default option. By simplifying the decision-making process can allow users to form sustainable habits more automatically. Furthermore, forming new habit may create a positive spillover effect where they make another positive change in future.

Prompts. Another technique to form positive habits is using prompts, messages that are given before the behavior occurs to remind the consumer what the desired sustainable behavior is (Lehman & Geller, 2004). Prompts work best when they are clear, easy to follow and received when the behavior will take place. It is easy to employ and cost-effective; therefore, it can be a good initial strategy.

2.4.2 Feelings and Cognition

The concepts of feelings and cognition are introduced together in this framework because consumers take one of two different routes to action: One that is driven by affect or one that is more driven by cognition (Shiv & Fedorikhin, 1999). Theory suggests that either an intuitive, affective route or a more deliberative, cognitive route can dominate in decision-making (Epstein, 2003; Kahneman, 2003, 2011), and it is highly relevant to information about ecological issues (Marx et al., 2007; White et al., 2019a, 2019b), as well as in the context of user experience. In this section, we first discuss about the feelings, followed by the cognition.

Emotions. Customers often consider the emotional consequences of either engaging or not engaging in sustainable behaviors (Rees et al., 2015). Negative emotions are usually related to fear, guilt, and sadness. For example, "fear appeals" is often used when highlighting the negative consequences of a certain action or inaction. As for positive emotions, research shows that people are more likely to engage in pro-environmental actions when they derive positive feelings from the behavior (Corral-Verdugo et al., 2009). Taking the actions has been shown to result in "warm glow" feelings that can lead to more favorable evaluations of the overall experience (Giebelhausen et al., 2016).

Information, Learning, and Knowledge. One top reason why people do not adopt a more sustainable lifestyle is because they do not have enough information (Deloitte, 2021), which can be due to the lack of exposure to that information (Gifford, 2011). Confusion and unfamiliar terms regarding sustainability issues can also contribute to low engagement in eco-friendly actions. Providing users with information that highlights why the desired behavior or product is sustainable can be effective as initial knowledge that influences pro-environmental actions.

3 Methodology

To encourage more conscious shopping when using ZALORA, this research will focus on making intervention on the user flow by first creating personas, mapping the current flows, identifying the barriers and opportunities, then using the SHIFT framework as guideline to create intervention on the user experience. All user flows will start at landing page (homepage) and will end at product detail page instead of payment page, due to some considerations: (1) The personas are created based on the customers who never made any purchase using Earth Edit feature; (2) there are many other factors that could influence the final decision (e.g., product size availability, color availability, detailed information, and comparison to other options), which at the same time will also limit the scope of this research.

3.1 Persona and User Flow

Based on the survey that has been discussed previously, the author creates three different personas that could represent the users who chose answer A or B, and then creates the current flow from each of them. According to the ZALORA Southeast Asia Trender Report 2021, many Indonesian customers have been purchasing sports equipment as well as other hobby-related goods during the Covid-19 pandemic situation; hence, all three personas are designed to search for new sportswear.

The above screen flows represent the user flow from the following three personas:

- Figure 3 First persona (P1), who does not recognize the Earth Edit feature and has low-to-zero interest or care about sustainability issues. P1 will immediately search for a new sports clothing, with or without using any filter. He may notice the Earth Edit filter or label but will ignore it completely.
- Figure 4 Second persona (P2), who does not recognize the Earth Edit feature but has some interest or a sense of care about sustainability issues; thus, there is still possibility of using the Earth Edit feature when noticing the filter or label. However, due to her limited knowledge about sustainability issues, and unfamiliarity with the terms used for the filter (e.g., "*Keterlibatan Masyarakat*" or community involvement), she might hesitate to use it, nor to try to understand the function.
- Figure 5 Third persona (P3), who is aware of the Earth Edit feature and has some interest or a sense of care about sustainability issues. She has noticed the feature from previous visit to the Web site and tried to use it but did not make any Earth Edit purchase at the end, because the product she was searching for was not included in the category. She is interested to retry using the filter sometime in future.

Based on the current user flows, we could identify some important information that will be considered when making the intervention, they are as follows:

- The user objective: The main objective of ZALORA users is usually to buy new product(s), whether they

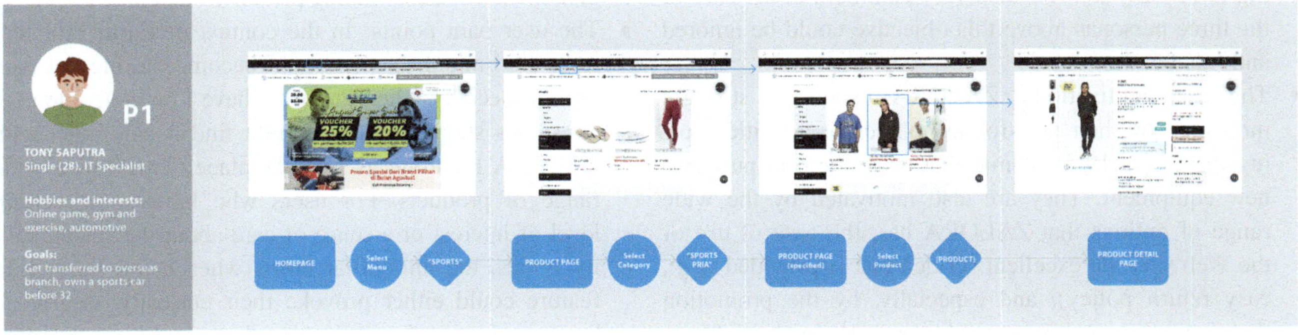

Fig. 3 User flow of P1

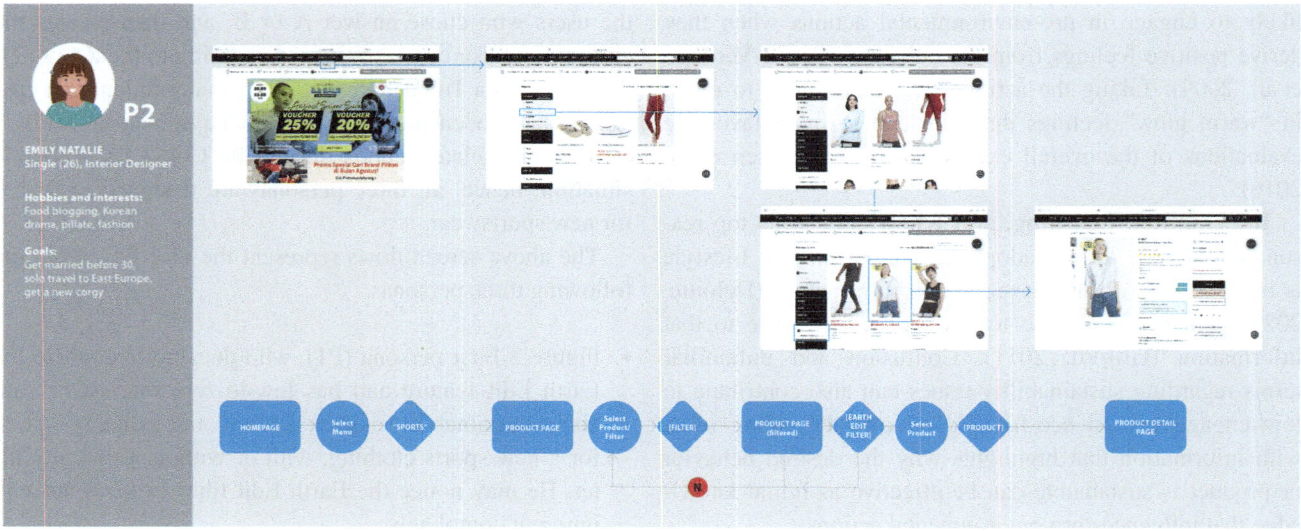

Fig. 4 User flow of P2

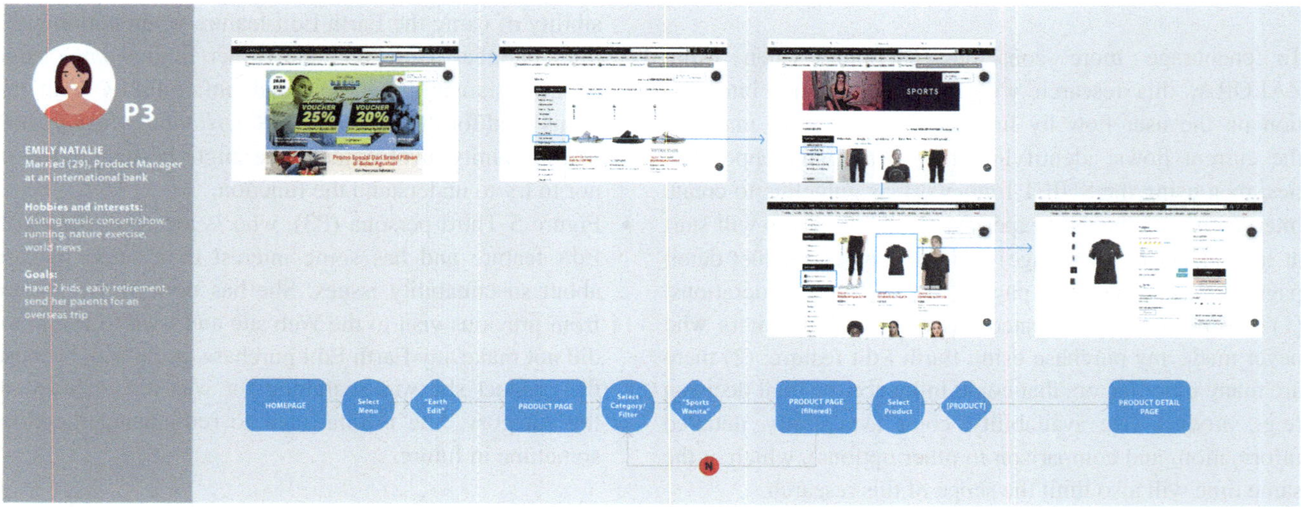

Fig. 5 User flow of P3

already had a particular item to buy or only general requirements (i.e., "sports clothing"). Although *window shopping* (browsing through the Web site without intending to buy any item) is also very common, in the case of the three personas above this objective could be ignored and ruled out.

- The user motivations: ZALORA users are strongly motivated by their new-found hobbies or activities during the Covid-19 pandemic that drive them to purchase new equipment. They are also motivated by the wide range of options that ZALORA has, the ease of use of the Web site, the excellent service that is provided (e.g., easy return policy), and especially, by the promotion that they offer, such as special price and vouchers. Users who have some interest about sustainability issues might become highly motivated by the Earth Edit feature since it could help them to feel less guilty when purchasing new item (regarding the fast fashion issue), or it would give them a sense of taking pro-environmental action.
- The user pain points: In the context of Earth Edit feature, the limited options could become the biggest pain point especially when the users have low motivation for conscious shopping. They might find it contradict with ZALORA branding as a marketplace that offers a wide range of products. For users who have some certain level of interest or a sense of care about the sustainability issues, the unfamiliar terms when using Earth Edit feature could either provoke their curiosity, or instead become another pain point that they would opt to ignore later. Outside of the Earth Edit feature context, the user

pain points may also include the Web site homepage interface that looks crowded with many texts, and the filter system that immediately updates the product page every time a filter is triggered.

3.2 Applying the SHIFT Framework

Making It Easy. Setting the sustainable action as default is the Earth Edit filter. We could be less aggressive by first offering this option to users while also ensuring them that this setting could be easily changed. It is important to keep in mind that there are still many users with low-to-zero interest or care about sustainability issues, such as our first persona. A recent survey by Deloitte found that the first main reason why people do not adopt a more sustainable lifestyle is because they are not interested. Making the Earth Edit as default may become irritating to them if it will limit their options; thus, it could result in bad experience. On the other hand, forming a new habit of shopping, the Earth Edit is hoped to create a positive spillover effect for the users, such as opting for a sustainable choice when shopping outside of ZALORA.

Prompts. It could be a simple reminder message to use the Earth Edit filter when the user is browsing through the product page. Regarding the Earth Edit feature using many terms that might be unfamiliar to some users, such as "community involvement" and "fair production processes," short and simple information could be provided by placing hover text on the filters. This hover text will help users to better understand the feature, hence supporting the prompts. Prompts can also be given as early as on the homepage, using encouraging message to shop consciously or by displaying Earth Edit campaign with button that directs to the Earth Edit menu.

Emotions. In consideration that ZALORA is not a 100% "sustainable" company, provoking negative emotions here could backfire. For example, using "guilt appeals," such as notifying the customers about the impact of fast fashion or the consequence of not using the Earth Edit, might reduce their intention to buy anything, which is not the result that ZALORA would desire. Consequently, we should focus on provoking positive feelings. According to the literature review presented, for ZALORA customers who are motivated by the availability of Earth Edit feature, making a purchase of any item under the category could give them a sense of taking sustainable action which would provoke positive emotion. We can use a literal approach to transform the "warm glow" feeling into a visual effect on the Web site interface, for example by applying hover color to every Earth Edit item's card on the product page, or by using different color for "order" button on product detail page.

Information, Learning, and Knowledge. Although the user objective and user motivation to shop on ZALORA could vary except to get educated or to learn something new, it is important to provide them with knowledge and information that could encourage conscious shopping. The current user flow of Earth Edit feature shows that customers would only find a short explanation about the particular filter on the product detail page (Fig. 6), which requires mouse scrolling, with the link "*selengkapnya*" ("more") underneath that pops-up a box with further information (Fig. 7).

As mentioned previously in the discussion about prompts, providing information could be as early as having them on the product page instead, using hover text on each Earth Edit filter. It can contain a short and simple information, as an initial knowledge that would encourage customers to turn on the filter. It might also provoke their curiosity to learn more about Earth Edit; therefore, we could place a text link, such as "more about Earth Edit," under the filter box that will guide them to a page that contains comprehensive information about the feature, or it could also direct them to ZALORA "Our Sustainability Journey" page.

Feelings and Cognition. Since each of these different routes could drive customers to take the desired actions, it is also possible to combine them together. Often people are hesitating to take a sustainable action because they do not believe that it makes a difference (Deloitte, 2019; White et al., 2019a, 2019b). To overcome this barrier, we could provide information that highlight possible consequences of purchasing Earth Edit items, such as the positive impact created by the brand or by the production process of that item. This information is then presented with wording that would invite customers to get "involved" by making the purchase, for example, "*By purchasing this item, you are supporting us to create an inclusive workplace...*". The feeling of taking part in a sustainable action could provoke positive emotions such as joy, pride, and optimism that we could make a difference. On the framework, it is also stated that this optimism can motivate the maintenance of sustainable behaviors over time (Peter & Honea, 2012), which would increase the use of Earth Edit.

4 Result and Discussion

The results that will be presented in this section are mock-up images created through modifying the real documentation of ZALORA Indonesia Web site pages. All original assets and elements, e.g., images, wordings, and buttons, are displayed with low opacity to distinguish them from newly created ones. Please note that this research only covers the type of intervention to recommend; therefore,

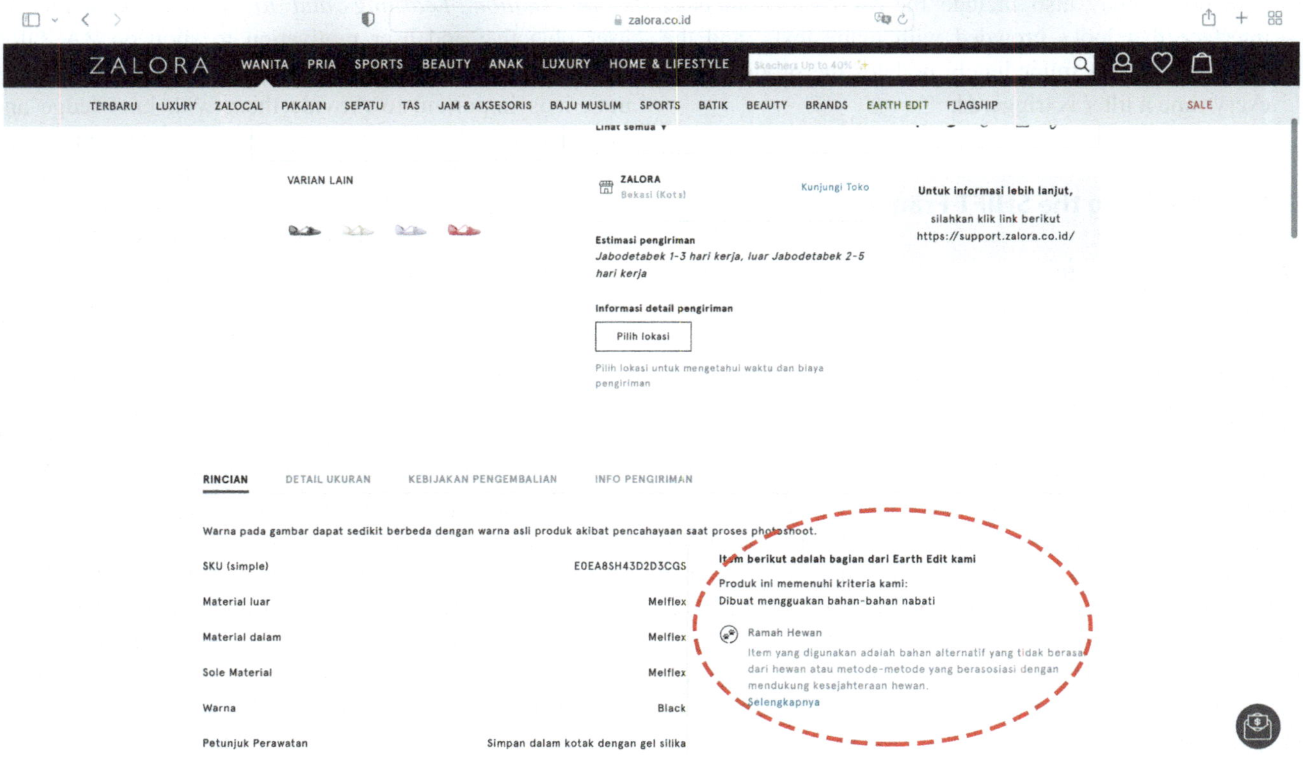

Fig. 6 Earth Edit filter information on product detail page

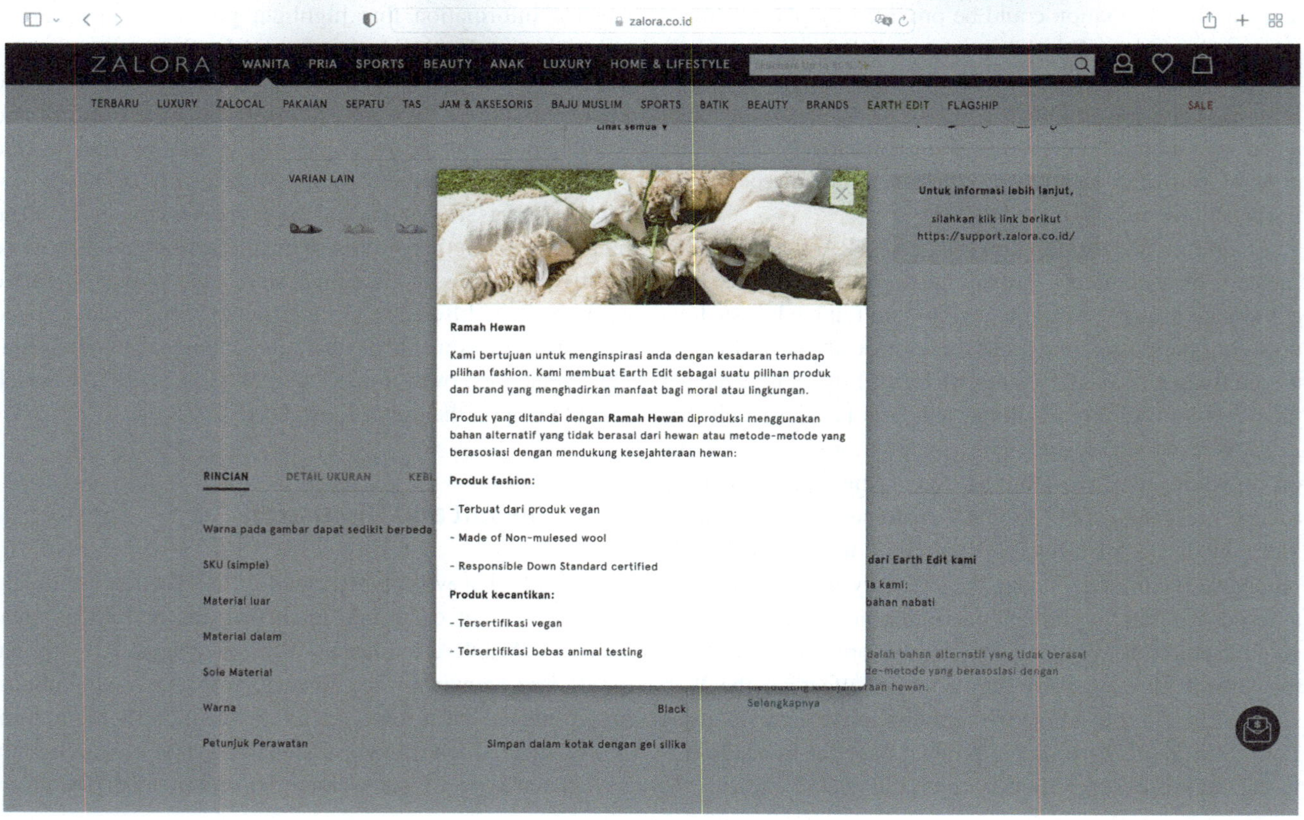

Fig. 7 "*Selengkapnya*" (more) link pop-up information

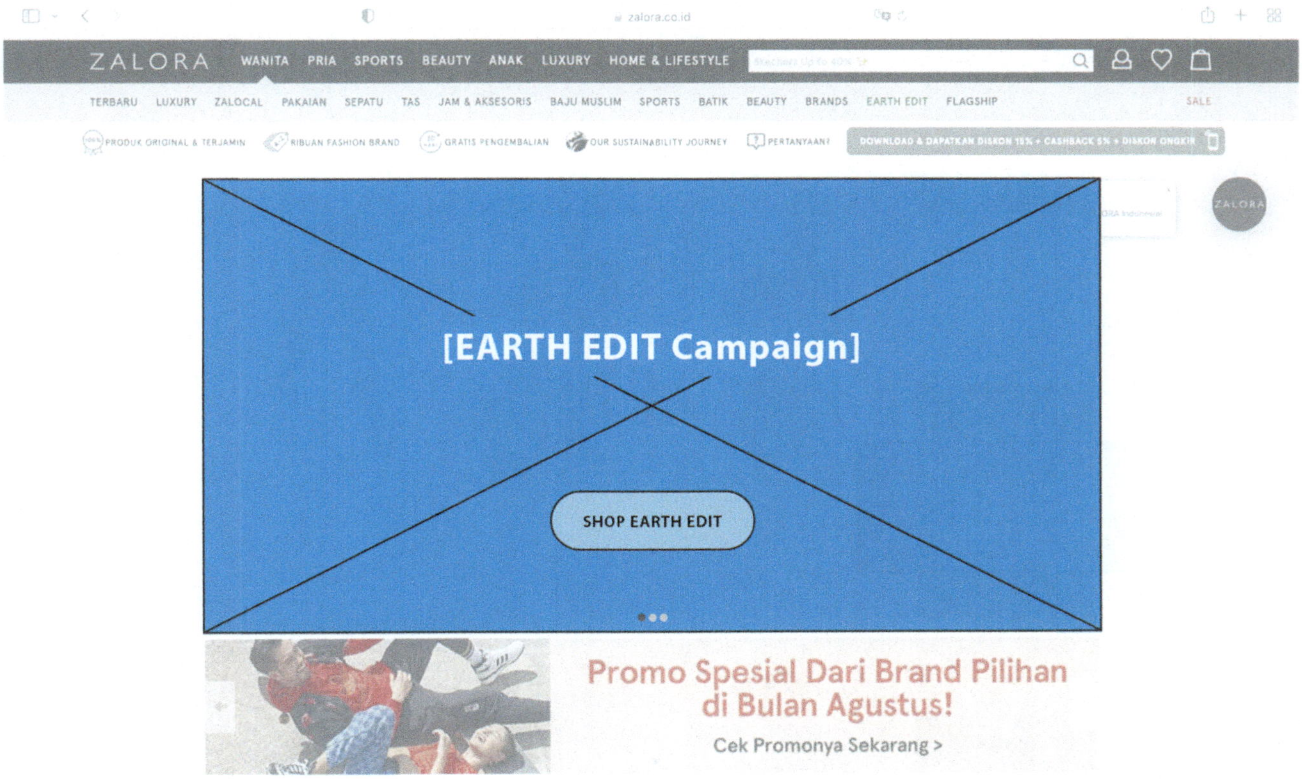

Fig. 8 Campaign picture and direct button on carousel

any UI element is chosen for the purpose of mock-up only (unless explained otherwise), with no regard to ZALORA brand visual identity.

4.1 Homepage

To raise user awareness about the Earth Edit feature, a campaign image is included in the homepage carousel (Fig. 8). Displaying it as one of the carousel images might be less powerful than as a single post, but it will avoid a sense of unfair treatment to products that are not part of Earth Edit. The campaign image will contain simple explanation that acts as a prompt to invite ZALORA customers to shop consciously. It will be equipped with an action button that directs user to Earth Edit page.

Posting a campaign image will not only tell customers about the existence of the Earth Edit feature or page, but it could also give them the initial knowledge about making sustainable choice. At the same time, it will show ZALORA's commitment to shaping sustainable fashion ecosystem and strengthen their brand image.

4.2 Product Page

When entering the product page, a small window will appear that displays prompt message (Fig. 9) to turn on the Earth Edit filter, with a check box option to make it as default filter every time the user is browsing a product page. By setting it as a default, customers are expected to form a new habit of shopping from sustainable options. If the user leaves the box unchecked, the pop-up message should only re-appear occasionally, for example every after 30 days. It is important to make customers feel that they are encouraged or reminded to use the filter, instead of pressured by repeated notification.

To give customers the basic idea of each Earth Edit category and to avoid confusion over unfamiliar terms, customers can move the cursor over one of the filters, and a pop-up window that contains short explanation will appear (Fig. 10). Since this explanation could also provoke their curiosity to learn more about the Earth Edit, there is a link at the bottom of the filter box that will direct users to another page with complete information.

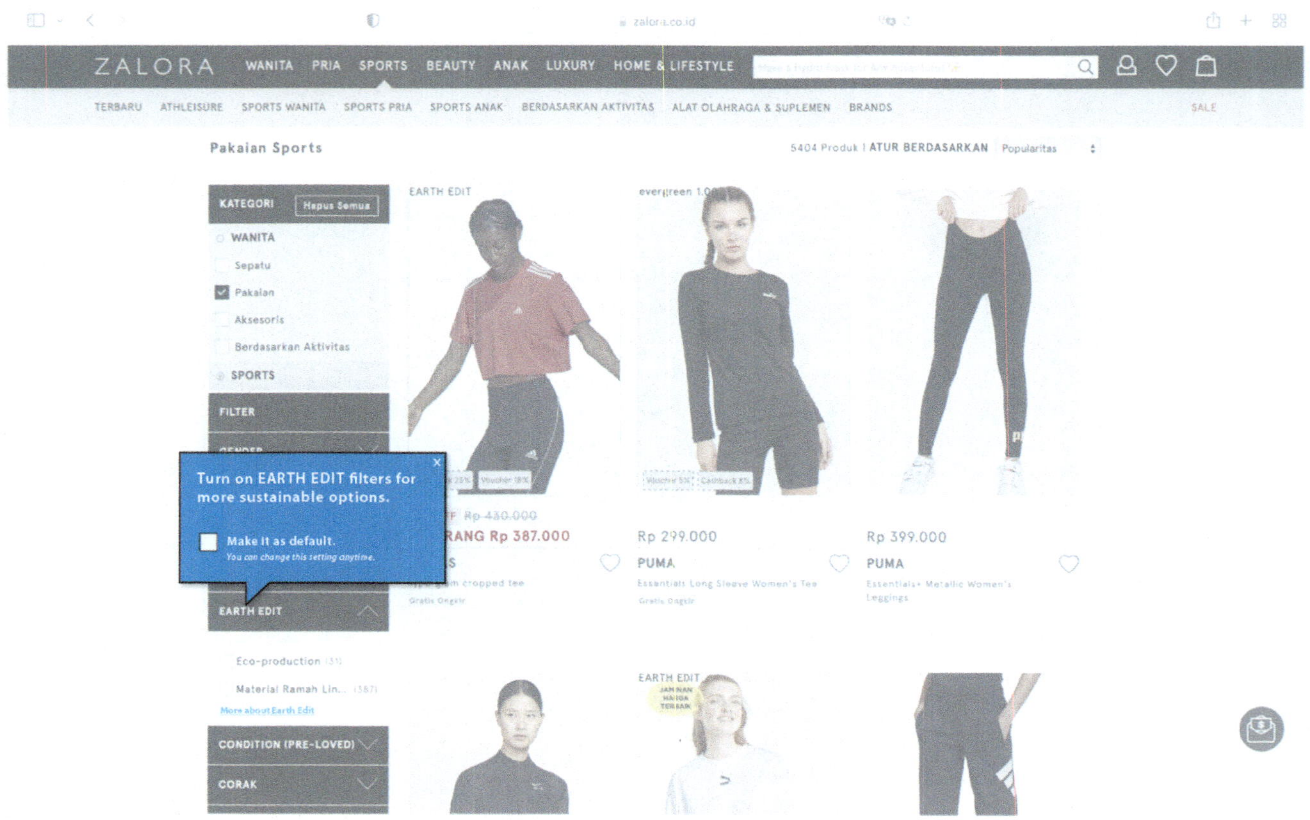

Fig. 9 Prompt to make Earth Edit as default

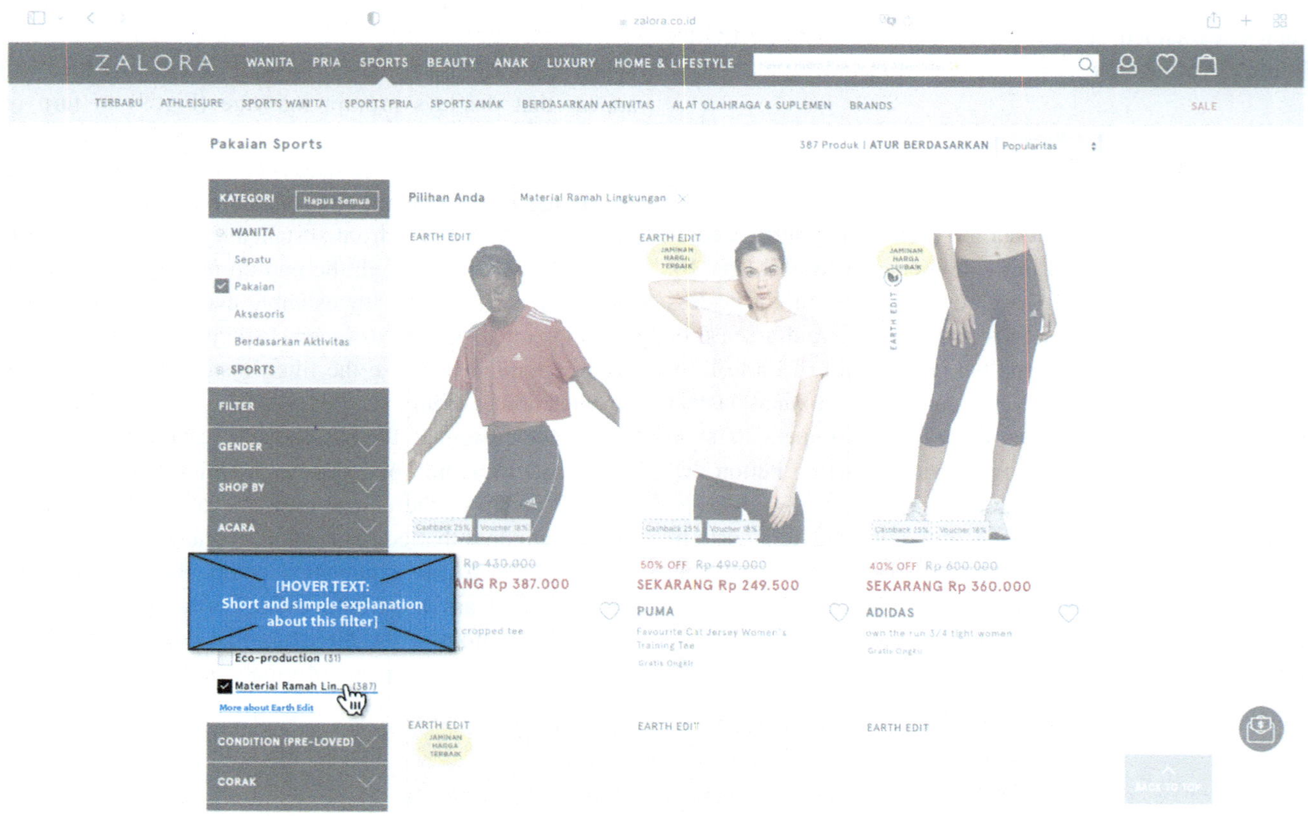

Fig. 10 Hover text with short explanation

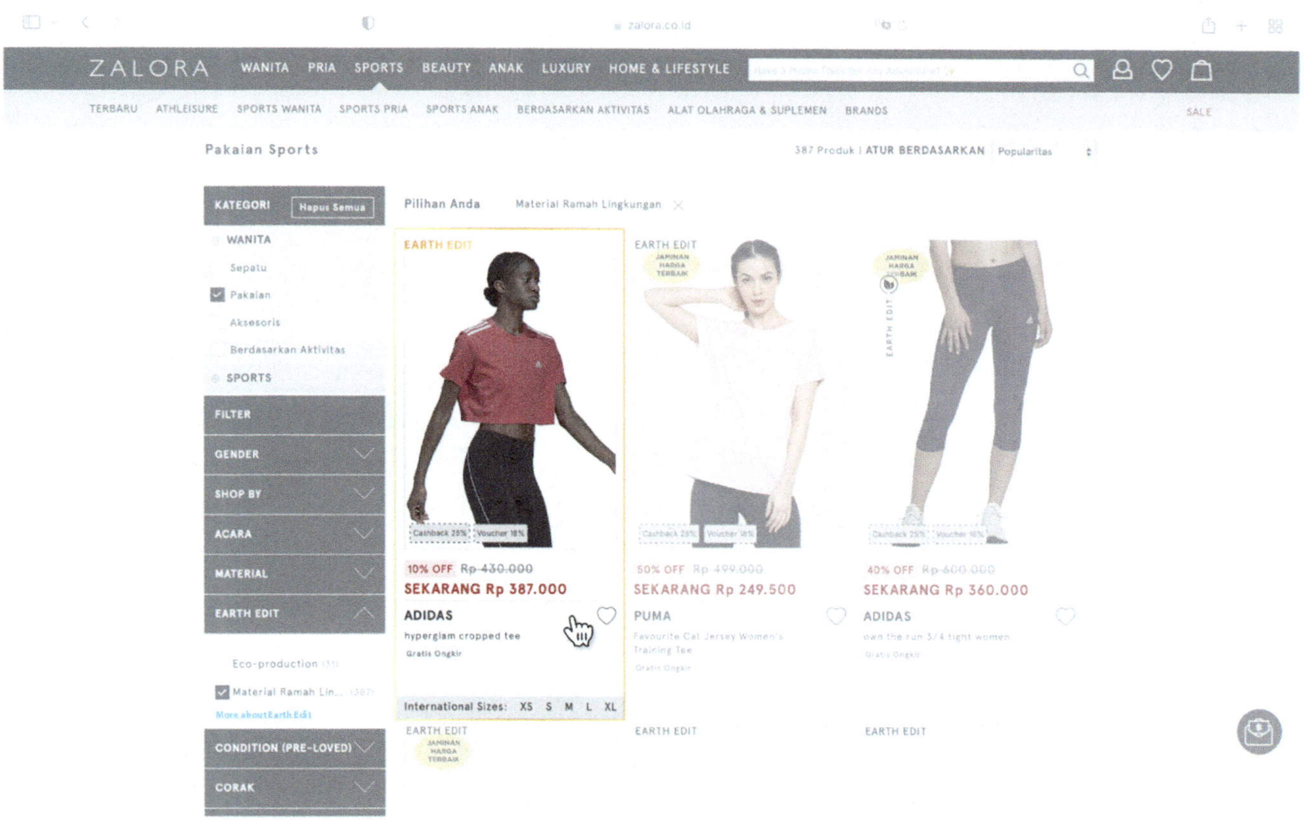

Fig. 11 Warm glow effect when hovering to Earth Edit item

As discussed previously about the "warm glow" feelings, which is the result of deriving positive feelings from taking pro-environmental actions, a hover visual effect is added to every Earth Edit item card (Fig. 11). A box with yellowish outline will appear to give the glow effect, and the "Earth Edit" text will also adjust its color.

4.3 Product Detail Page

The "warm glow" effect is also applied on the "Order" button (Fig. 12). Instead of the original black box that will hover into white box with black outline, Earth Edit item's "Order" button will turn into yellowish color. Other than that, a pop-up window will also appear with a message about the positive impact that both customer and the brand are making through the product. This message would encourage customers to finalize the purchase, by giving them a sense of making contribution through making sustainable choice.

5 Conclusion

ZALORA has taken a huge step in running their business while being responsible to the environment and society by creating the Earth Edit initiative, along with their other sustainability strategies. Customers, as the top stakeholder, are expected to take part in the initiative by shopping sustainable items. However, the Earth Edit feature seems to be rather passive and not engaging, while also there is a lack of information or message that could encourage customers to use it. Considering that these issues are highly related to the user experience aspect of the Web site, the authors propose to create some interventions by combining several methods; they are as follows:

- Creating personas based on ZALORA customers who never made any Earth Edit purchase, then generating their user flows
- Analyzing the user objective, motivation, and pain points, to understand the existing barriers and opportunities
- Adopting SHIFT framework as a guideline, in particular the factors "Habit Formation" and "Feelings and Cognition" to design the interventions that would encourage the intended sustainable action, which is shopping the Earth Edit items.

The interventions then applied and presented as mock-up images of the Web site, containing the homepage, product page, and product detail page.

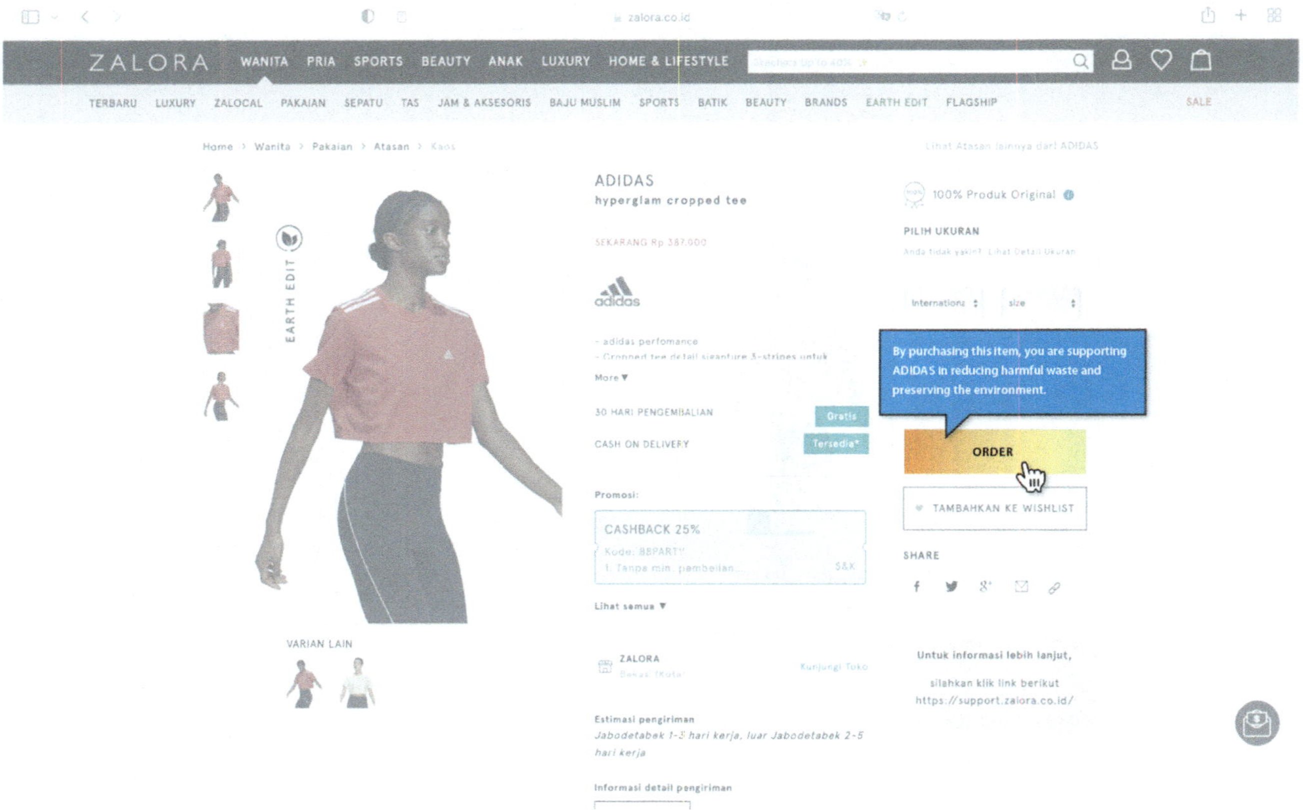

Fig. 12 Warm glow colored button; positive message to provoke emotion

Created using a framework that covers various fields, such as behavioral science, business, and psychology, these practical recommendations are expected to improve the user experience in using ZALORA Earth Edit. Through forming a simple conscious shopping habit, obtaining more information and knowledge about the issues, and building positive emotions from taking pro-environmental actions, customers could experience a behavior change and thus making more sustainable choices. Further research is needed to test these ideas and turn them into prototype for a more comprehensive user test. For future research, the SHIFT framework can be reused to create more extensive interventions, including ones outside the context of user experience.

References

Aurell. (2021). *The impact of Indonesia's growing digital economy towards e-commerce industries*. Bright Indonesia. Retrieved from https://brightindonesia.net/2021/10/28/the-impact-of-indonesias-growing-digital-economy-towards-e-commerce-industries/

Botha, C. (2020). Customer journey mapping. *Goodreads*. https://www.goodreads.com/book/show/55594876-customer-journey-mapping

Dixon, M., Freeman, K., & Toman, N. (2010). Stop trying to delight your customers. *Harvard Business Review, 88*(7). Retrieved from https://hbr.org/2010/07/stop-trying-to-delight-your-customers

Garrett, J. J. (2011). The strategy plane: Product objectives and user needs. In *The elements of user experience* (2nd ed.). New Riders.

Graham, D., Cromwell, E., Perkins, B., Walton, Dr. B., & Heagren, P. (2020). *Shifting sands: The unchanging consumer landscape*. Deloitte LLP.

Kathleen, A. (2021). Analisis perbandingan user flow dari aplikasi E-catalogue ifurnholic. *Jurnal DKV Adiwarna*.

Kramer, K.-L. (2012). *User experience in the age of sustainability: A practitioner's blueprint*. Morgan Kaufmann.

Lai, O. (2021). *What is fast fashion?* Earth.Org. Retrieved August 1, 2022, from https://earth.org/what-is-fast-fashion/

Maini, D. S. (2022). Search pulse insights: What's top of mind for consumers in Indonesia now. *Think With Google*. Retrieved August 3, 2022, from https://www.thinkwithgoogle.com/intl/en-apac/consumer-insights/consumer-trends/indonesia-search-trends-consumer-behavior/

Sinha, M., & Fukey, L. (2020). Web user experience and consumer behaviour: The influence of colour, usability and aesthetics on the consumer buying behaviour. *Test Engineering and Management., 82*, 16592–16600.

Statista Research Department. (2021, May 20). *Online retail users in Indonesia 2017–2025*. Statista. Retrieved from https://www.statista.com/forecasts/251635/e-commerce-users-in-indonesia

Watulingas, E. B., & Permana, D. (2020). The influence of user interface, user experience and digital marketing toward purchase intention. *International Humanities and Applied Sciences Journal, 3*(2), 35–39.

White, K., Habib, R., & Hardisty, D. J. (2019a). How to SHIFT consumer behaviors to be more sustainable: A literature review and guiding framework. *Journal of Marketing, 83*(3), 22–49. https://doi.org/10.1177/002F0022242919825649

White, K., Habib, R., & Hardisty, D. J. (2019b). Consumer behavior: The elusive green consumer. *Harvard Business Review*, 124–133. Retrieved from https://hbr.org/2019/07/the-elusive-green-consumer

Wolff, H. N. (2021, June 2). *Leading fashion e-commerce sites Indonesia Q1 2021, by monthly visits*. Statista. Retrieved August 3, 2022, from https://www.statista.com/statistics/1012489/indonesia-leading-b2c-ecommerce-sites-fashion/

Xia, V. (2018, October 17). User journey VS user flow—differences and similarities. *Medium*. Retrieved from https://uxplanet.org/user-journey-vs-user-flow-differences-similarities-c970e1dd58c7

Zalora. (2020). *Zalora sustainability strategy 2020–2025*. Retrieved June 3, 2022, from https://worldwide.zalora.com/docs/ZALORA_Sustainability_Strategy_March2020.pdf

Zalora. (2020). *Zalora annual sustainability report*. https://zalora-mktg.s3-ap-southeast-1.amazonaws.com/Sustainability+Report+2020.pdf

Zalora. (2021). *Zalora Southeast Asia trender report 2021*. https://zalora-mktg.s3.ap-southeast-1.amazonaws.com/zalora-mktg/trender/ZALORA-Southeast-Asia-Trender-Report-2021.pdf

Zalora.co.id. (n.d.). *Tentang zalora Indonesia: Destinasi fashion Indonesia*. Retrieved August 1, 2022 from https://www.zalora.co.id/about/

GPSR Compliance

The European Union's (EU) General Product Safety Regulation (GPSR) is a set of rules that requires consumer products to be safe and our obligations to ensure this.

If you have any concerns about our products, you can contact us on ProductSafety@springernature.com

In case Publisher is established outside the EU, the EU authorized representative is:

Springer Nature Customer Service Center GmbH
Europaplatz 3
69115 Heidelberg, Germany

The manufacturer's authorised representative in the EU is Springer Nature Customer Service Centre GmbH, Europaplatz 3, 69115 Heidelberg, Germany. If you have any concerns regarding our products, please contact ProductSafety@springernature.com

Printed and bound by CPI Group (UK) Ltd, Croydon, CR0 4YY
01/10/2026
02225885-0003